D1157824

Roland Kalb

The Pathways of Mind

A Neural Theory of Mental Processing

Mathematical Principles, Empirical Evidence, and Clinical Applications

SpringerWienNewYork

Prof. Dr. Roland Kalb
Erlangen, Germany

Typesetting: Composition Design Services, Minsk, Belarus
Printing and binding: Manz, A-1050 Wien

Printed on acid-free and chlorine-free bleached paper

SPIN: 10784965

With 192 Figures

CIP data applied for

ISBN 3-211-83565-2 Springer-Verlag Wien New York

Preface

The model for this work was the description of the physical world by mathematical laws.

It were always the simplest phenomena which were treated by this scientific method. Physicists studied simple motions in order to find the mathematical laws. Astronomists observed the orbits of planets in order to find the laws of gravity.

One of the simplest measurable phenomenon in the brain is the stimulus-response task.

Such tasks have been known since the last century by psychiatrists and psychologists (v. Helmholtz). There exists a vast literature about the measurement and theory of simple reaction tasks and various choice reaction tasks, visual or auditory. They have been measured and have been described mathematically. One of the first models for the reaction times used a logarithmic function. But many intriguing questions remained open about reaction tasks especially the neural explanation of the findings.

The new tool to investigate the neural structure of stimulus-response sequences was the computer. Now it was possible to measure the reaction times by using special programs, to compute the elementary times and the pathway structures from these reaction times, to evaluate the results statistically, to simulate the results, and to write this text. It was this instrument which permitted to save large amounts of data and evaluate them by special software written for this purpose. Thus it was possible to compute the time quanta and the pathways and to understand each reaction time as an integer multiple of this time quantum (plus a constant value). Computers were also used to record the event-related potentials. Two of the both fundamental parameters of this work (the time quantum called elementary time and the length of the linear pathway) have been confirmed by event-related potentials.

Eight pathways have been investigated: the visual and the auditory pathways in each hemisphere and the pathways of one stimulus and two stimuli tasks. These eight pathways have been described by mathematical equations and confirmed by event-related potentials.

All empirical findings have been correlated to the histological structure of the human cortex. The correspondence between the results and the neuro-anatomical substrate gives strong evidence for the reality of these processes. Nevertheless all these findings have to be replicated by independent researchers.

Every new medical knowledge has to serve our patients, therefore the instruments of this work were used to explore the deformations of mental pathways caused by monohemispheric brain lesions (tumors etc.) or by schizophrenia. This work is dedicated to all our patients which participated in this study in order to find new knowledge about their disease. I was deeply encouraged by my patients who asked me to investigate their illness in order to find some new knowledge. Three groups have been defined in order to compare their pathways with each other: a control group with healthy subjects, a group with patients suffering from chronic schizophrenia, and a group of patients with monohemispheric organic brain lesions. I hope that the findings of this work will contribute to some new therapy in the future.

At last, there was the deep belief that the spatiotemporal structure of mental activity can be found and understood, which implies that the human mind has the power to understand itself, at least in some respect.

The belief that physical, chemical, biological, and mental processes have common features has been an inspiring factor for this investigation. One of these common principles is the search for some elements and the selection of one. This is a Darwinian process within the mind which has unique properties: its search and selection processes do not need millions of years but occur within milliseconds. The existence of such fast Darwinian processes during reaction tasks was the great surprise of this investigation. To prove the existence of these material processes by gathering empirical evidence had the first priority for this work.

The book "The Pathways of Mind" investigates the spatiotemporal basis of the mind without asking: "what is mind, what is this which moves along the pathways". Only the physical parameters of these pathways have been measured. Fortunately the investigated mental processes are of great simplicity and can be described by simple mathematical laws. Furthermore they are of great beauty and show similarities with certain natural processes outside the mind.

All proposed structures of this work are hypothetical, but evidence is gathered to support these structures. The book has four parts: in the first part the hypotheses are developed. In the second part, the methods are introduced and in the third part the results of 72 subjects are shown. In the fourth part the findings are confirmed by event-related potentials and by simulation programs. In this part the shortcomings and possible errors are discussed. There will be mistakes in the calculations, results and simulations of this work. I apologize for these shortcomings but hope, that some basic principles of the mind have been found.

For data safety reasons the initials of the subjects have been changed into a code like PNN, HNN, ONN or SNN with P=preparatory, H=healthy, O=organic, S=schizophrenic, and NN=natural number. The tasks are denotated as xNNy with x=v (visual) or x=a(auditory), N = number of stimuli = number of responses, y=r (right-handed task) or y=l (left-handed task).

This work has been approved by the local ethics commitee at the University of Erlangen- Nuremberg.

Acknowledgments

I thank God, who has allowed me to finish this book. I think of my late mother who has loved me all her life. She looked forward to this book but did not see its final realization. I thank her for her love, she gave to me and my family. I thank my wife and my children who will be happy that the work is finished. I will not forget all the people taking interest in my endeavor, and helping me to proceed. I thank all the scientists who worked without public recognition. I thank all the writers of books and papers I had the privilege to read. I got inspired by the work of (in alphabetical order) Calvin, Churchland, Damasio, Donders, Eichert, Gazzaniga, Kandel, Pinel, Posner and Raichle, Pöppel, Spitzer, Zigmond and many others with their work forming the unconscious fundament of our thinking. I want to thank my patients who encouraged me and my students who supported this work by active help and many discussions. Some persons who supported this work by giving me the opportunity to pursue it or talked to me about its themes or accompanied it with their sympathy, shall be acknowledged explicitly:

Strategic support

Prof. Bogerts, Director of the Department of Psychiatry, University Magdeburg
Prof. Dr.Joachim Demling, Acting Director of the Department of Psychiatry (1998-2000), University Erlangen- Nuremberg
Prof. Fahlbusch, Director of the Department of Neurosurgery, University Erlangen-Nuremberg
Prof.Dr.Max-Josef Hilz, Department of Neurology, University Erlangen-Nuremberg
Dipl.Inf.Renke Hobbie,
Dr.Norbert Hofmann, Department of Mathematics, University Erlangen-Nuremberg
Dr.Huth (TÜV Nürnberg)
Prof.Dr.Kaiser, Department of Psychology, University Erlangen-Nuremberg
Prof. Dr.Johannes Kornhuber, Director of the Department of Psychiatry and Psychotherapy, University Erlangen-Nuremberg
Prof.Dr.Kühner, Director of the Department of Neurosurgery, Nuremberg
Prof. Dr.Eberhard Lungershausen, former Director of the Department of Psychiatry (1983-1995), University Erlangen-Nuremberg
Dr.rer.nat.Dipl.math.Martin Mayer, Department of Medical Statistics, University Erlangen-Nuremberg
Dipl.Ing.M.Schwind, Nicolet Biomedical

Prof.Dr.Singer, Director of the Max-Planck-Institute for Brain Research, Frankfurt
Dr.Th.von Stockert, Director of the Department of Neurological Rehabilitation, Klinikum am Europakanal, Erlangen
Prof.Dr.Vieth, former Director of the Dept. of Experimental Brain Research, University Erlangen-Nuremberg

Active support

Dr. Gerhard Bauer, Jörg Beer, Maresi Berry, Dr. Gertrud Fricke-Kalb, Joachim Höfig, Jens Huber, Michael Müller, Ulrike Pöllath, Udo Reulbach, Gesine Raydt, Dagmar Schöpplein, Christian Schorr.

I am especially grateful to Raimund Petri-Wieder and Susanne Mayr for providing me with the opportunity to present this research in a publication of Springer, Wien New York.

Contents

Part I
The time measurement of stimulus-response pathways

Part II
The spatiotemporal structure of stimulus-response pathways

Part IV
Critical evaluation of the results and the model

Part I

The time measurement of stimulus-response pathways

In Part I some notions and hypotheses are formulated which have to be proven in the Parts II – IV. The data of this Part I are not intended to give empirical evidence but to show the development of the theory. They have stimulated the investigations of Part II – IV. This part may be omitted because of its preparatory character. The empirical evidence and the theoretical model are presented in the Parts II – IV.

Measurement of reaction times in healthy subjects

Bihemispheric visual reaction tasks

Summary

The number of alternative stimuli plus the number of alternative responses and the reaction times are strongly related. But the exact relation is questionable. The law of Hick postulates a logarithmic fit and has dominated the discussion. The question arises whether new data fit Hick's law or some other law. Five subjects with nearly 10,000 measurements were used to answer this question. A linear relation was found between reaction time and the number of alternatives when the number of alternatives is not too high. With more alternatives, the relationship is clearly non-linear.

Introduction

In a discrimation task, one has to decide which stimulus of a set of stimuli is present and which key of a set of keys is to be pressed in response of this stimulus. The coordination of the stimuli to the keys is given by an instruction before the task begins. The subject has therefore only to decide, which stimulus is present and which key is to be pressed in each single trial.

The measurement of reaction times has a long history. Helmholtz (1850) used it to measure the time of neural conduction from the foot to the head (20 ms). Wundt (1863) wrote about it in his lectures. Donders (1868) distinguished three types of reaction: A-reaction with one signal and one response, B-reaction with n signals and n responses and C-reaction with n signals but respond to only one of them. He subtracted the reaction time of the A-reaction from the reaction time needed for the C-reaction and found 50 ms difference time. He took these 50 ms for the discrimination time between the different signals. Merkel (1885) found that reaction time increased as the number of alternative signals was increased from one to ten. Hick (1950,1952) , Hyman (1953) and Crossman (1953) tried to interpret these results theoretically. Hick used 10 lamps and ten corresponding Morse keys and measured the mean reaction times for several series of equiprobable signals with number of alternatives from two to ten. He fitted his results to the equation $t(n)= b \log (n+1)$ where $t(n)$ is the reaction time to one of n equiprobable signals. The above equation provided a slightly better fit than the equation $t(n)= a+b \log n$ (the last equation seems to me more realistic because there are input and motor times which do not correspond to the number of alternatives). This idea of a logarithmic relation between reaction times was first mentioned by Blank (1939). Of late, Kvalseth (1996) showed that the Hick-Hyman law is a special case of another law (power law) where a square root formula predicts reaction times, when the number of equiprobable stimulus-response pairs ranges from 1 to 10.

Sternberg (1966) introduced his additive factors method which provides a procedure for isolating subtasks which are serially performed within a reaction task. In Sternberg's task of responding to a single target number, which is tested to be member of an previously shown list of numbers, the reaction time correlated linearly with the list length. He found 40 ms for every number in the list and took this as an evidence against a self-terminating but exhaustive mental process. However, there are tasks to be performed in a parallel fashion (eg letter recognizing), in these cases one cannot apply the additive factors method.

One can ask whether the logarithmic fit is the adequate function or only an elegant mathematical approximation to a more complex reality. The only way to answer this question is to make new and more measurements of reaction times in tasks with changing number of stimuli and response alternatives.

Pöppel stimulated the start of this work with his observation that a visual reaction task with 2 lamps and 2 keys produced on average a 70 ms longer reaction time that the 1lamp-1key- task, the 2lamps-1key-task needed a 50 ms longer reaction time and the 3lamps-3keys-task needed 105 ms longer than the 1lamp-1key-task. If one takes the 1lamp-1key-task as the reference task, the plus of the other times can be computed by this formula:

RT(mlamp-nkey-task)= (m+n)* 17.5 with 17.5 being a personal constant of Pöppel. The computed value for the 2lamp-1key-task is 52.5 ms, for the 2lamp-2key-task 70 ms and the 3lamp-3key-task 105 ms. The following chapter tries to generalize and explain this findings.

Methods

Subjects

The experiments were conducted with 15 persons. The subjects were healthy and did not take any medication except ovulation inhibitors. There were no musculoskeletal complaints or vision problems that might interfere with reaction time measuring. 15 subjects performed all the tasks vMN with M=N and 3 subjects performed all tasks vMN (v=visual task, M=number of stimuli, N=number of responses).

Tasks

The subject faced a PC screen, located approximately 40–50cm from their eyes and had their fingers on distinct keys of the keyboard:

left little finger	a
left ring finger	s
left middle finger	d
left index finger	f
left thumb	space
right thumb	–>
right index finger	4 (at the numeric keyboard)
right middle finger	5 "
right ring finger	6 "
right little finger	+ "

For each task there were only so many fingers on the keyboard as were needed for the task

A white fixation arrow was always visible in the center of the display. The stimulus configuration of the visual task was a vertical line of five "O"s under a row of numbers

The area on the display used by the stimuli was 10 cm x 5 cm. The subject was instructed to press one key as rapidly as possible as soon as the stimulus appeared on the display. The attachment of a stimulus to a key was told to him by a brief instruction on the display before the beginning of each task. By pressing any key the task started. The sequence of visual stimuli was randomized, all stimuli had about the same probability to be shown, the mean interstimulus interval was 2.5 sec. Each task consisted of a training sequence of 10 trials and then a sequence of 100 trials. Only the latter were used as results. The subject performed 100 trials with the possibility to pause 10–60 seconds after every 10 trials.

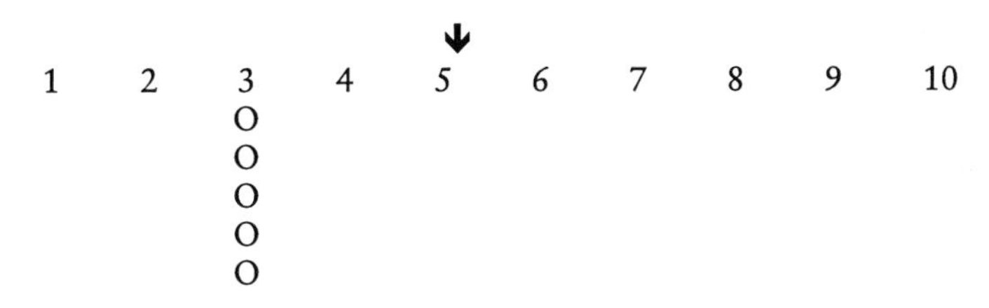

Fig. 1. Example of a trial with 10 stimuli and 10 keys used when the third stimulus was on

There were different kind of tasks examined ("v" stands for visual):

v11, v21, v22, v32, v33, v43, v44, v54, v55, v65, v66, v76, v77, v87, v88, v98, v99, v109, v1010. The first number stands for the number of stimuli, the second number for the number of keys used in this special task. For example v11 means one stimulus, one key. v21 means two stimuli, one key, one unresponded stimulus. v43 means four stimuli, three keys, one unresponded stimulus.

Apparatus

The PC used for these tasks had a Pentium processor. The system timer was not used because of its discontinuous time measuring. Instead, loops were used to measure time with ms accuracy. The number of loops which are performed during a distinct period has been gauged by help of the system time. All tasks have been programmed in Basic.

Procedure

The testing took place in an unlighted, dimmed, sound-attenuated room. The subject sat in a comfortable chair in front of the PC, both arms resting on the desk. The computer program was started and the subjects *read the instruction on the screen*, telling them to start the task by pressing any key. Then they got the instruction which stimulus should be responded to by which key. Before the visual tasks started, the subjcts were asked to fix their eyes on the central arrow. Responded stimuli vanished after pressing the key. All mistakes were recorded by the program. Reaction times below 100 ms were considered as premature and omitted (Ito 1997). The time of the experiments was mostly in the afternoon, the subjects had their last meal some hours before testing. The reaction times are influenced by many factors like season, daytime, last meal and so on. Therefore each subject has its own reaction times. Moreover the reaction times of each subject vary in every trial. For that reason one has to calculate mean or median values from a set of trials.

Design

The experimental session consisted of 19 visual tasks of different complexity with 100 trials per task. The program computed the mean reaction time, the standard deviation, the mean interstimulus time, the number of mistakes, the stimulus frequencies and the mean reaction times for each key. The results were printed and saved at the hard disk for further computing.

Data analysis

First one had to assign the number of decisions for each kind of task: in this paper the number of decisions is the number of occurring stimuli + the number of occurring keys. For the task v11 the number of decisions has been assumed as 0 because no decisions have to be made if there is only one stimulus to respond to. The tasks have the following number of decisions:

v11 (0 decisions), v21 (3 decisions), v22 (4 decisions) , v32 (5 decisions), v33 (6 decisions), v43 (7 decisions), v44 (8 decisions), v54 (9 decisions), v55 (10 decisions), v65 (11 decisions), v66 (12 decisions), v76 (13 decisions), v77 (14 decisions), v87 (15 decisions), v88 (16 decisions), v98 (17 decisions), v99 (18 decisions), v10 9 (19 decisions), v10 10 (20 decisions).

Comparison of the computer program with a standard device

One subject compared his results with a commercial device for measuring reaction time. The results were nearly the same.

Results

Illustration of linear relationship between the number of basic elements (stimuli and responses) and the mean reaction time

There are some measures of special importance. The *slope* of the regression line, the *intercept* of the regression line (intersection of regression line with the y-axis), the end of serial processing *ESP* (defined by the first task > v55 with falling median reaction time in the next task).

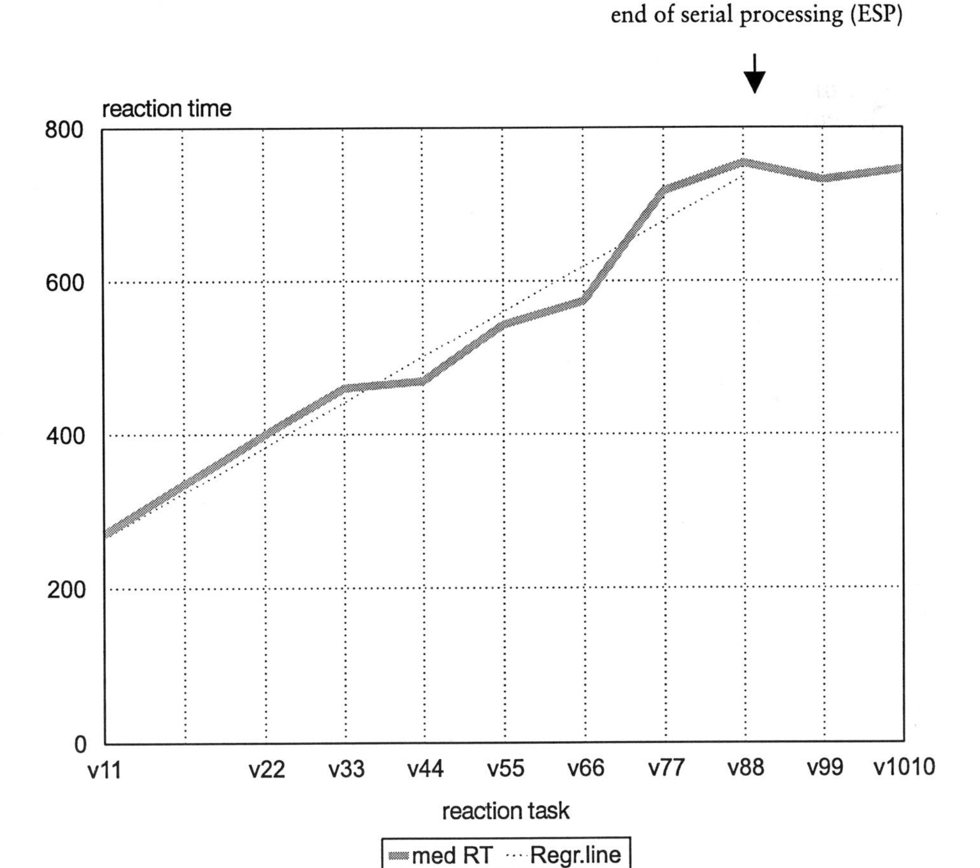

Fig. 2. Bihemispheric visual median reaction times of the tasks v11 to v1010 of subject P15 with regression line (broken line). The task v11 has a non linear position relative to the other tasks

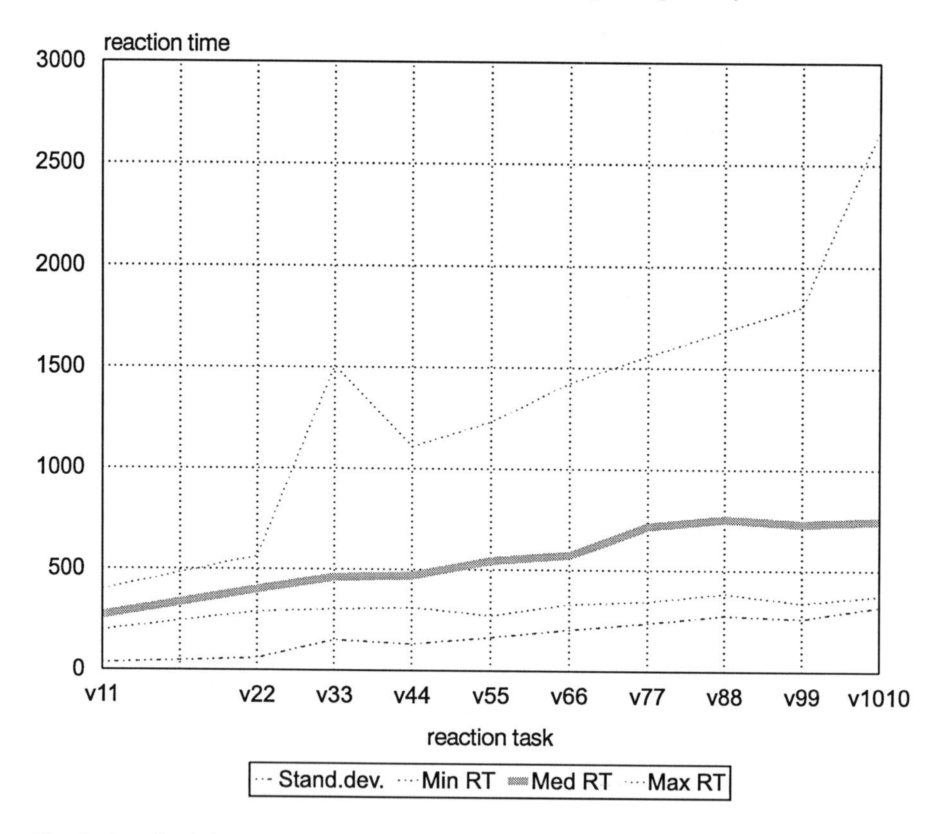

Fig. 3. Standard deviation, minimal reaction times, median reaction times, maximal reaction times (from below) of the tasks v11 to v1010 of subject P15

Fundamental data

The next sections of this chapter only use the bihemispheric visual tasks v11, v22, v33, v44, v55, v66, v77, v88, v99, and v1010. The slope and the intercept of this series of tasks are given below for each participating subject.

Table 1. Basic data, slope, and intercept from the bihemispheric visual tasks. Subjects who have repeated the tasks get a small letter behind their initials

Sub	Sex	Age	Hand	Slope=vCT	Intercept
P01	fem	24	R	19.8	230.8
P02a	male	26	R	24.9	236.4
P02b				14.7	248.0
P03	fem			20.8	248.9
P04	fem	45	R	39.9	243.0
P05	male	25	R	28.3	363.4
P06	male			32.7	235.7
P07	male	30	R	24.2	240.7
P07a				24.5	218.0

Table 1. Continued

Sub	Sex	Age	Hand	Slope=vCT	Intercept
P08	male			56.1	237.7
P09	fem	25	R	21.9	257.2
P10	male			33.6	194.7
P11	fem	27	L	31.7	258.5
P12	male	48	R	22.7	217.2
P13	male	24	R	27.3	235.1
P14	male			24.4	208.7
P14a				19.7	193.8
P15	male	25	R	29.5	264.8
P15a				22.3	244.5
P15 b				18.8	251.5

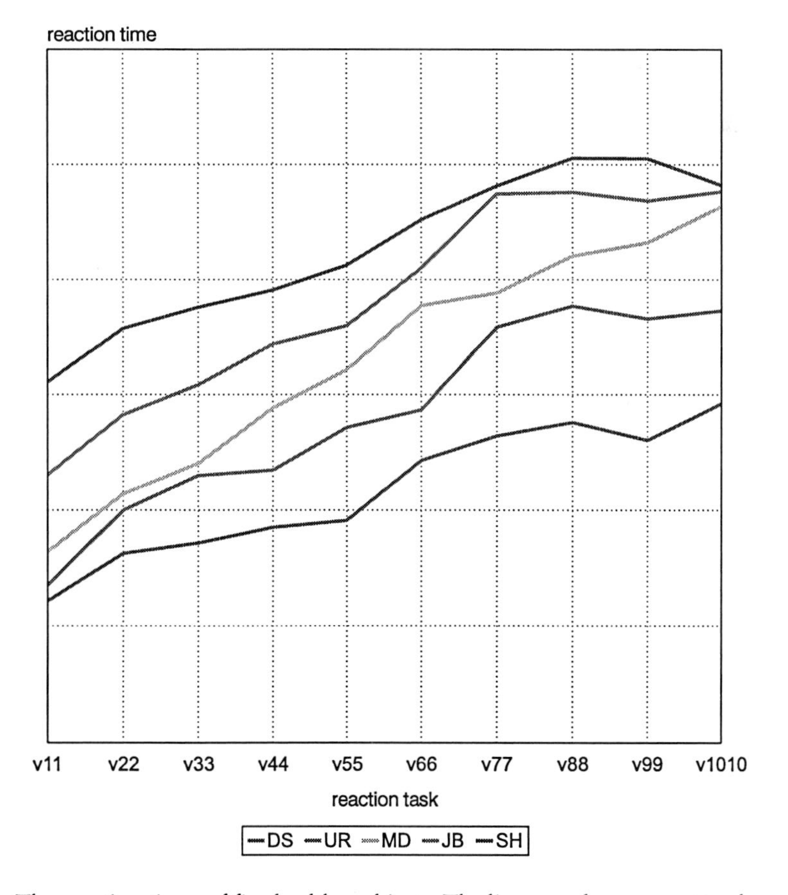

Fig. 4. The reaction times of five healthy subjects. The lines are drawn apart to show each single line. The v11 is set on equal distance to the other tasks. One sees the relative large increase from v11 to v22, the linear section between v22 and ESP and the horizontal section beyond ESP

The time differences between the tasks

Table 2. The differences between the tasks for each subject. Subjects who have repeated the tasks get a small letter behind their initials

sub	v22-v11	v33-v22	v44-v33	v55-v44	v66-v55	v77-v66	v88-v77	v99-v88	v1010-v99
P01	82	18	27	12	104	42	24	–31	63
P02a	90	35	69	14	71	62	62	–15	32
P02b	64	31	23	38	33	16	30	–2	56
P03	99	48	–20	46	70	74	20	–14	21
P04	122	76	118	74	–19	219	–33	41	74
P05	84	71	16	90	37	110	28	–20	–43
P06	104	52	71	32	100	129	3	–15	16
P07	59	41	67	33	61	79	21	–33	30
P07a	77	44	37	58	43	101	–33	38	77
P08	188	47	92	205	123	141	31	–82	–44
P09	102	60	62	19	46	49	33	19	–12
P10	87	66	43	80	109	81	–94	49	76
P11	98	182	–2	96	45	82	25	14	–35
P12	111	17	34	60	69	24	66	18	–3
P13	99	28	23	66	100	81	–1	39	20
P14	93	37	30	43	79	59	48	–1	–46
P14a	61	34	30	39	28	44	62	72	–50
P15	129	61	9	74	30	144	37	–22	14
P15a	94	77	–7	63	26	77	34	43	–25
P15 b	72	65	18	29	77	38	1	36	37

The quotient (v(N+1)(N+1) – vNN)/dvCT

Subsequently the slope is called vCT.

The dvCT is the "directly observed cycle time" of a subject. It is observed in the distributions of monohemispheric visual reaction tasks.

After gathering empirical evidence for the number of cycles between v11 and v22 one can ask for the number of cycles between v22 and v33, v33 and v44 and so on. The theory says that in untrained subjects these differences are always two cycles and in trained subjects one cycle.

To compute the empirical number of cycles one first needs the empirical cycle time (directly observed cycle time). The measuring of this period is described in a later chapter. The empirical observations of cycle numbers in the tasks v22, v33 etc. are therefore delayed to this chapter. Table 4, with the directly observed cycle time as the divisor, shows that the value of (v22 – v11)/dvCT lies between 2.0 and 6.5.

The maximal difference is obtained if v11 needs a minimal number of cycles (0 cycles) and v22 needs a maximal number of cycles (4.5 cycles and more). The minimal difference demands a v11 with a maximal number of cycles (eg 2.5 cycles) and a v22 with a minimal number of cycles (eg 3.5 cycles). Medium values result

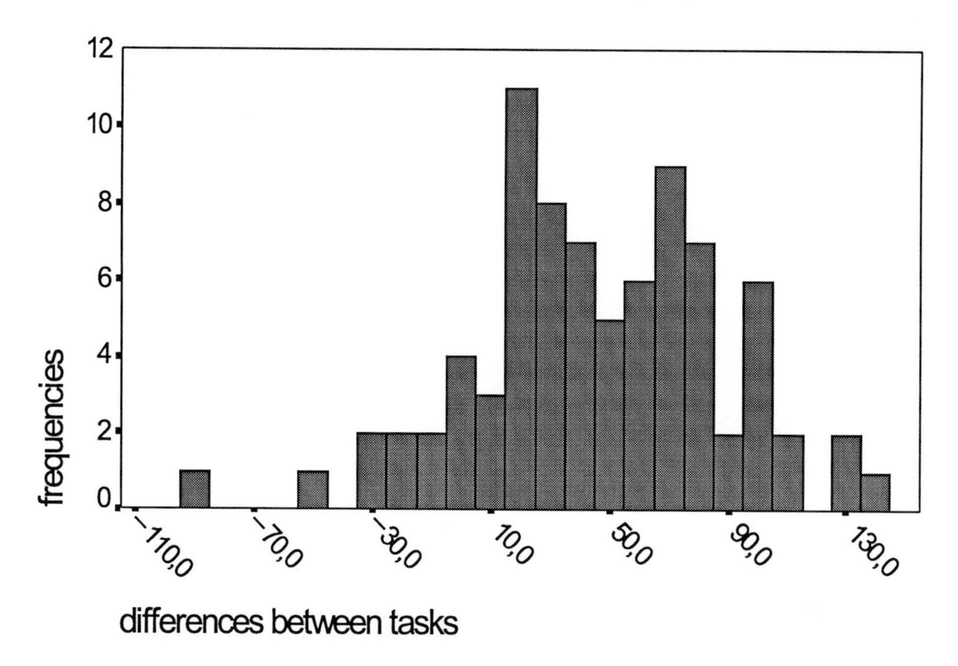

Fig. 5. The differences between the reaction times of two tasks (see Table 2) prefer certain values

Table 3. The reaction time differences between the tasks for each subject, divided by the individual slope. Subjects who have repeated the tasks get a small letter behind their initials

	vCT	(v22-v11) /vCT	(v33-v22) /vCT	(v44-v33) /vCT	(v55-v44) /vCT	(v66-v55) /vCT	(v77-v66) /vCT	(v88-v77) /vCT
P01	19.8	4.1	0.9	1.4	0.6	5.3	2.1	1.2
P02a	24.9	3.6	1.4	2.8	0.6	2.9	2.5	−0.6
P02b	14.7	4.3	2.1	1.6	2.6	2.2	1.1	2.0
P03	20.8	4.8	2.3	−1.0	2.2	3.4	3.6	1.0
P04	39.9	3.1	1.9	3.0	1.9	−0.5	5.5	−0.8
P05	28.3	3	2.5	0.6	3.2	1.3	3.9	1.0
P06	32.7	3.2	1.6	2.2	1.0	3.1	3.4	0.1
P07	24.2	2.4	1.7	2.8	1.4	2.5	3.3	0.9
P07a	24.5	3.1	1.8	1.5	2.4	1.8	4.1	−1.3
P08	56.1	3.4	0.8	1.6	3.7	2.2	2.5	0.6
P09	21.9	4.7	2.7	2.8	0.9	2.1	2.2	1.5
P10	33.6	2.6	2.0	1.3	2.4	3.2	2.4	−2.8
P11	31.7	3.1	5.7	−0.1	3.0	1.4	2.6	0.8
P12	22.7	4.9	0.7	1.5	2.6	3.0	1.12	2.9
P13	27.3	3.6	1.0	0.8	2.4	3.7	3.0	0.0
P14	24.4	3.8	1.5	1.2	1.8	3.2	2.4	2.0
P14a	19.7	3.1	1.7	1.5	2.0	1.4	2.2	3.1
P15	29.5	4.4	2.1	0.3	2.5	1.0	4.9	1.3
P15a	22.3	4.2	3.5	−0.3	2.8	1.2	3.5	1.5
P15 b	18.8	3.8	3.5	1.0	1.5	4.1	2.0	0.1

Table 4. Comparison of the directly observed cycle times in the various tasks in 15 healthy subjects. A small letter behind the initials means repetition of the tasks. vCTl for example means the "directly observed *v*isual *c*ycle *t*ime *left-hand*ed task"

	vCT	dvCTl	dvCTr	dvCT	(v22-v11)/ dvCT	(v33-v22)/ dvCT	(v44-v33)/ dvCT	(v55-v44)/ dvCT
P01	19.8	33.5	27	30.3	2.7	0.6	0.9	0.4
P02a	24.9	not	made					
P02b	14.7	25	40	32.5	2.0	1.00	0.7	1.2
P03	20.8	28.5	31	29.8	3.3			
P04	39.9	27.5	32.5	30	4.1	2.5	3.9	2.5
P05	28.3	32	25	28.5	2.9	2.5	0.6	3.2
P06	32.7	32	32.5	32.3	3.2	1.6	2.2	1
P07	24.2	34.5	27	25.8	2.3	1.6	2.6	1.3
P07a	24.5	26	28	27	2.9	1.6	1.4	2.1
P08	56.1	31.5	26	28.8	6.5	1.6	3.2	7.1
P09	21.9	30.5	31	30.8	3.3	1.9	2.0	0.6
P10	33.6	27	33.5	30.3	2.9	2.2	1.4	2.6
P11	31.7	28	30	29	3.4	6.3	-0.1	3.3
P12	22.7	32.5	27	29.8	3.7	0.6	1.1	2
P13	27.3	28	31.5	29.8	3.3	0.9	0.8	2.2
P14	24.4	23	31	27	3.4	1.4	1.1	1.6
P14a	19.7	28	25	26	2.3	1.3	1.2	1.5
P15	29.5	23	28.5	25.8	5	2.4	0.3	2.9
P15a	22.3	27.5	30.5	29	3.2	2.7	0	2.2
P15 b	18.8	33	26.5	29.8	2.4	2.2	0.6	1

from a v11 with a maximal number of cycles (eg 2.5 cycles) and a v22 with a maximal number of cycles (eg 4.5 cycles) or a v11 with a minimal number of cycles (0 cycles) and a v22 with a minimal number of cycles (eg 3.5 cycles).

See the real values for the 15 subjects in the column (v22–v11)/dvCT in Table 4.

Surprisingly the most frequent value is 3.5 with a v11 of 0 cycles and a v22 of 3.5 cycles.

There are some subjects (eg P12) who use the fast mode in simpler tasks, which have fewer stimuli and keys to be connected but use the slow mode in more complex tasks because of their abundance of stimuli and keys and therefore many more connections. The subjects which have repeated the bihemispheric visual tasks (small letter after the initials in the above table) show a reduction of cycle numbers in the repetition.

Beyond the end of sequential processing (ESP): the begin of parallel processing

The existence of an upper limit for the reaction times is a unique feature of the bihemispheric visual reaction tasks. In complex tasks where the searching set is too large, the brain changes it's strategy to solve these tasks and save time. Without this new strategy the fast speaking and understanding of human language would not be possible.

In some subjects the parallel processing becomes visible in the tasks v88, v99 and v1010. In these subjects the median reaction times of more complex tasks do not increase any more. By using the model of the searching set, one can say that the first searching set is full and an other searching set has to be opened which is scanned in parallel by a searching mechanism.

Bihemispheric visual intermediate reaction tasks

If the linear correlation between the number of task elements (stimuli plus responses) and the reaction time of this task is true, then intermediate tasks with new combinations of stimuli and responses must follow this correlation, too.

A good test of the linear correlation theory is, if tasks with the same sum of stimuli and responses have the same reaction times. So the meanRT(v32) = meanRT(v41) or meanRT(v33) = meanRT(v42) = meanRT(v51) and so on.

All the possible combinations are shown below. The sum of stimuli and responses in each column is the same. The 0 represents the 10.

v11 v21 v22 v32 v33 v43 v44 v54 v55 v65 v66 v67 v77 v87 v88 v98 v99 v09 v00
v31 v41 v42 v52 v53 v63 v64 v74 v75 v76 v86 v96 v97 v07 v08
v51 v61 v62 v72 v73 v83 v84 v85 v95 v05 v06
v71 v81 v82 v92 v93 v94 v04
v91 v01 v02 v03

This similarity of reaction times in tasks with equal numbers of task elements but different composition of stimuli and response has to be shown.

Table 5. The mean values from 22 visual tasks for 3 subjects

Subject 1(P12)						Subject 2 (P16)				
Task	min. RT (ms)	meanRT (ms)	St.-Dev. (ms)	meanIS (ms)	false	min. RT (ms)	meanRT (ms)	St.-Dev. (ms)	meanIS (ms)	false
V11	182	212	27.4	3189	0	173	220	23.1	2636	0
V21	222	272	37.3	2538	0	207	266	48.5	2859	4
V22	230	293	35.1	2435	7	212	293	59.7	2605	8
V32	243	311	40.2	2568	6	224	309	59.1	2886	2
V33	255	320	40.0	2406	5	232	340	74.8	2696	2
V43	249	324	49.2	2560	2	244	367	86.1	2701	1
V44	255	360	58.3	2436	6	238	399	84.6	2706	7
V54	276	394	85.9	2497	6	230	408	83.7	2777	5
V55	277	406	67.2	2391	4	260	420	84.8	2736	2
V65	314	465	93.5	2691	4	316	478	93.5	2761	4
V66	307	463	84.1	2434	8	323	491	98.5	2764	3
V76	326	481	84.7	2667	6	264	519	122.9	2864	5
V77	331	511	126.6	2430	6	348	587	128.1	2716	7
V87	310	512	113.1	2652	6	277	568	150.4	2680	3
V88	312	546	133.5	2423	5	319	614	191.1	2748	2
V98	338	552	98.0	2660	4	256	626	192.9	2748	2
V99	343	596	127.4	2514	9	318	675	202.6	2655	2
v109	338	618	130.3	2609	4	322	657	218.9	2671	4
v1010	427	614	122.1	2513	5	336	666	218.2	2646	4

The results of some intermediate visual tasks are shown in Table 5 for some subjects.

Table 5. Continued

Subject 3 (P17)

Task	min. RT (ms)	meanRT (ms)	St.-Dev. (ms)	meanIS (ms)	false
V11	195	251	35.5	2691	0
V21	218	297	46.1	2828	2
V22	261	346	49.1	2847	2
V32	284	384	78.1	2782	0
V33	262	379	69.6	2829	0
V43	253	439	111.9	2812	5
V44	295	451	108.7	2763	4
V54	316	452	61.3	2797	3
V55	345	535	140.1	2694	2
V65	356	560	110.9	2844	3
V66	357	564	147.7	2709	3
V76	381	601	147.2	2786	6
V77	373	682	164.6	2790	3
V87	373	661	168.3	2808	2
V88	377	659	162.3	2771	3
V98	398	711	195.3	2721	3
V99	459	678	151.0	2785	5
v109	431	707	197.3	2731	0
v1010	463	696	209.3	2776	2

Now the mean reaction times for each number of task elements (=number of decisions) is displayed graphically for three participating subjects:

One can see that the reaction time for every additional decision increases by about the same amount of time. This difference time which is needed for each additional decision is the *individual cycle time* of each subject. As one can see the cycle time for the task v11 has to be v11=0. Because of this regularity one can compute the decision time for every task if the number of decisions and the personal cycle time is known.

In turn, one can compute the individual cycle time by using every two tasks of the same modality and divide it through the difference of the number of decisions of both tasks (in v11 the sum of decisions is zero):

$$\text{cycle time 1} = \frac{\text{reaction time(vMN)} - \text{reaction time(vKL)}}{(M+N) - (K+L)}$$

where M and K are the numbers of stimuli and N and L are the numbers of responses for the tasks.

For example:

$$\text{cycle time} = \frac{\text{reaction time(v22)} - \text{reaction time(v11)}}{(2+2)-(0+0)}$$

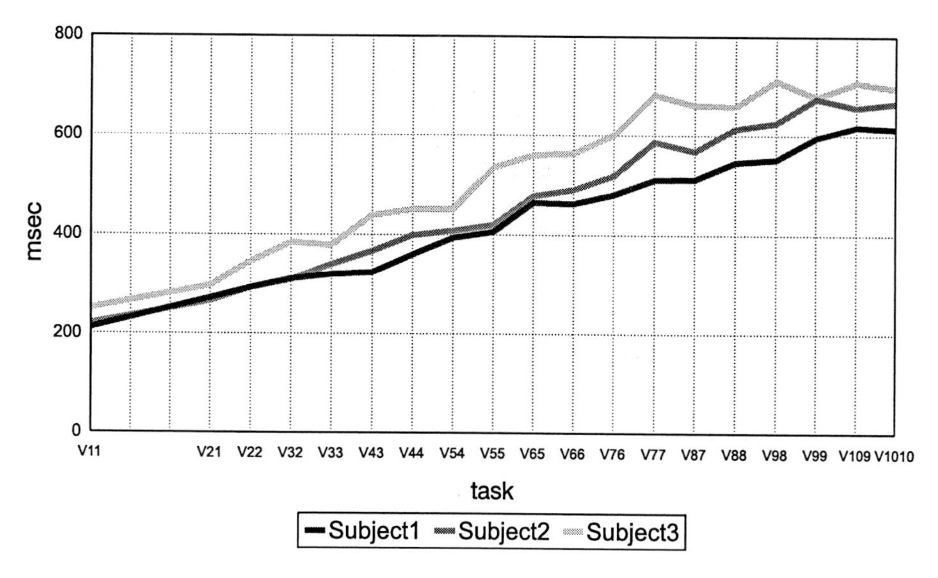

Fig. 6. Mean reaction times for subjects1–3 (P12, P16, P17) in relation to the number of decisions (=number of task elements)

The *decision time* is the product of the number of cycles needed to decide a task and the cycle time, for example:

decision time(v22)= cycle time * (2+2)

If one has computed the cycle time of a subject from two reaction times, one can compute the reaction times of all other tasks for this subject by the above formula.

Another method to compute the cycle time from all the known mean reaction times is given by the following formula:

$$\text{cycle time 2} = \frac{(RT1 + RT2 + \ldots RTn) - n * v11}{(stim1 + key1) + (stim2 + key2) + \ldots + (stimn + keyn)}$$

Having determined the cycle time of a subject, one can compute each reaction time of this subject by

RT(vMN)=v11+(M+N)*CT

or

RT(vMN)=RT(vKL)+(M+N)–(K+L))*CT

There is still a *third method* to compute the cycle time from the measured reaction times:

cycle time 3 = ((v22–v11)+(v33–v22)+(v44–v33)+...+(v99–v88))/18 v1010 is not used because there is no linear correlation between the number of alternatives and the reaction time in this range. 18 is the sum of steps between two tasks, with four steps between v22 and v11 and two steps between all other tasks.

It is an open question whether there exists a more convenient method to compute the gradient of the line approaching the reaction times of the different tasks. It is better not to compute the cycle time from only two measured values because of insufficient measurements. The more measured values one uses the more precise is the computed value.

Minimal reaction times (also linear growing with number of alternatives)

For each subject the minimal reaction time which was greater than 100 ms for every task was taken and drawn into a coordinate system to show its course. As one can see, the minimal reaction time is in a linear relation to the number of alternatives until some point where it flattens (see next figure).

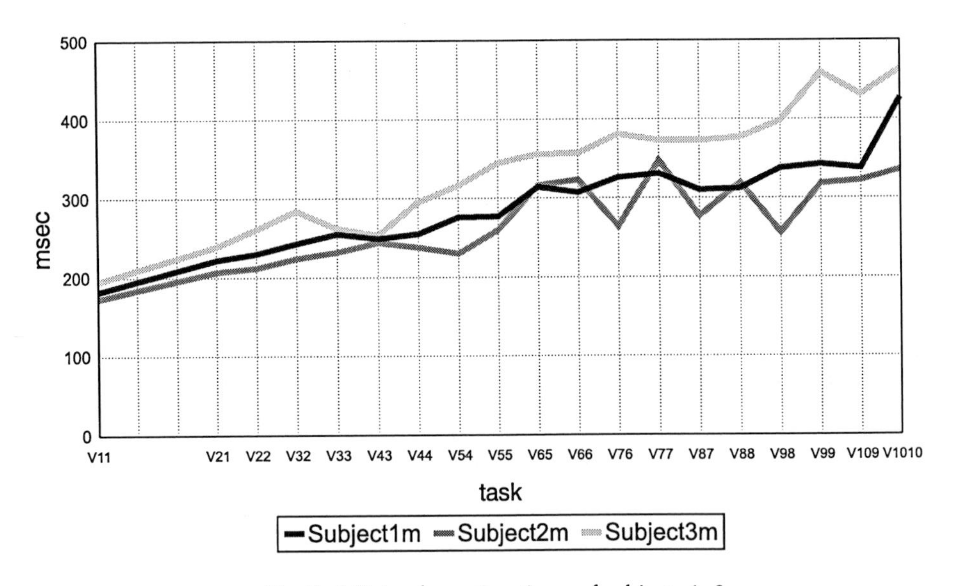

Fig. 7. Minimal reaction times of subjects 1–3

Direct observation of the cycle time and comparison with the computed cycle time values

The 70-Hz monitor with a formation of a picture each 14 ms does prevent a high time resolution. Therefore the direct visualisation of visual tasks demands the use of light bulbs, LEDs or a laptop plasma display. Therefore a TOSHIBA laptop (T3100 or T4400) was used which has a plasma display. TOSHIBA (Technical Support Hotline) asserted that this display builds up a picture faster than 14ms.

Discussion

Discussion of the method

The succession of tasks can be reversed without fundamentally changing the values of the mean reaction times. That means learning is not responsible for the slower growth of the curve at v99 and v1010.

Comparison of the linear with the logarithmic relation

How great are the differences for the above formulas and how great is the difference for a logarithmic formula used by Hick?

What does this mean that the (mean) decision process needs as much decision times as there are stimuli and responses to select from? The only answer until now can be that the decision mechanism needs as much processing steps as there are stimuli and responses. This does not mean, that the decision process looks at each of the stimuli and responses one after the other. Because we know that there is a wide range of possible reaction times within one task. The above is only valid for the mean reaction times.

It is well known for a long time, that more complex stimuli need more time to respond to than simple stimuli.The difference times between the simple and complex tasks were known too. There were many explanations and mathematical theories which relate the reaction time to the number of alternative stimuli-response pairs.

New is the idea that the reaction times are related to both: the number of stimuli and the number of respones as one can show when these two numbers are different.

The second idea is not new but under contention: is there linearity of mean reaction time relative to the sum of all stimuli and all responses or not? To answer this question the tasks with unequal stimulus and response numbers should fill the gaps in the functions which prevent the right answer to this old question. The more complete functions look like being composed of three parts: v11 with no decision at all, v21 to v88 (etc.) with a linear ascent and the tasks more complex than v88 with a flat course.

The present findings suggest that the brain needs the same time for one decision, be it sensory or motor and adds these cycle times over a wide range. This addition suggests a serial processing within this range. After this range, the brain operates in parallel similar to letter tasks where it decides in parallel, too.

Comparison with reaction times given by other authors

Falkenstein et al. (1993) have given more than two values of reaction times, therefore the cycle time can be computed from two of them and tested by help of the third:

- visual cycle time vCT = (v22–v11)/4 = (323ms–217ms)/4 = *26.5ms*
  v44 = v11+8*vCT = 217ms+212ms = 429ms. The measured value is 420 ms.

Table 6. Comparison with some reaction times given by other authors

Author	v11	v21	v22	v33	v44	v55	v66	v77	v88	v99	v00
Falkenstein et al. (1993)	217		323		420						
Matsuoka et al. (1996)	253	335									
Merkel (1885) in Woodworth	187		316	364	434	487	532	570	603	619	622
Damon(1966) in Kvalseth				400					600		
Eichert et al. (1996)		261									
Falkenstein et al. (1993)	187		328		435						

- auditory cycle time aCT = (a22–a11)/4 = 328ms–187ms/4 = *35ms*
  a44 = a11+8*aCT = 187ms+280ms = 467ms. The measured value is 435 ms.

- The visual cycle time of Matsuoka et al. (1996) is:
  vCT = (v21–v11)/3 = (335ms–253ms)/3 = *27.3ms.*

This value is comparable with the vCT of Falkenstein et al. It still has to be explained why the aCT differs from the vCT. *These calculations shall not prove the correctness of the favorite theory. They are presented because they have stimulated the investigations of Part II – IV.*

Some theoretical considerations

At the next pages a hypotheticel neural representation of reaction tasks is proposed.
The end of sequential processing (ESP) and the beginning of parallel processing only emerge when the task set is greater than a definite size. Below this size *one* searching set is possible. Beyond this size no unique integrating searching set exists so that *many* searchings sets have to work in parallel to do the job. The searching sets compete with each other for the task set.

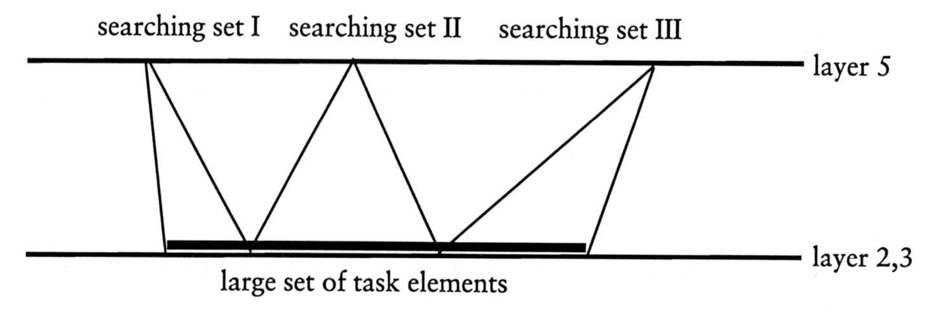

Fig. 8. Large sets of task elements (= task set) need more than one searching set

The task set determines its searching set by converging onto one or more elements of layer5. Subsets in layer5 are omitted due to competition between the layer5 elements of the searching sets. How does the searching set determine the cycle time? The representation of the searching set in layer5 (= top element) doesn't contain only one neuron but many. If this number of neurons is reduced in layer5, the cycle time between the two layers 2,3 and 5 is reduced, too. The same is true if the number of connections between the layer5 and the layer2,3 is reduced. Then the activation of one layer by the other costs more time.

We have, therefore, two parallel effects of reducing the layer5 neurons: the cycle time between the two layers is prolonged and the searching set does not fit to the task elements as exactly as usually. This implies inappropriate elements in the searching set.

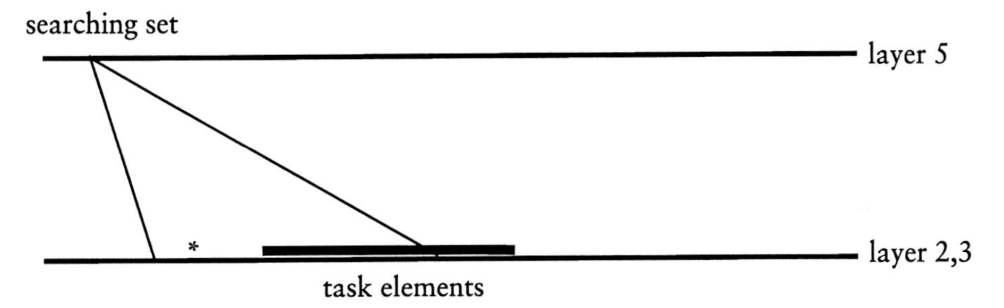

Fig. 9. The searching set does not fit to the task elements

Such a situation has many consequences: the search covers inappropriate elements and omits appropriate elements.The slowing down of searching sets is only the beginning. If the loss of layer5 cells continues, the emergence of inappropriate searching sets becomes possible.

Minimal visual reaction time

If one excludes the possibility of guess or accidentally pressing the key, but takes these short times as real reaction times then they would give us the minimal possible pathways of this task. Therefore the minimal visual reaction time would be:

reaction time with *n* cycles in a subject with a cycle time CT = 20 ms and an elementary time ET = CT/2 = 10ms:
= (Input time + linear time1 + n*cycle time + linear time2 + corticospinal time + motor execution time)
= *50 ms* + 20 ms + n*20 ms + 10 ms + *20 ms* + *50 ms* = 170 ms (Minimum with n=1)

The underlined times can be summed up as the non-cortical time. This would mean a minimal visual reaction time of 170 ms. If there are lower times, the pathway either has some shortcuts or it does not use the cortex at all but passes only *subcortical areas.*

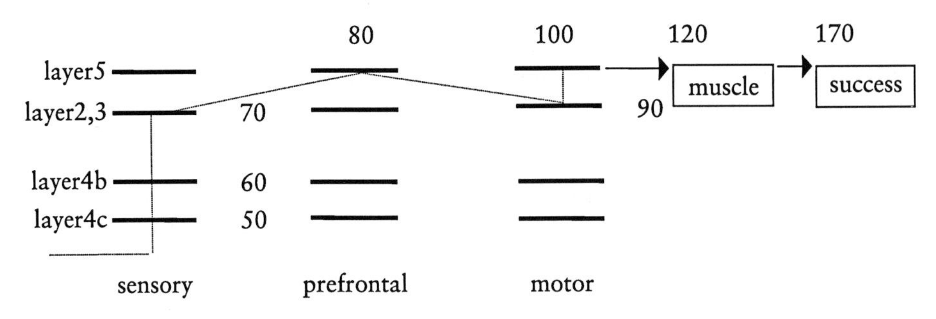

Fig. 10. Hypothetical minimal visual pathway with an elementary time of 10ms. The layers of the cerebral cortex are designated by their histological names. At the right side are the times in milliseconds

Brain imaging of reaction task pathways

Kotrla (1997) describes inYudofsky and Hales (eds.) the areas which are involved in simple visual and auditory tasks. In fMRI the right motor cortex is activated primarily during left-hand movements; the left motor cortex is somewhat activated even during ipsilateral finger movements (Kim et al 1993, Rao et al. 1993). During sustained visual attention, fMRI localizes signal changes in primary and secondary visual cortex, the dorsolateral prefrontal, inferior parietal, and anterior cingulate cortices (Simpson et al. 1995). If participants must discriminate between auditory stimuli, PET reveals metabolic changes in the frontal, parietal, and cingulate cortices (Cohen et al. 1988).

A first mathematical theory of bihemispheric visual reaction tasks

Fundamentals

If one identifies one cycle time with the slope of the linear regression than one may ask: why is the additonal average number of cycle times, necessary to decide the task v(N+1)(N+1), approximately the difference v(N+1)(N+1) – vNN? The task v22 needs 4 cycle times on average more than the task v11. The task v33 needs 6 cycle times on average more than v11 etc. What do these numbers tell us? Is there any theory to explain these numbers?

The mechanism which (unconsciously) solves these tasks uses the following strategies in untrained subjects:

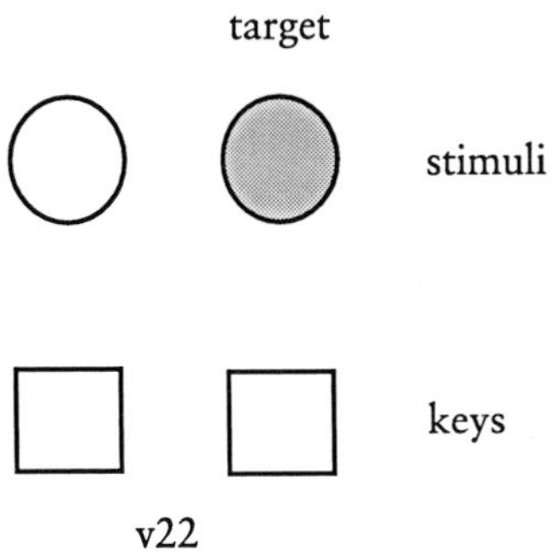

target

stimuli

keys

v22

first searching in 4 elements:	4 cycles
then searching in 2 elements:	2 cycles
sum:	6 cycles

1. A scanner is looking for the target stimulus by chance. This procedure is similar to find a black ball within a box with otherwise white balls. If n is the number of all balls in the box, the mean number of searching acts (grips into the box) is n if the balls are laid back (re-selection possible).
2. If the target stimulus has been found, the scanner is also looking for the target key by chance. The target stimulus and the target key form an own subset, the so called sensorimotor set. It has two elements. Therefore the accidental search on this set for the target response takes 2 cycle times on average.

The two procedures are performed sequentially, therefore the sum of searching acts is n + 2. In the case of v22 the sum is 4+2=6.

If the number of alternatives (stimuli+keys) equals four, than 4+2 cycles are needed on average.

If the number of alternatives (stimuli+keys) equals six, than 6+2 cycles are necessary on average to find the target stimulus and the target key.

target

v33

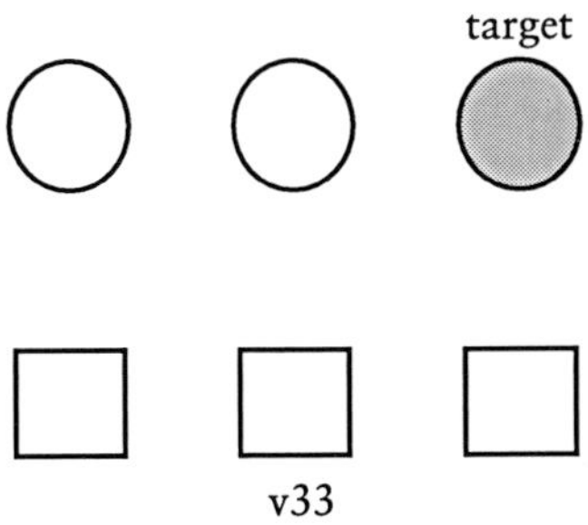

first searching in 6 elements:	6 cycles
then searching in 2 elements:	2 cycles
sum:	8 cycles

That means the difference between two tasks is always 2 cycles.

Difficult is the structure of v11. Using the above ideas one would get the following situation:

First search on two elements, cycles = 2 (search for the target stimulus)
Second search on two elements, cycles = 2 (search for the target key)
Sum, cycles = 4

But there is still another possibility. There could be no search at all because there are fixed connections between the elements of this task which direct the way (see implicit learning). Which of these two possibilities, the maximal, the minimal or some intermediate solutions have been chosen must be proven by the empirical cycle numbers.

The states lying between the extremes could also exist, so we can meet a variety of cycle numbers from the maximum ones over intermediate states to the minimum ones.

Implicit learning may decrease the number of cycles

There is empirical evidence that v22 needs 4 cycle times more than v11. But the search for the target stimulus and then for the target response would take 6 cycle times on average. Is there the possibility of a faster mechanism?

The number of 6 cycles in the task v22 is only valid if there is no learned connection between a stimulus and its attached key. With such a connection the number of cycles is reduced:

v22 —> 4 cycles to find the target stimulus + 1 cycle to take the connected key —> sums up to 5 cycles

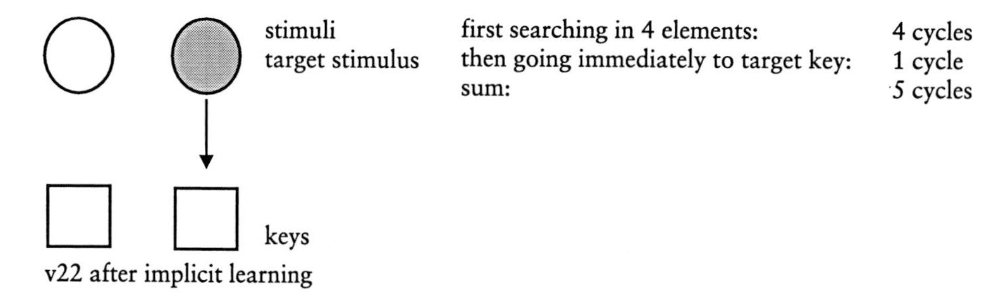

v22 after implicit learning

The connection (arrow) between stimulus and attached key allows the scanner to go directly from the found target stimulus to the target key without any new search. This saves 1 cycle per trial in v22. If the arrow alone is capable of activating the target key, the random search mechanism is not necessary.

Can the number of cycles be reduced furthermore?

There is still a difference between the empirical value of additional 4 cycle times in v22 and the theoretical 5 cycle times in the above calculations. This may be due to some parallel processes which save another cycle time (see later chapters).

The bihemispheric visual median finger reaction tasks

Summary

There is the astonishing effect that each finger alone performing the task v11 has nearly the same reaction time but many fingers together (mostly from v55 to v1010) show a different median finger reaction time for each finger. There are two questions to be answered: is the cycle time or the cycle number responsible for the different reaction times and, secondly, what could be the cause of this differentiation? This chapter shows that the cycle times of the different fingers of the left hand are rather constant and the mean cycle number is the cause of the increased reaction time of the finger LMF in the subject P12.

The second question for the cause of this differentiation may be answered by the many arguments against lateral inhibition as the cause of different median finger reaction times. Rather, differentiating influences (from layer 5) are supposed to be the cause.

Introduction

The median reaction times for each finger in tasks with more than four fingers are different: the fingers which are at the rim have lower reaction times than the fingers which are amidst others. This is not due to the finger *because each finger yields similar reaction times when tapping for example at the first key (a)*. The different reaction times result from the competition within the stimulus and the response row: when the competition is greater, the reaction time is longer and vice versa (number of indirect and direct neighbors). The neighbors in the stimuli row are slightly different from the neighbors in the response row, because the latter consists of two hands with the two thumps and some distance between them while the stimulus row is more homogene. If one takes a distinct finger and observes its mean finger reaction time for the different number of alternatives one sees that the mean finger reaction time grows linearly relative to the number of alternatives.

Method

The method is the same as in the first chapter. But in the results, the trials for each finger are computed separately. In the task v22 for example there are two fingers which are active: the left little finger, LLF, and the left ring finger, LRF. Therefore in about one half of the hundred trials of task v22 the left little finger presses the "a" key, in the other half the left ring finger has to press the "s" key. Now the median of the fifty reaction times of the LLF is computed and the same is done with the LRF. These are the median finger reaction times, medFRT, of LLF and LRF. It can be observed that the medFRT deviates from the median reaction times, medRT, of the various tasks. Moreover, it looks like the median reaction times are the mean of the medFRT.

There are two possibilities why the medFRT deviates from the medRT. First, the cycle time of each finger is different from the other. Second, the cycle time is constant but the cycle number of each finger deviates from the cycle number of the other fingers.

The cycle time of a finger is the time difference of this finger in two successive tasks greater or equal v22 divided by two (=one stimulus more and one finger more).

$$CT = (v(N+1)(N+1) - vNN)/2$$

The cycle time of a finger can also be observed directly.

The other possible cause for an increased median finger reaction time could be an increased number of cycles performed by this finger in the search process. The number of cycles necessary to find the target stimulus and the target key may depend on the pre-activation of the target elements. Target elements with low pre-activation may be found not so fast as target elements with a high pre-activation.

Results

Each finger yields similar reaction times when tapping at the same key

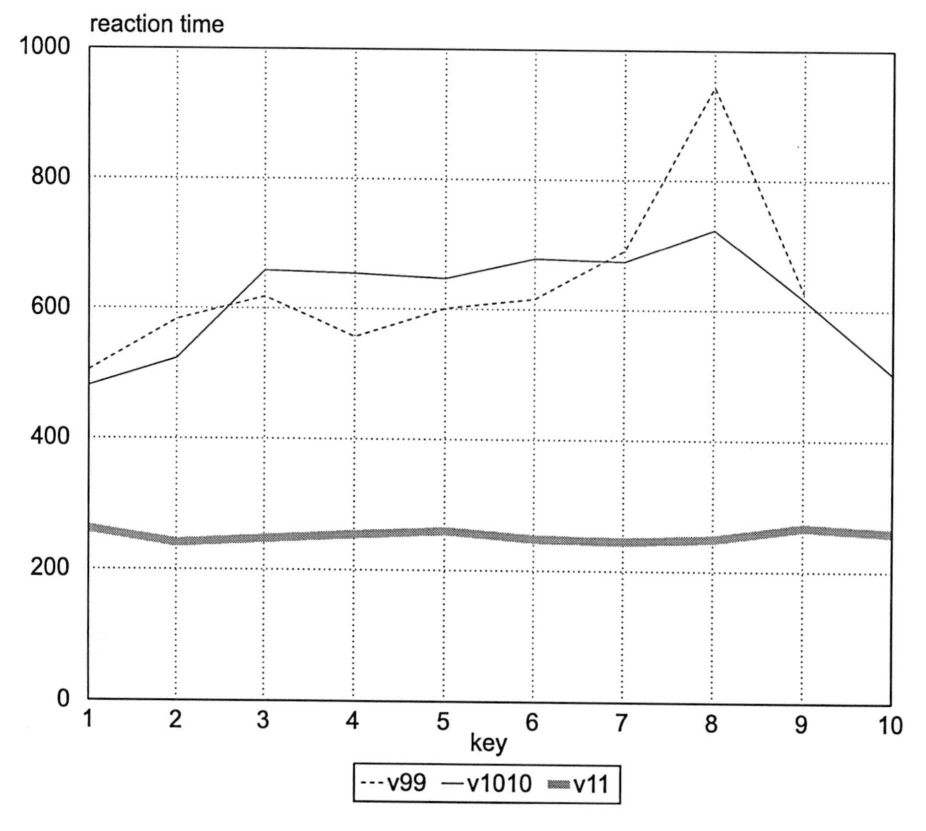

Fig. 11. Bihemisperic visual median finger reaction times in the tasks v99 and v1010 (upper two lines) and the task v11 for each single finger (line below) of subject P12

Each finger yields different reaction times within the tasks v99 or v1010

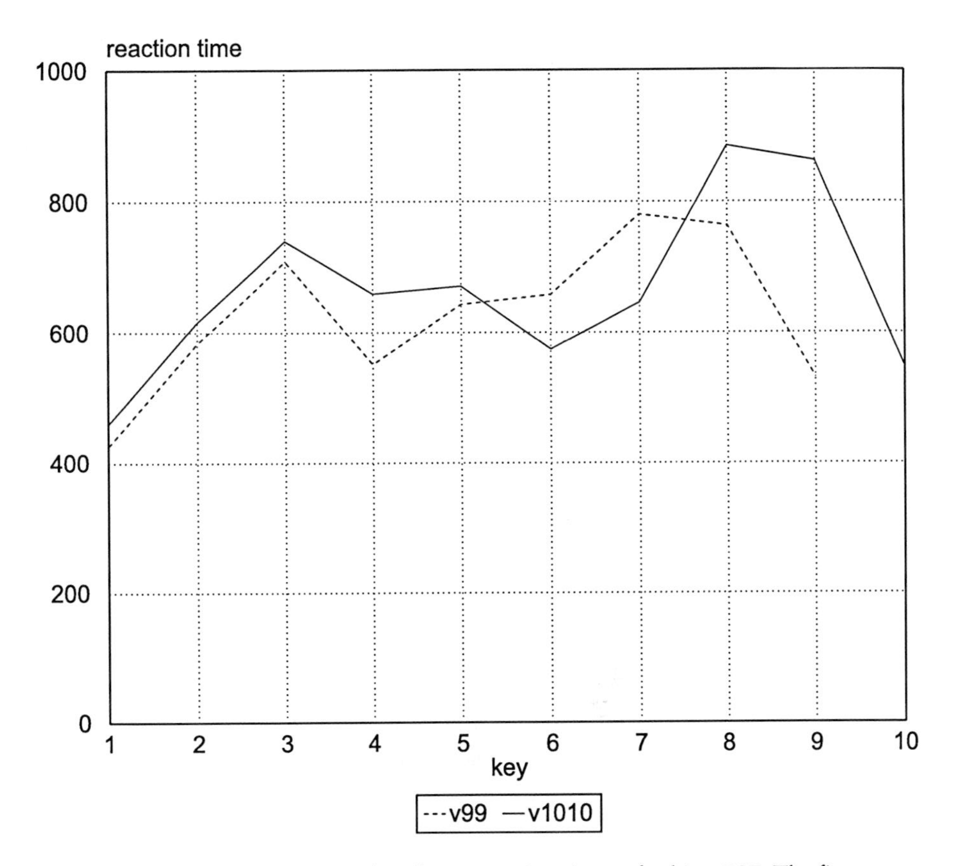

Fig. 12. Bihemispheric visual median finger reaction times of subject P07. The fingers are numbered from left to right

This appearance of median finger reaction times is seen frequently: the reaction times for the fingers 1 (=LLF) and 10 (=RLF) are smallest, rise to the middle and decline to a local minimum at the fingers 5 (=LTH) and 6 (=RTH). The differences between finger1 and finger3, for example, can be up to 400 ms or smaller.

The median finger reaction time of a specific finger has its own slope from the task v11 to the task v1010

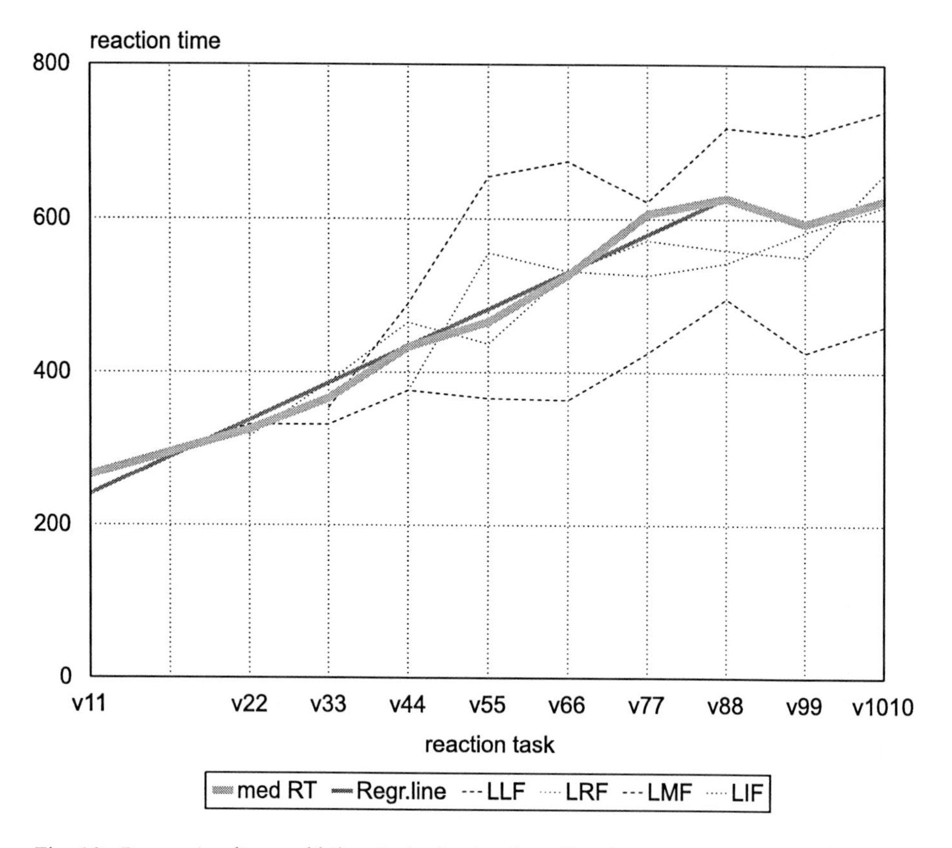

Fig. 13. Regression line and bihemispheric visual median finger reaction times of the four left fingers (LLF, LRF, LMF, LIF) of the subject P07

The greatest difference of slope is between the LLF (lowest line) and LMF (highest line). This appearance is typical for many subjects but the difference of slopes must not be as great as observed in this subject.

Discussion of the bihemispheric visual median finger reaction times

The discrepancy between the finger reaction times in the tasks v99 and v11 and the reaction time for each single finger in the tasks v11LLF etc.

This observation is remarkable that the reaction times for each single finger are nearly constant and the finger reaction times in the tasks v99 and v1010 show such differences between the fingers. The v11 findings tell us that the differences in v99 and v1010 are not qualities of the fingers themselves, but result from the fingers concurring within the searching set. Here the fingers must be influenced to produce the different finger reaction times.

One known interaction between neural cells (and assemblies) is *lateral inhibition*. This would explain why marginal elements are favored and middle elements are disadvantaged. Another possibility would be a *differentiating influence* giving the different fingers different support.

There are two questions: first, *which influence* causes the differentiation of median finger reaction times second, *does this influence change the cycle time or the the cycle number of the fingers* to produce the finger reaction times of, for example, the task v1010? To answer this question the simpler task v55 has been taken. See the next section for the results (finger LLF and LMF).

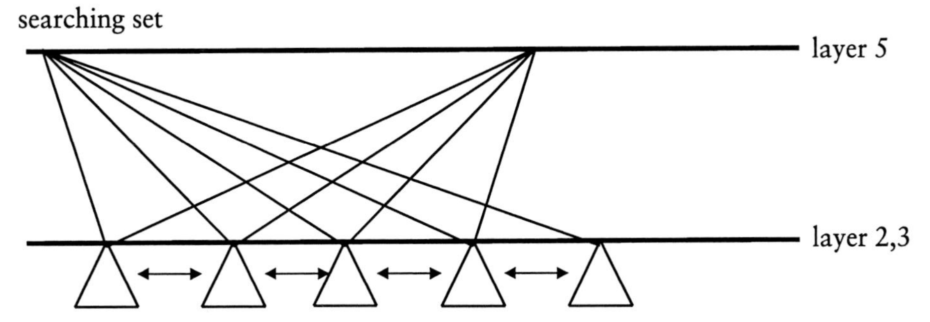

Fig. 14. The task v55 with representation of the five fingers in layer 2,3 and the searching set in layer 5. There can be other searching sets in layer 5, influencing the level of activity in the layer 2,3. These additional sets may contribute to the overall activity of the finger representations. Representations with high activity may be searched first and vice versa. The symmetrical arrows between the finger representations mean hypothetical lateral inhibition. The different acitivating elements of layer 5 may be started by a set of elements in layer 2,3. If two hands are used, for example, new elements may be recruited in layer 5 concerning bimanual behavior

The fact that the differentiation of fingers begins mostly with the task v55 (seldom with v44 or v66) contradicts the hypothesis of lateral inhibition being the cause of differentiation. These would be present in the tasks v33 and v44 too, but *in these tasks no differentiation can be detected*. Finally the lateral inhibition cannot explain why the *marginal* finger5 in the task v55 of some subjects has an *increased* median finger reaction time.

The finger reaction times in the tasks v22 to v44 are not very different from each other. *There is no evidence for lateral inhibition in these tasks*, otherwise the marginal fingers would be faster and the middle fingers slower. The first greater deviation appears in task v55 where the LTH is much slower than the other fingers. Therefore it has to be discussed *whether the different recruitation of layer 5 elements*, dependent on different sets of layer 2,3 elements, is responsible for the differentiation of median finger reaction times.

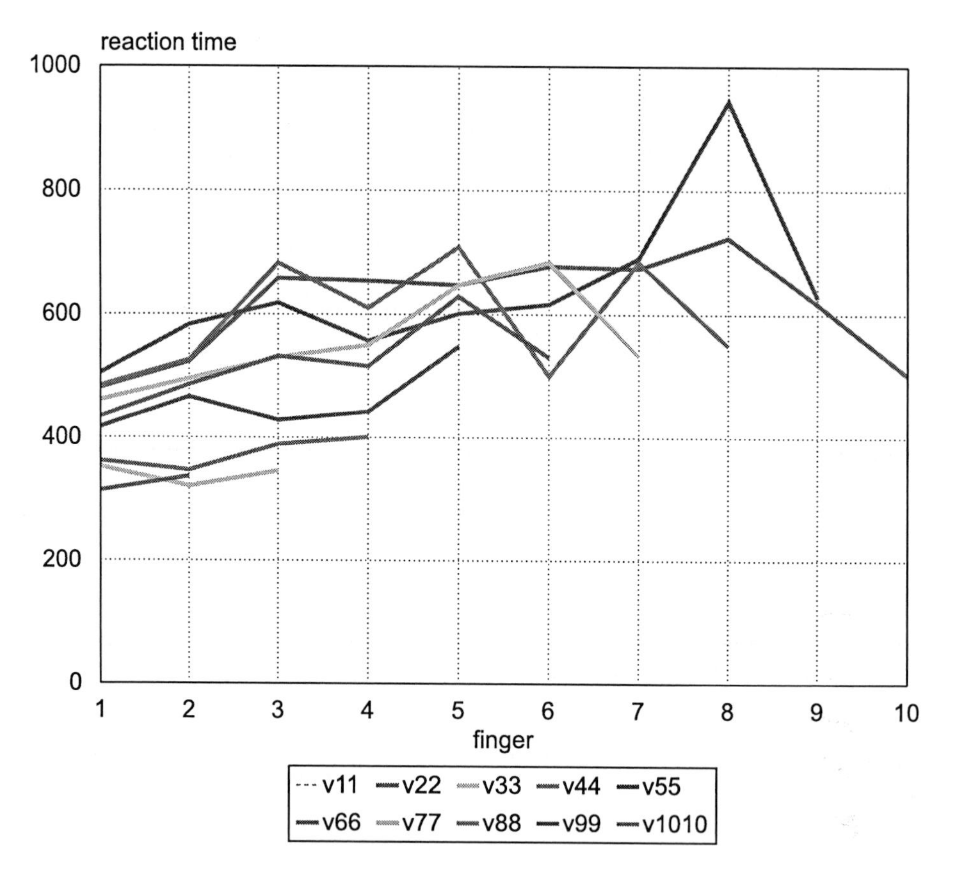

Fig. 15. Bihemispheric visual median finger reaction times of subject P12 in the tasks v22 to v1010

Is cycle time or cycle number responsible for different finger reaction times?

Computing cycle numbers and cycle times may produce false results, the correct way is to observe these sizes by *direct observation*. How can these observations clarify the cause of the different finger reaction times? The next figure shows some differences of finger reaction times in the task v55 of subject P12:

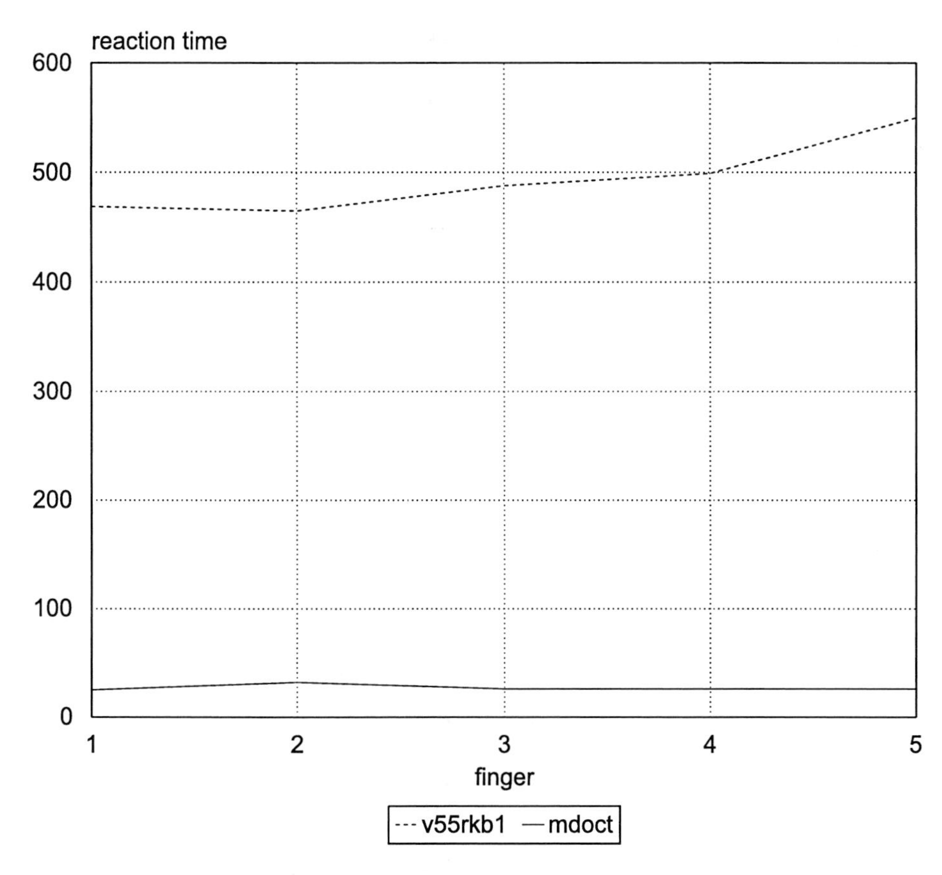

Fig. 16. Median finger reaction times in task v55 (500 trials) and below the mean directly observed cycle time (meandoCT) for each finger of the left hand in subject P12

It is obvious that the cycle times cannot be the cause for the different median finger reaction times in task v55. The next figure shows the distributions of the reaction times of the fingers LLF and LMF in subject P12.

The program DRUCK96p gives nearly the same cycle time for both fingers. This can be expected because the length of one searching cycle (=cycle time) cannot be influenced by the target element. The reason is, the target element is not known until it is found, therefore it cannot determine the length of the previous actions.

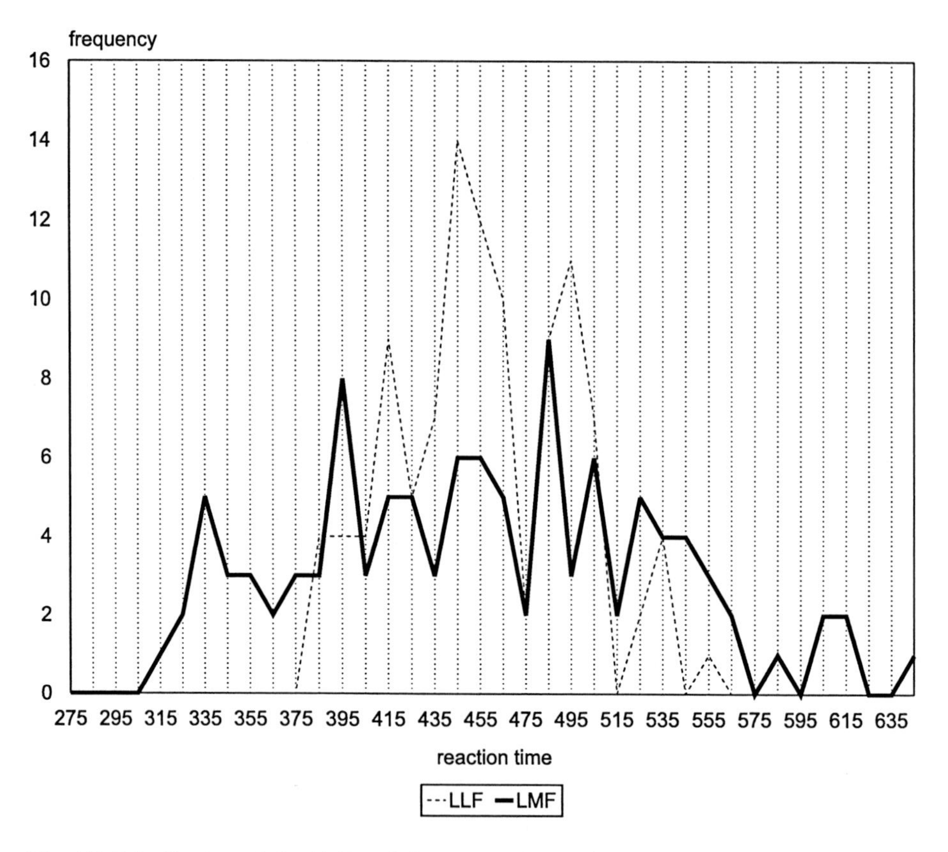

Fig. 17. Distribution of the LLF and the LMF subsets of v55P12b1–5 (n=500). The two fingers possess nearly the same median finger reaction times, nearly the same cycle times and similar medians of distributions

The most different fingers in the above figure are the LLF and the LTH. If one looks at the distributions of these two fingers, one sees that the distribution of the LTH is shifted to the right. This right shift, not the cycle time, is the reason for the increased median finger reaction time of the left thumb (LTH).

Why is the distribution of the left thumb shifted to the right? That means the number of trials with more searching cycles is increased in this finger. Therefore one can reformulate the question: why does the brain need more searching cycles for the left thump? We know that the search before detecting the target element cannot be determined by this element. But if the brain has difficulties to find the left thumb as the target finger, it has to apply more searching actions than usual. Why then, may some elements be difficult to detect? The lateral inhibition may be one factor but not the only one as the task v55P12b1 shows. Other factors could be different attention or the outstanding position of the left thumb within the row of fingers and some subsequent difficulty to activate it.

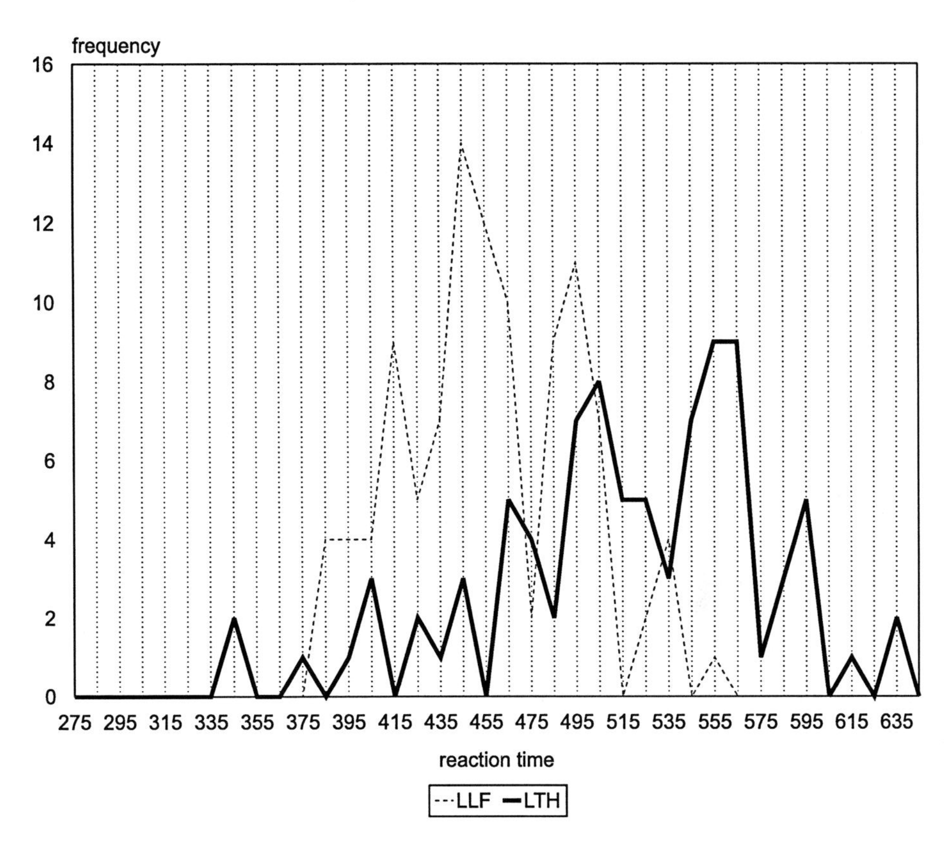

Fig. 18. One possible measure of this right shifted distribution is the "median cycle number" computed by the program DRUCK98p. The five values for the five fingers are from left to right: 4, 4, 4, 3, 6

Observing or computing the different cycle numbers?

If the different slopes of the finger reaction times are due to different cycle numbers then these cycle numbers may be computed.

The subject P18 has the following cycle numbers in the task v44 for his left four fingers:

LLF: 7.84, LRF: 8, LMF:10, LIF: 7.07, mean CN: 8.2
with cycle number of LLF in task v44 = (RT(v44)–RT(v11))/vCT

Because the direct observation of cycle time confirms that the cycle number and not the cycle time accounts for the different decision times of the fingers, there remains the question:

are these cycle numbers interindividual constant?

The line of the median reaction times is the mean of the lines of median finger reaction times

The line of the median reaction times is the mean of the lines of the median finger reaction times. Every factor influencing the probability of one element of the searching set changes all the other probabilities at the same time. Therefore, the sum remains constant and the median reaction time is the sum of the median finger reaction times.

Monohemispheric visual reaction tasks

Summary

The previous reaction tasks used pathways without regarding the hemisphere in which the process took place. By changing the design, monohemispheric pathways can be investigated. In this chapter, the visual pathways v22r and v22l are compared to each other. The intention was to look for asymmetries between the right and left hemisphere in processing these tasks.

Introduction

The reason for this design was the observation that some subjects have side differences in their reaction times. The righthanded subject P15 eg has a slight *left-hand (i.e. right hemisphere) advantage of cycle time* and a *right-hand (i.e. left hemisphere) advantage of cycle number*. Perhaps a right hemisphere advantage in visual cycle time can be explained by a general right hemisphere visual processing advantage. Do other subjects show interhemispheric differences, too?

Method

In auditory reaction tasks the tone is given to one ear and the hand which is used lies on the same side as the ear, so the opposite hemisphere mainly performs this task. In visual tasks the stimuli are confined to one half of the visual space by looking at an arrow in the middle of the screen reacting to peripheral stimuli in either the right or the left half of the visual space. The reaction hand lies again on the same side as the stimuli so that the opposite hemisphere will be active in performing this task.

Examples of monohemispheric visual reaction tasks

v11r			v22r			v33		
↓			↓			↓		
1			1	2		1	2	3
O			O			O		
O			O			O		
O			O			O		
O			O			O		
O			O			O		

Fig. 19. Appearance of the visual task on the screen. The arrow is to be looked at. The visual stimulus consists of a colomn of circles if the stimulus is activated. The number above identifies the stimulus

The keys were symmetrically used: the j,k and l key by the index finger, the middle finger, and the ring finger of the right hand and the f,d and s key by the index finger, the middle finger, and the ring finger of the left hand. The subjects has been informed to lay only these fingers on the keyboard which were to be used in a task.

Direct observation of visual cycle times

The direct observation of cycle time in distributions of monohemispheric visual reaction tasks needs some precautionary remarks:

1. The 70-Hz monitor with a formation of a picture each 14 ms does prevent a high time resolution. Therefore the direct observation of visual cycle times demands the use of light bulbs, LEDs or a laptop plasma displays. For the observation of mean doCT (mean directly observed cycle time) TOSHIBA laptops (T3100 or T4400SX) were used with plasma displays. TOSHIBA (Technical Support Hotline) asserted that this displays build up a picture in parallel.
2. One has to observe one finger separately because the single fingers have shifted distributions (in the time axis) and therefore sum up without showing clear peaks. To get approximately 100 trials for one finger, one has to do 200 trials of the task x22y, 500 trials of the task x55y and so on. The distribution of one finger is then computed and drawn separately and the peaks recognized automatically by a computer program (DRUCK52p or DRUCK98p). The program measures the distances between the peaks and computes the mean value of these distances. This is called the mean interpeak time or the mean directly observed cycle time (mDOCT or dvCT).

Direct observation of visual cycle numbers

There is another way to get the cycle number of a task by direct observation of the distribution of reaction times.

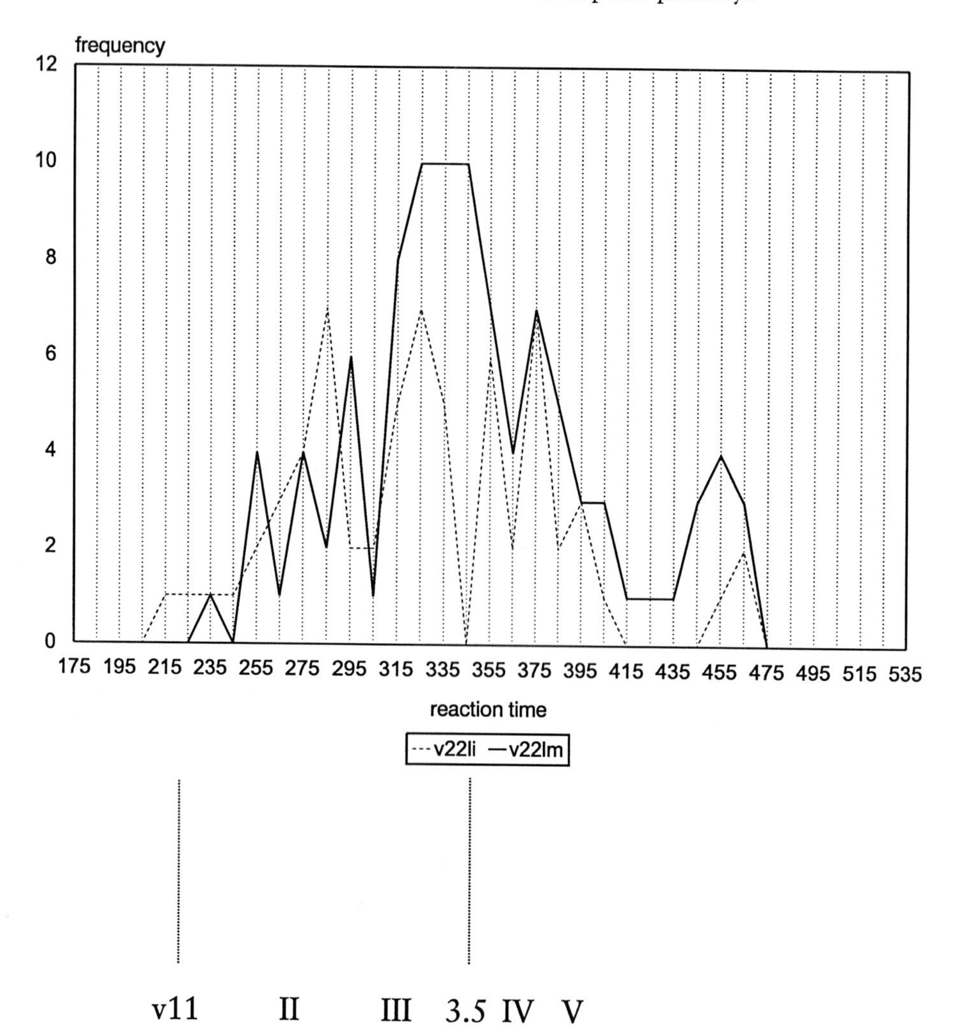

Fig. 20. Distribution of reaction times of the task v22lP18 (index finger and middle finger)

We consider the *left index finger* (v22li) of the above figure. The subject P18 has an intercept of 222 ms. Therefore the first peak at 285 ms means two cycles with a cycle time of 32 ms. The third cycle lies at 325 ms, the fourth at 355 ms and the fifth at 375 ms. *The directly visualized mean cycle number* is therefore

$$\begin{aligned} mCN &= (2 + 3 + 4 + 5)/4 \\ &= 3.5 \end{aligned}$$

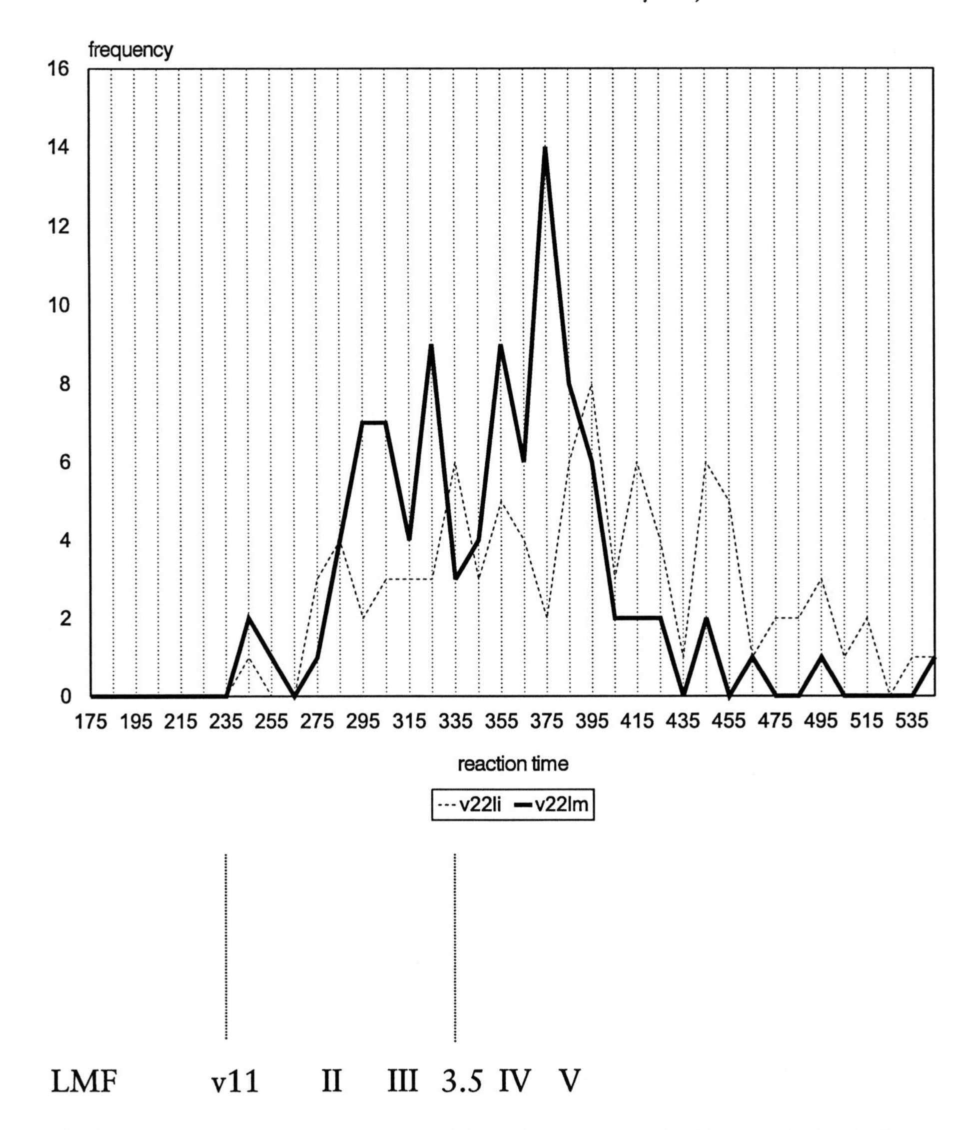

Fig. 21. Distribution of reaction times of the task v22lP06 (index finger and middle finger)

The subject P06 has a directly visualized cycle time of 25 ms for the left middle finger and 27.5 ms for the left index finger in the above figure.

The computed cycle number for the left middlefinger is: (v22lm–*v11li*)/25 =
359–235/25 = 4.96

with *v11li* instead of *v11lm* because the latter is not known

and for the left index finger: (v22li–v11li)/27.5 =
398–235/27.5 = 5.9

That means in the subject P06 and the task v22l direct visualisation tells us: the left index finger needs 5 cycles, each 27.5 ms, and the left middle finger needs 4 cycles, each 25 ms. The decision time of v22li is therefore 5*27.5ms = 137.5ms and the decision time of v22lm 4*25ms = 100 ms.

The problem is where to begin with counting the cycles. One could take the median reaction time of v11l or v11r as the beginning and use the special cycle time of this task to count the cycles correctly. The minimal number of cycles is two: one to detect the target stimulus, the other to detect the target key. So the next peak after the "zero" times of v11l or v11r and within the next cycle time must be the 2-cycle-peak. This could help to count the following peaks correctly. This works only if all cycles occur in the distribution at hand. Otherwise some cycles which are skipped have to be counted as if they did occur.

With this arrangement the directly visualized median cycle number of the left middlefinger is 5, the directly visualized median cycle number of the left index finger is 6 in task v22l of subject P06. The right middlefinger needs 5 cycles too.

Can this directly visualized parameters be replicated in the same individual at different times?

No. A replication of v22l by subject P06 failed. The directly visualized cycle time of the left index finger was now 33 ms, the cycle time of the middle finger was 45 ms with the possibility that some cycles did not occur in the replicated distribution so that the observed cycle time is too high.

The newly observed cycle time of the left index finger is similar to the bihemispheric visual cycle time of subject P06 which is 32.7 ms. Summing up, the directly observed cycle times and cycle numbers can fluctuate considerably. Therefore, only the mean values computed by help of many single measurements should be considered as intra-individual constants.

Results

The monohemispheric visual median reaction times

There is the same tendency of linear increase as in the bihemispheric visual tasks. The above figure shows the advantage of vNNr relative to vNNl in the subject P15. The healthy control group contains until now the following 11 subjects: P06, P19, P20, P12, P14, P10, P03, P07, P01, P15, P21 (P22, P23, P24).

The following subjects are faster in the task v22l than in the task v22r: P01, P21, P19, P20, P10, P12, P14
The following subjects are faster in the task v22r than in the task v22l: P03, P07, P15
The following subject has nearly the same reaction time in v22l and v22r: P06
That means most of the subjects have a faster v22l.

As in the bihemispheric tasks one can estimate how many cycles the task v22y has more than the task v11y by CN(v22y)–CN(v11y) = (v22l – v11l)/vCT

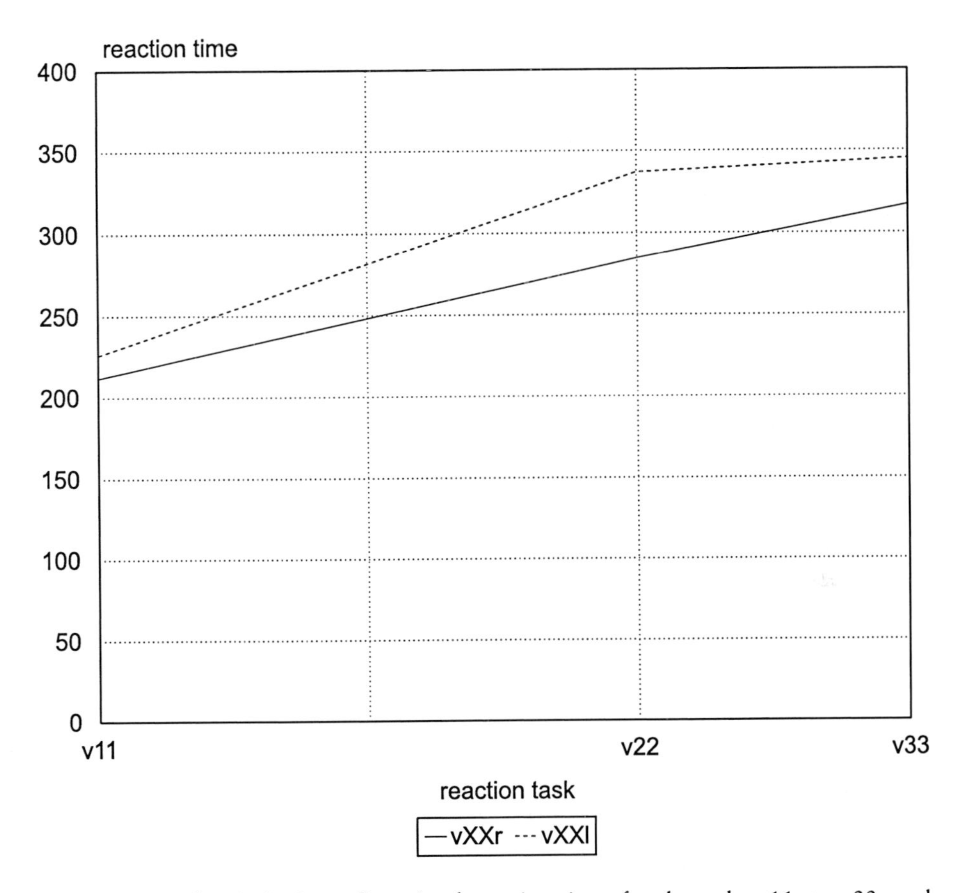

Fig. 22. Monohemispheric median visual reaction times for the tasks v11r to v33r and v11l to v33l (broken line) in subject P15

This estimation presumes that v22y and v11y differ only in the number of searching cycles. The later sections will show that this is not correct, but may be used as a first estimation. vCT is the cycle time from the bihemispheric visual tasks (slope of the regression line). The values of subject P15 are 2.5 for the right hand task and 3.8 for the left-hand task. This means if the pathway would differ only in the numer of searching cycles, then the tasks v22y would need 2.5 to 3.8 cycles more than the v11y tasks.

The monohemispheric visual median finger reaction times

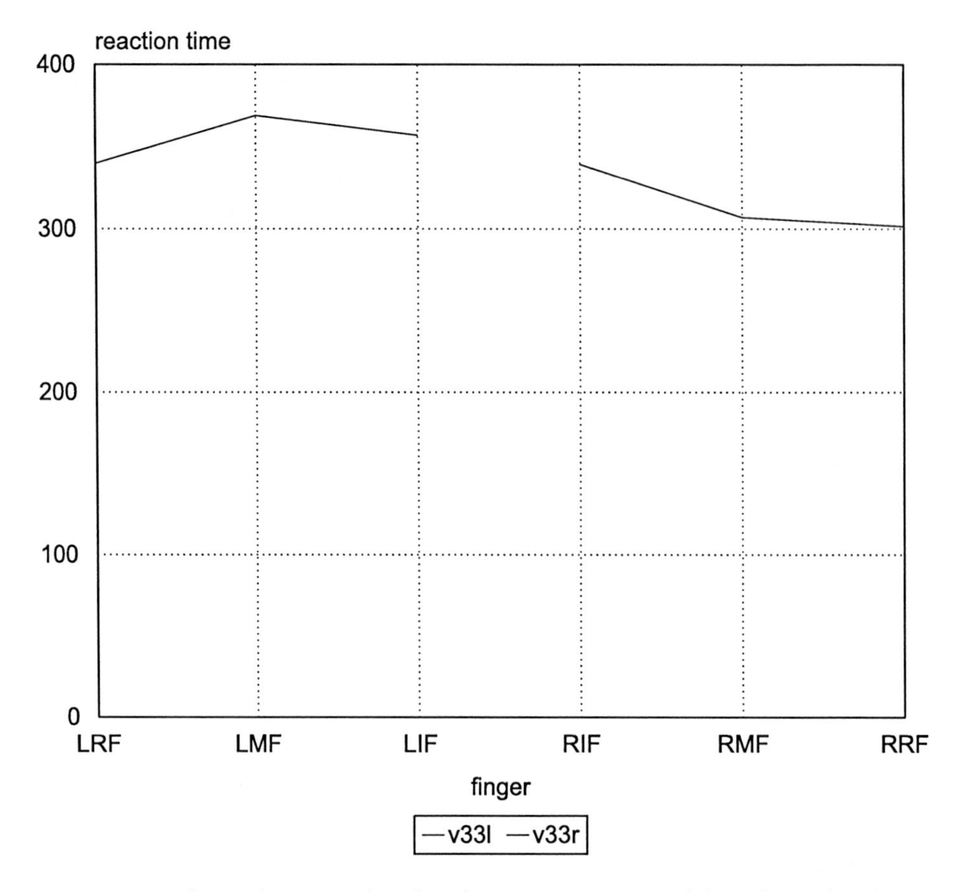

Fig. 23. Monohemispheric visual median finger reaction times of the tasks v33l and v33r in subject P15. These results show no clear tendency of difference between the three fingers. In both above figures the right visual reaction times are slightly faster than the left. The subject P15 is righthanded

In the healthy control group (subjects: P06, P19, P20, P12, P14, P10, P03, P07, P01, P15, P21) the following subjects have *no clear tendency of differentiation* within the finger reaction times of one hand in the task v33y: P01, P06, P20, P10, P14, P15

and the following subjects have *a tendency of differentiation* within the finger reaction times of one hand (with disadvantage of the left resp. the right middlefinger) in the task v33y: P03, P21, P07, P19, P12

Altogether, there is no uniform tendency to a differentiation within the fingers of one hand in monohemispheric visual finger reaction times.

But if one compares the finger reaction times of both hands with each other, one sees the side difference of the previous chapter: most subjects show an advantage of the left-hand tasks relative to the right hand tasks.

The directly observed cycle times in monohemispheric visual reaction tasks (v22y)

The question is whether there is a difference between the left-hand cycle times (i.e. the right hemisphere) and the right-hand cycle times (i.e. the left hemisphere)?

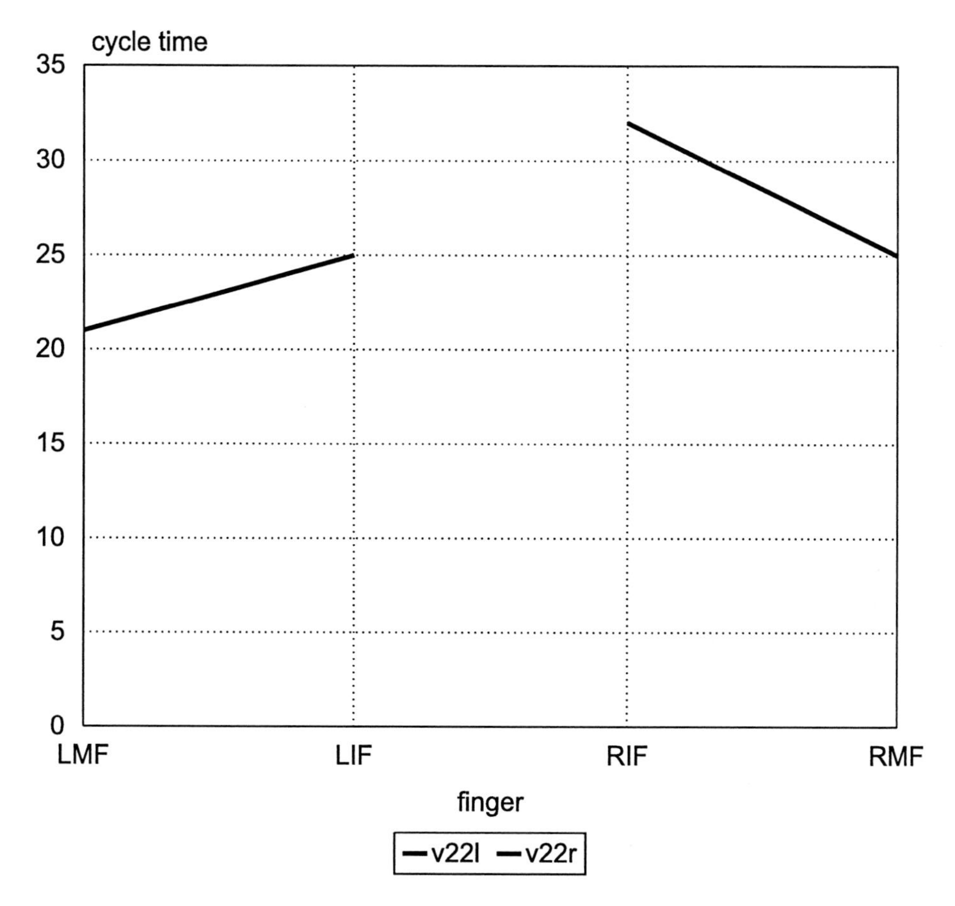

Fig. 24. Directly observed cycle times for the tasks v22l and v22r of the subject P15. The left cycle times are shorter than the right cycle times

The other subjects show the following directly observed cycle times of single fingers in the tasks *v22l and v22r*:

Table 7. Comparison of the directly observed cycle times in the tasks v22l and v22r in healthy subjects. vCT is the bihemispheric visual reaction time i.e. the slope of the linear regression. A small letter behind the initials means repetition of the tasks

Sub	vCT	dvCT LMF	dvCT LIF	Mean dvCTl	dvCT RIF	dvCT RMF	Mean dvCTr	dvCT	dvCTr-dvCTl
P01	19.8	34	33	33.5	20	32	27	30.3	−6.5
P02b	14.7	25	25	25	30	50	40	32.5	15
P03	20.8	30	27	28.5	37	25	31	29.8	2.5
P04	39.9	32	23	27.5	30	35	32.5	30	5
P05	28.3	31	33	32	25	25	25	28.5	−7
P06		32	32	32	25	40	32.5	32.3	0.5
P06a		45	33	39					
P06b		34	26	30					
P07	24.2	32	37	34.5	26	28	27	25.8	−7.5
P07a	24.5	25	27	26	28	28	28	27	2
P08	56.1	35	28	31.5	28	24	26	28.8	−5.5
P09	21.9	30	31	30.5	32	30	31	30.8	0.5
P10	33.6	24	30	27	35	32	33.5	30.3	6.5
P11	31.7	26	30	28	35	25	30	29	2
P12	22.7	40	25	32.5	30	24	27	29.8	−5.5
P12a		30	30	30	28	40	34	32	4
P12b		20	28	24	26	30	28	26	4
P13	27.3	26	30	28	33	30	31.5	29.8	3.5
P14	24.4	21	25	23	30	32	31	27	8
P14a	19.7	28	26	28	20	30	25	26	−3
P15	29.5	21	25	23	32	25	28.5	25.8	5.5
P15a	22.3	25	30	27.5	31	30	30.5	29	3
P15 b	18.8	43?	23	33	25	28	26.5	29.8	−6.5
Male									6.5
Fem									3.5

There is no reproducible side difference between the directly observed cycle times of task v22l and those of task v22r.

The subjects who have repeated the monohemispheric visual reaction tasks change the difference (dvCTr-dvCTl) from positive to negative and vice versa. Either the method is not precise enough to measure stable side differences or the side differences can change after implict learning. *This would mean that one hemisphere is better in implicit learning than the other.* This hypothesis can only be confirmed if in successive repetitions there is no more than one change of side preference. The four subjects who have repeated the monohemispheric visual tasks until now, show the following sequences of side preferences. Only two of them have performed more than two exercises. They both show only one change of side preference. To test this hypothesis more subjects with more than two exercises are needed.

Against the implicit learning hypothesis of changing interhemispheric differences in repeated tasks stands the fact that the cycle times of the repeated tasks are not generally shortened relative to the naive task. In reality, one hemisphere has a

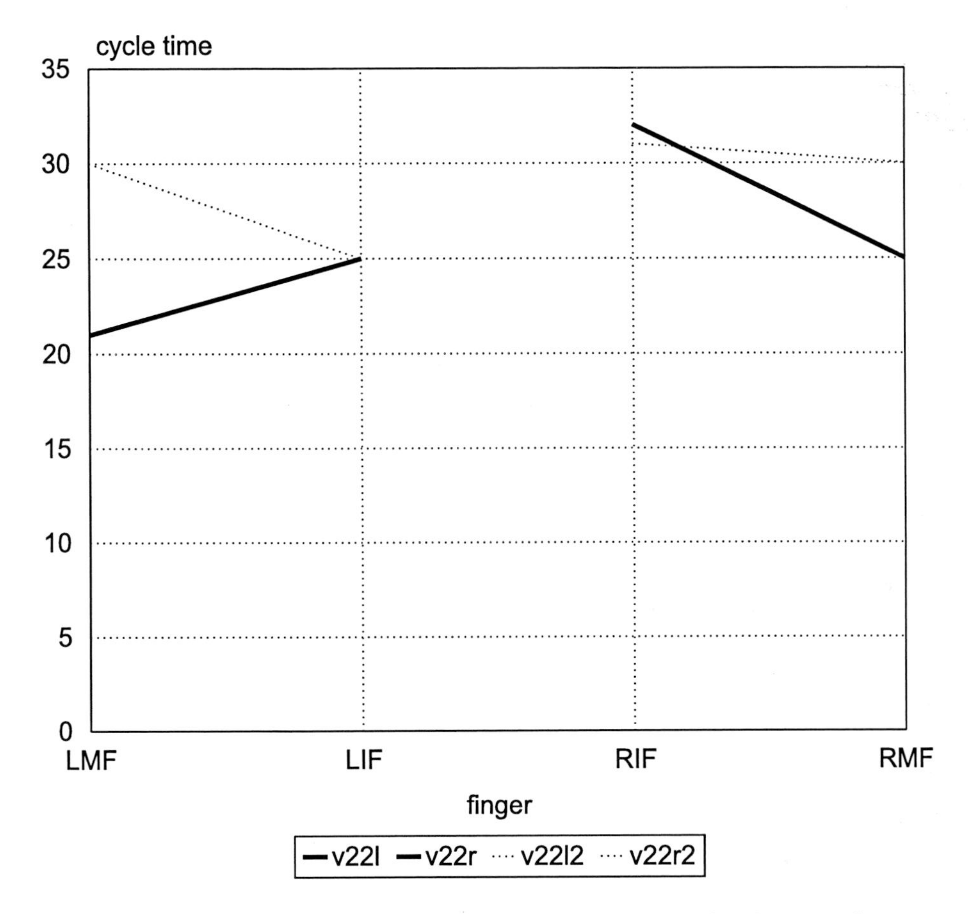

Fig. 25. Replication of directly observed cycle times in the tasks v22l and v22r (with n=200 trials i.e. each finger approximately 100 trials) in the subject P15. The slight advantage of left cycle times is replicated. The differences between the cycle numbers could not be replicated

shortened cycle time and the other a prolonged cycle time in repetition. Is there another explanation than random? The hemisphere which becomes faster may be the one which learns implicitly. But why does the other hemisphere become slowlier than in the naive state?

It is striking that in the below table always the slower side becomes faster while the faster side does not. This could suggest a limit to cycle time. Only the cycle time slower than this limit can change. The increase in cycle time can only be explained by uncontrolled variables like time of day etc.

Table 8. The change of side preference in repetitions of monohemispheric visual reaction tasks

Subject Subject	First exercise Naive task	Second exercise First repeat	Third exercise Second repeat	Fourth exercise Third repeat
P07	−7.5	+2		
P12	−5.5	+4	+4	
P14	+8	−3		
P15	+5.5	+3	−6.5	

Differences between right and lefthanded subjects (hypothesis: right-handed subjects show a slower visual cycle time at the left hemisphere, i.e., the right hand dvCTr > dvCTl). This hypothesis cannot be supported by the present method of direct observation of cycle time.

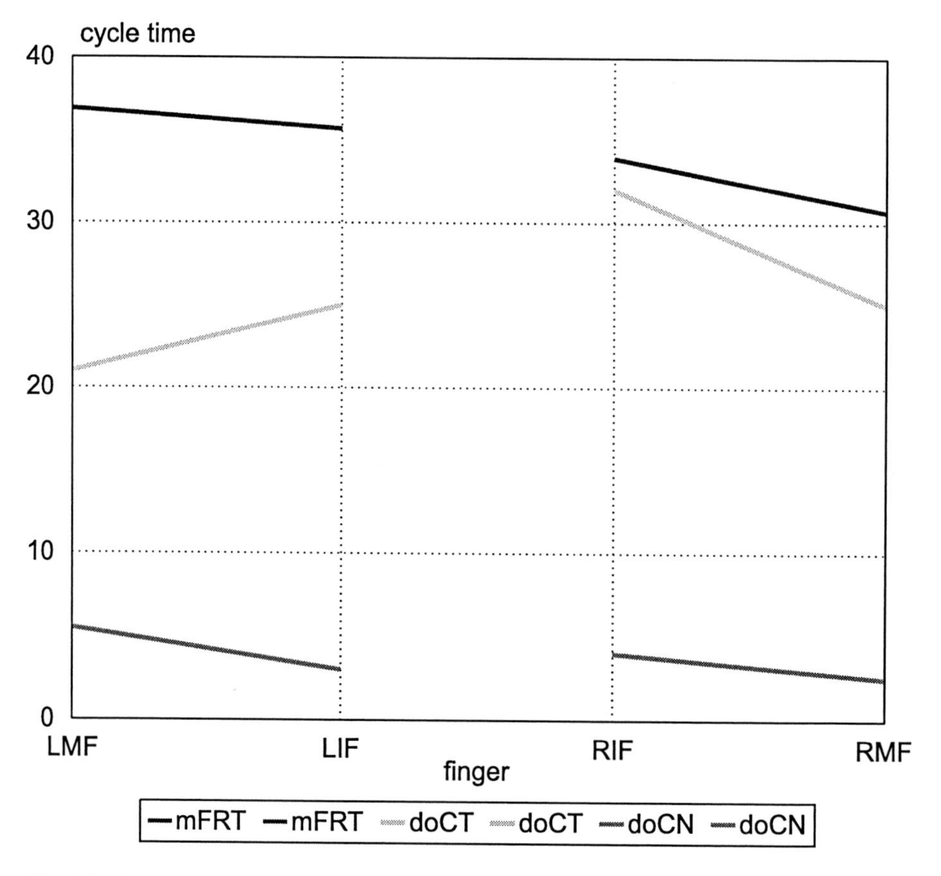

Fig. 26. Common display of median finger reaction times (above, divided by 10), dvCT (middle), and dvCN (below) in subject P15. The left cycle times are lower than the right ones, the left cycle numbers are higher than the right ones. In the sum, the left finger reaction times are increased relative to the right ones

Differences between male and female subjects (hypothesis: female subjects are more symmetric than male subjects). If one computes the mean of the absolute values (i.e. the size of the side difference independent from its location), then the female subjects show a mean side difference of 3.5 ms and the male subjects of 6.5 ms.

The directly observed cycle number in monohemispheric visual reaction tasks

Apparently, there is no safe correlation between the directly observed cycle time and the directly observed cycle numbers of a finger. That means that these two variables are independent from each other.

The cycle number has something to do with the target elements. If the task elements are differentiately influenced, the less pre-activated ones need more cycles to be found. In the sections before, such an influence was only found in tasks with more than four fingers.

The cycle time is independent from the target elements because during the search- which needs cycle times- the target element is not known.

If one computes the difference of cycle numbers difference(CN)= (v22r–v11r)/CT, one gets 2.5 for the right cycle number and difference(CN) = (v22l–v11l)/CT = 3.8 for the left cycle number. These cycle numbers are mean values for both fingers. The respective mean values for both fingers computed from the directly observed

Table 9. Comparison of the directly observed cycle numbers in the tasks v22r and v22l in healthy subjects. A small letter behind the initials means repetition of the tasks

Sub	CN (LMF)	CN (LIF)	Mean dvCNl	CN (RIF)	CN (RMF)	Mean dvCNr
P01	4	2	3	2	3	2.5
P02b	2	2	2	2	1	1.5
P03	4	4	4	3	2	2.5
P04	3	4	3.5	3	5	4
P05	5	5	5	4	3.5	3.8
P06	4	4	4	5	5	5
P06a	2	4	3			
P06b	3	4	3.5			
P07	3	3	3	4	3.5	3.8
P07a	2	3	2.5	5	5	5
P08	3	4	3.5	5	3	4
P09	5	3.5	4.3	4	2	3
P10	4	5	4.5	3	3	3
P11	2	3	2.5	2	4	3
P12	3	4	3.5	3	5	4
P12a	3	4	3.5	2.5	3	2.75
P12b	3	3.5	3.25	5	4	4.5
P13	4	3	3.5	3	4	3.5
P14	4	3	3.5	3	3	3
P14a	3	3	3	3	2	2.5
P15	5.5	3	4.25	4	2.5	3.25
P15a	4	3	3.5	4	3	3.5
P15 b	5	4	4.5	3	3	3

cycle numbers of single fingers would be (4+2.5)/2 = 3.3 for the right side and (5.5+3)/2 = 4.1 for the left side.

One can directly see this reduced cycle number in the distribution of the task v22r:

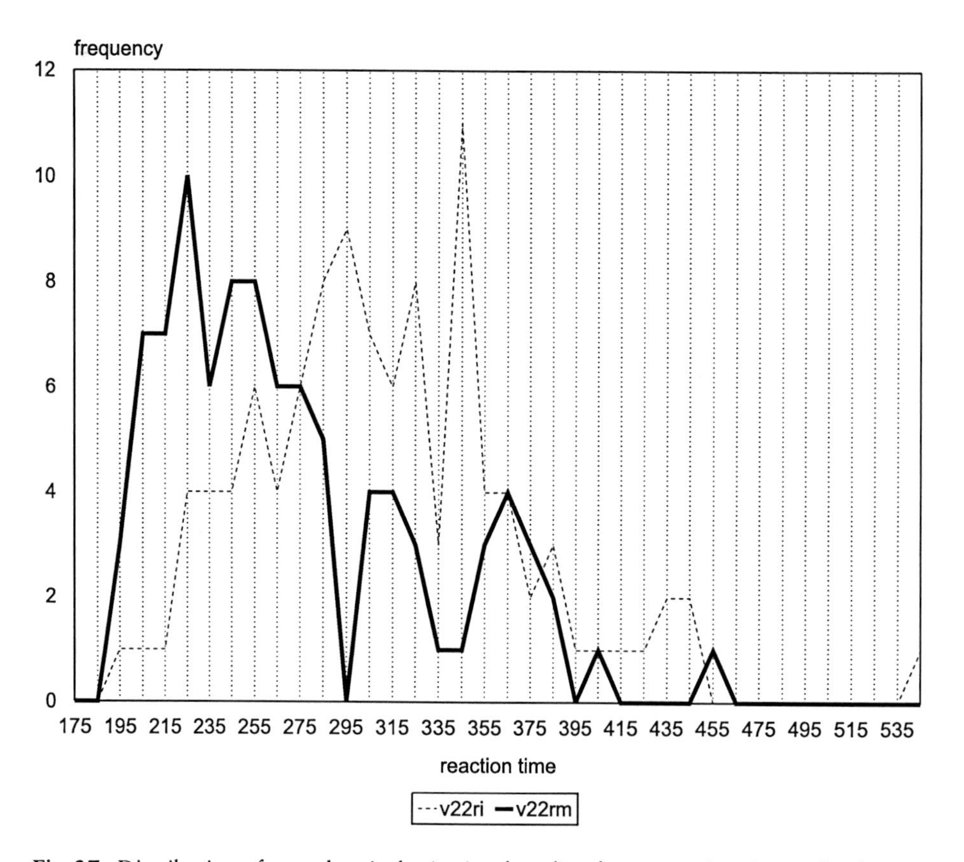

Fig. 27. Distribution of monohemispheric visual median finger reaction times of task v22r in the subject P15. Especially the right middle finger shows a distinct left shift

Because on the average the task v22y needs only four cycles more than the task v11y, *all* cycles may be observable in the distribution.

The side differences of directly observed cycle numbers show a tendency of replication. This must be further examined.

Until now, the methods to measure the cycle time and the cycle number were not so precise to be able to show replicable lateral differences. That means: if there are lateral differences they are not detectable by the applied methods. The fewer the cycle number, the more uncertain is the mean value of the cycle time. The repetition of v22y proves the limited accuracy of this methods.

Is there any correlation between the side differences in directly observed visual cycle times (dvCTr–dvCTl) and the side differences in directly observed visual cycle numbers (dvCNr– dvCNl)? Until now, in 9 subjects both differences behave differently and in 5 subjects, similarly.

If one asks for the cause of this side differences between v22l and v22r in some subjects one has to consider the following:

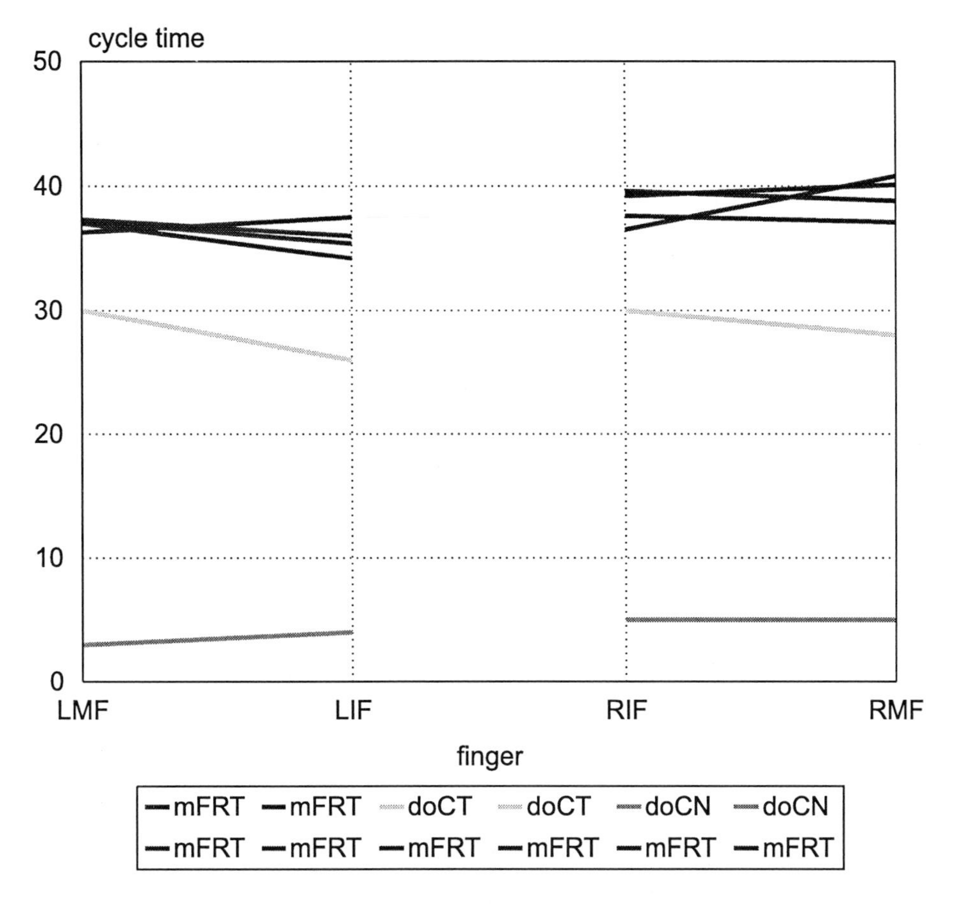

Fig. 28. Median finger reaction times (divided by 10), directly observed cycle times and directly observed cycle numbers of subject P12, n=4*100 (v22lP12–1,2,6,7 and v22rP12–1,2,6,7). The doCT and doCN are measured by help of the sum file, the median finger reaction times (mFRTs) are depicted for each single file

The monohemispheric visual median reaction times of this subject show an advantage for the left-hand tasks, this advantage continues in the left finger reaction times, the left directly observed cycle numbers (and the dvCT of the left index finger).

Important is that these differences are personal traits. There are subjects with a left side advantage, a right side advantage or symmetric conditions.

The number of cycles in monohemispheric visual reaction tasks computed by dvCT

There is one value which has to be considered once again: the computed number of cycles in monohemispheric visual tasks. The idea of computing these cycle numbers is the following: if one knows, that the decision time of a task and the cycle time of a subject area nearly constant, one could compute the number of cycles (i.e. number of elementary actions) which are necessary to perform this task. Simple tasks would be performed by a lower number of cycles, more complex tasks by a higher number of cycles. Optimistically, this number of cycles should be interindividually constant. Instead of this, three groups of subjects can be found: The first group having nearly the same cycle numbers which approximates the number of task elements within the searching set (number of stimuli plus number of keys). The next two groups show either distinctly higher or lower cycle numbers. See the three groups in the table below:

If only the decision times of a task are known, it is not possible to clarify whether the cycle time of a singular task, its cycle number or both are changed. Because the cycle time is the same in these tasks as the direct observation proves, the extended monohemispheric reaction times must be caused by an increasing number of cycles. These numbers are computed in the table below.

Table 10. Computed cycle numbers in monohemispheric visual tasks. In this table the expression mRT(vNN) is abbreviated by vNN. The abbreviation vCT means bihemispheric visual cycle time (=slope of the bihemispheric regression line) and dvCT = directly observed visual cycle time. The direct observation of CT is taken from v22yy (v22li, v22lm, v22ri, v22rm). The repetitions are omitted because of their changed vCT

Sub	vCT	(v22l-v11l)/ vCT	(v33l-v11l)/ vCT	(v22r-v11r)/ vCT	(v33r-v11r)/ vCT
P01	19.8	3.4	7	5.1	7.3
P02b	14.7	3.5	5.8	5.0	5.9
P03	20.8	3.4	7.1	2.6	5.6
P04	39.9	3.9	6.8	4.8	7.8
P21	19.4	5.7	10.2	6.8	7.9
P05	28.3	3.3	4.2	3.8	5.5
P06	32.7	3.6	6.3	3.8	6.4
P07	24.2	3.4	6.3	3.2	5.4
P07a	24.5				
P08	56.1	3.4	4.3	3.5	4.7
P19	29.5	2	5	2.1	4.7
P09	21.9	5.2	9.4	3.8	6.0
P10	33.6	4.7	6.7	4.6	7.2
P11	31.7	2.5	4.7	2.7	4.3
P12	22.7	4	8.5	6.4	10
P13	27.3	3.1	5.2	5.0	5.7
P14	24.4	3	5.9	2.8	6.1
P14a	19.7				
P15	29.5	3.8	4	2.5	3.6
P15a	22.3				
P15b	18.8				

The use of vCT to compute the cycle numbers in repetitional tasks is inappropriate, as the examples P15 and P15b show. vCT is not constant but depends on the cycle numbers between tasks which can be changed by implicit learning. This comes from the method to determine vCT: this method uses a distance of 4 between v11 and v22 and a distance of 2 between all subsequent tasks. However, this is only true in naive subjects. In practiced subjects, the differences are smaller (see the explanation of fast mode and slow mode in the corresponding chapter).

It is striking that the subjects with higher results possess a relativ low vCT and the subjects with lower results possess a relative high vCT. Perhaps in these subjects, the vCT and the "real" dvCT differ from each other in order to give the deviating results. In the first group vCT and dvCT should correspond. It is also possible that in the first group with the correct cycle numbers the vCT is a more accurate value than the dvCT!

If the same subjects show a too high or too low auditory cycle number, computed by (aNNy–a11y)/vCT, this would support the vCT being too low or too high. Persisting deviations should be due to deviating cycle numbers (advantage) or random deviations.

The question of intra-individual replication of median reaction times and computed cycle numbers (cvCN) in the pathways v22l and v22r:

The median reaction times of the tasks v22l and v22r were repeatedly measured in some subjects:

Table 11. Replication of median reaction times of the tasks v22r and v22l in subject P12 v22l and v22r replicated also in subject P15 etc

Task	Median RT	Task	Median RT
v22lP121	366.5	v22rP121	390
v22lP122	358.5	v22rP122	393.5
v22lP126	365.5	v22rP126	373.5
v22lP127	358.5	v22rP127	398.5

The side differences in computed cycle numbers are due to side differences in median reaction times. These differences are due to different cycle numbers.

The intercepts of monohemispheric visual reaction tasks

Table 12. The intercepts are listed in the table. The values of nearly *all* subjects are larger in the left hand. In v11y the non cortical and the non decision periods are prevalent. That means the right hemisphere is handicapped relative to these periods. One subject P15 has repeated this tasks, the intercepts were reduced, the difference remains the same

Subject	v11r	v11l	v11l-v11r
P01	213.5	223	+9.5
P02b	168	198	+30
P03	245.5	247	+1.5
P04	222	241	+19
P05	302	329	+27

Table 12. Continued

Subject	v11r	v11l	v11l-v11r
P06	237	235	−2
P07	219	225	+6
P08	224	265	+41
P09	232	234.5	+2.5
P10	194	198.5	+4.5
P11	188	213	+25
P12	234	243	+9
P13	217	226	+9
P14	183	216	+33
P15	212	226	+14
P15b	202	217	+15

The number of cycles in monohemispheric visual reaction tasks computed by dvCTy

There are some interesting changes in the vCT of intra-individual repeated tasks. In the subject P15, for example, the vCT is reduced from 29.5 (P15) to 22.3 (P15a) and 18.8 (P15b). Because the directly observed cycle time does not show this reduction (see table below), the cause must be a reduction of cycle number.

Table 13. Comparison of the directly observed cycle times in the tasks v22l and v22r in healthy subjects. vCT is the bihemispheric visual reaction time, i.e., the slope of the linear regression. A small letter behind the initials means repetition of the tasks. dvCTl for example means "directly observed visual cycle time left hand"

Sub	vCT	Mean dvCTl	Mean dvCTr	(v22l-v11l)/ dvCTl	(v22r-v11r)/ dvCTr	(v33l-v11l)/ dvCTl	(v33r-v11r)/ dvCTr
P01	19.8	33.5	27	2	3.7	4.1	5.4
P02b	14.7	25	40	2.1	1.8	3.4	2.2
P03	20.8	28.5	31	2.5	1.8	5.2	3.8
P04	39.9	27.5	32.5	5.6	5.9	9.9	9.5
P05	28.3	32	25	2.9	4.3	3.8	6.2
P06	32.7	32	32.5	3.9	3.7	6.5	6.3
P07	24.2	34.5	27	2.4	2.9	4.4	4.8
P08	56.1	31.5	26	6	7.5	7.7	10.2
P09	21.9	30.5	31	3.7	2.7	6.8	4.2
P10	33.6	27	33.5	5.9	4.6	8.3	7.3
P11	31.7	28	30	2.8	2.8	5.3	4.6
P12	22.7	32.5	27	2.8	5.4	5.9	8.4
P13	27.3	28	31.5	3.0	4.3	5.1	4.9
P14	24.4	23	31	3.1	2.2	6.2	4.8
P15	29.5	23	28.5	4.8	2.5	5.2	3.7
P15 b	18.8	33	26.5	3.3	1.9	4.1	4.3

Because the supposed cycle numbers are (v22y–v11y)/dvCTy = 4 and (v33y–v11y)/dvCTy = 6, one sees a dvCTy dependent deviation from the expected cycle numbers. That means, that the dvCTr and dvCTl show an error which influences the quotient (=cycle number). If, for example, dvCTr is too large, the quotient will be too small.

Another observation is that the empirical cycle numbers differ slightly from the expected cycle numbers. That means that v11y does not need 0 cycles, but does need some cycles.

The third observation is that in some elderly subjects, the cycle numbers are increased.

The fourth observation is that practiced subjects (P02b and P15b) need fewer cycles than unpracticed subjects.

Discussion

The reason of the asymmetry between the two monohemispheric visual pathways v22l and v22r

The cycle numbers of the visual monohemispheric tasks are not as constant interindividually as the cycle numbers of the auditory monohemispheric tasks. There is an important difference between the two groups: in the visual monohemispheric tasks motions of eyes are hardly to prevent. This could be the reason of the great interindividual differences in the visual tasks. This reason is absent in auditory tasks.

Differences of cycle time could be provided by different number of neurons being involved in a monohemispheric pathway. It is known that visual words start foci of electrical activity in the occipital brain with an assymetry favoring the right hemisphere (Posner and Raichle). This could mean a faster passage within the right hemisphere.

In the above table, the difference (v22l–v11l) represents the decision time of the task v22l relative to the task v11l. If one states a constancy of cycle number, the left-hand cycle times could be computed by the expression:

left-hand cycle time = (v22l–v11l)/4
left-hand cycle time = (v33l–v11l)/6

It should be observed that the left-hand cycle time of the task v22l is the same as that of the task v33l, otherwise the theory is false. It has to be noted that a *left-handed task* is processed within the *right hemisphere* and vice versa.

Subject P18 has a left-handed visual cycle time (vCTl) of 32 ms in task v22r and 32.7 ms in task v33r but a right-handed visual cycle time (vCTr) of 25.59 in task v22l and 25.14 in task v33l.

Why is the vCT measured by help of v11 to v1010 31.5 ms in this subject? In the subject P18 the vCT is similar more to vCTl than to vCTr. Can this be replicated for other subjects?

Perhaps these numbers can be explained by surveying the fingers used in the single tasks. In the measurement of the right-handed cycle times only the right index finger, middlefinger and ringfinger are used, in measuring the left-handed cycle times the respective left fingers are used.The bihemispheric visual cycle time vCT is generated by all ten fingers. If one uses only the tasks v11 to v33 then the cycle time vCT is similar to the left-handed cycle time. Therefore one must ask, whether less data produce a greater deviation because of the differences in median key reaction times (v11 to v33, v11l to v33l, on the contrary v11 to v1010).

This argument is supported by the observation that *the left-handed cycle time of subject P18, computed from v11 to v55* (these tasks are only processed by the left hand) has a value of 33 ms but the left-handed cycle time of subject P18, computed from v11 to v33, has a value of 25 ms. This value may also be seen in the median key reaction times of the left little finger and the left ring finger for this subject. For further clarification the direct visualisation has to be observed.

The question is: is the laterality caused by the selected fingers or by a real difference in cycle time between hemispheres? Is there any experiment to decide this question? In the direct visualisation, the cycle time of single fingers is computed. If one compares the task v11ri (that means the right index finger was used to press the single key) and the subset of task v22ri (that means these trials in which the right index finger was used) and computes (v22ri-v11ri)/4 – is this the *cycle time of the right index finger*?

It is possible that often used fingers are represented in cortex by a greater set of neurons than fingers which are used rarely. Therefore the velocity and cycle time of the frequently used fingers may be faster.

As a consequence only the reaction times of equal sets of fingers can be compared to each other. If one compares sets with different fingers the *key reaction times of these fingers influence the results* of computing the cycle time.

Try to compute (a22ri-a11ri) and (a33ri-a11ri). Is there a difference to (a22r-a11r) and (a33r-a11r) respectively?

To say it frankly: the differences between the key reaction times of three fingers can be responsible for the deviances of decision times in monohemispheric visual tasks. But the same fingers are used in monohemispheric auditory tasks with rare deviances.

The external task elements have a slight advantage against the internal elements. This effect could explain why the system needs more searching actions (i.e. a higher cycle number) to detect the middle finger than the little finger. The difference of cycle number and the indifference of cycle time have to be confirmed. The hypothetical reason for the slightly different probabilities of being scanned may be the number of neuron which represent an object. If the object is often used, it has a greater set of representing neurons and vice versa. This number of representing neurons could directly account for its availability by the scanning layer.

If the mean number of cycles (mCN) is slightly different for all fingers, why does the sum of all key reaction times result in the nearly linear function between the number of alternatives and the reaction time of vNN? Because the sum of all probabilities is 1, if one finger has less, the other finger must have more. The subject P18 has in the task v44 for his left four fingers the following cycle numbers:

LLF: 7.84, LRF: 8, LMF:10, LIF: 7.07, mean CN: 8.2
with cycle number of LLF in task v44 = (RT(v44)–RT(v11))/vCT

The hypothesis that the cycle number and not the cycle time accounts for the differences in the decision time of the fingers has to be proven.

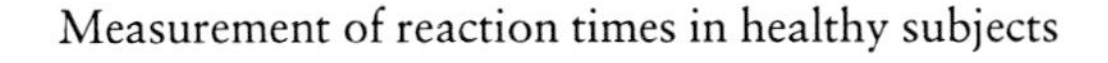

Fig. 29. An element represented by a greater subarea possesses a greater probability of being detected by the searching layer. A second reason could be the lack of lateral inhibition in elements with marginal position

Why is the mean cycle number reduced in some fingers ($n>4$) and in some sides?

If the number of fingers is smaller than four, there are no certain differences in median finger reaction times, cycle times or cycle numbers within these fingers. However if one compares the both sides with each other one sees side differences in median reaction times, median finger reaction times and cycle numbers but not in cycle times.

The results of subject P15 in Fig. 27 show a shift to the left in the distribution of the middle finger in the task v22r as the explanation for the v22r advantage relative to v22l. A similar observation is made in subject P12, here with an v22l advantage.

If the cycle number is different between the two sides and between fingers (if their number is greater than four) in a replicable manner, one has to ask why.

A reduction of cycle number means the occurrence of trials with fewer cycles. If the median cycle number of v22 is four, than the trials with two or three cycles would become more frequently.

Monohemispheric auditory reaction tasks

Introduction

The pathways of the auditory tasks a22l and a22r are the objects of this chapter. Similar to the visual tasks a possible asymmetry of cycle time and cycle number between the two hemispheres has to be examined. The cycle times and the cycle numbers are again directly observed. Of special interest is the computed cycle number and its relative interindividual constancy.

Method

Collecting the data

The monohemispheric auditory reaction tasks used an earphone in one ear and the reacting hand at the same side as the earphone. Altogether 6 different monohemispheric auditory reaction tasks were examined: a11r, a22r, a33r, and a11l, a22l, a33l. The "a" means auditory, the first number the number of different auditory stimuli that can be heard and the second number the number of different keys which can be pressed. The fourth letter means left or right. The tasks were constructed in the following ways:

a11l:
low tone
f
left index finger

a22l:

middle tone	low tone
d	f
left middle finger	left index finger

a33l:

high tone	middle tone	low tone
s	d	f
left ring finger	left middle finger	left index finger

The right hand tasks were symmetrical to the left-hand tasks and used the keys j = right index finger = low tone, k = right middle finger = middle tone, l = right ring finger = high tone. The interstimulus interval lay about 2500 ms with random deviations to prevent guessing. Each task comprises 100 trials.

Evaluating the data

To observe the cycle times directly the tasks a22l and a22r had 200 trials each. The further procedure is given on page 33. The danger of this direct observation is that cycles can be omitted, with the consequence that the distance between two peaks becomes too large. To prevent this kind of mistake, one has to use a limit: distances between two peaks which were larger than 50 ms have been omitted in computing the mean.

The second rule consists in defining a peak. If the following expression was valid

((freq(n)–freq(n–1)) + ((freq(n+1)–freq(n)) > 2

then the freq(n) was called a peak at the reaction time n

A second kind of peak was defined by the expresssion:

((freq(n)–freq(n–1)) >1 and ((freq(n)–freq(n+1))>1

Results

The monohemispheric auditory median reaction times

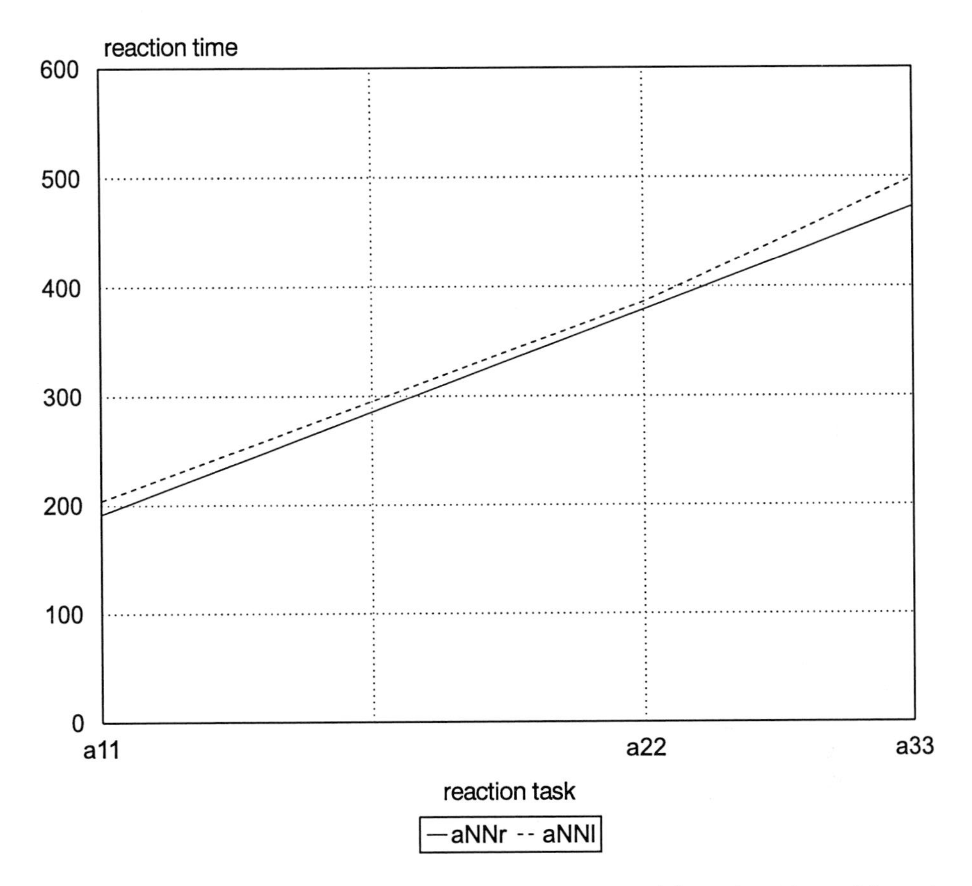

Fig. 30. Monohemispheric median auditory reaction times of the tasks a11r to a33 r and a11l to a33l in subject P15. In this subject the auditory reaction times are highly symmetric

The 11 healthy subjects show the following interhemispheric differences in auditory tasks:

The following subjects are in the task a22l faster than in the task a22r: P07, P10, P14,
The following subjects are in the task a22r faster than in the task a22l: P03, P21, P06, P20,
The following subjects have nearly the same reaction time in a22l and a22r: P15, P01, P19, P12, P15

That means in auditory tasks there is no clear preference of one side.

The monohemispheric auditory median finger reaction times

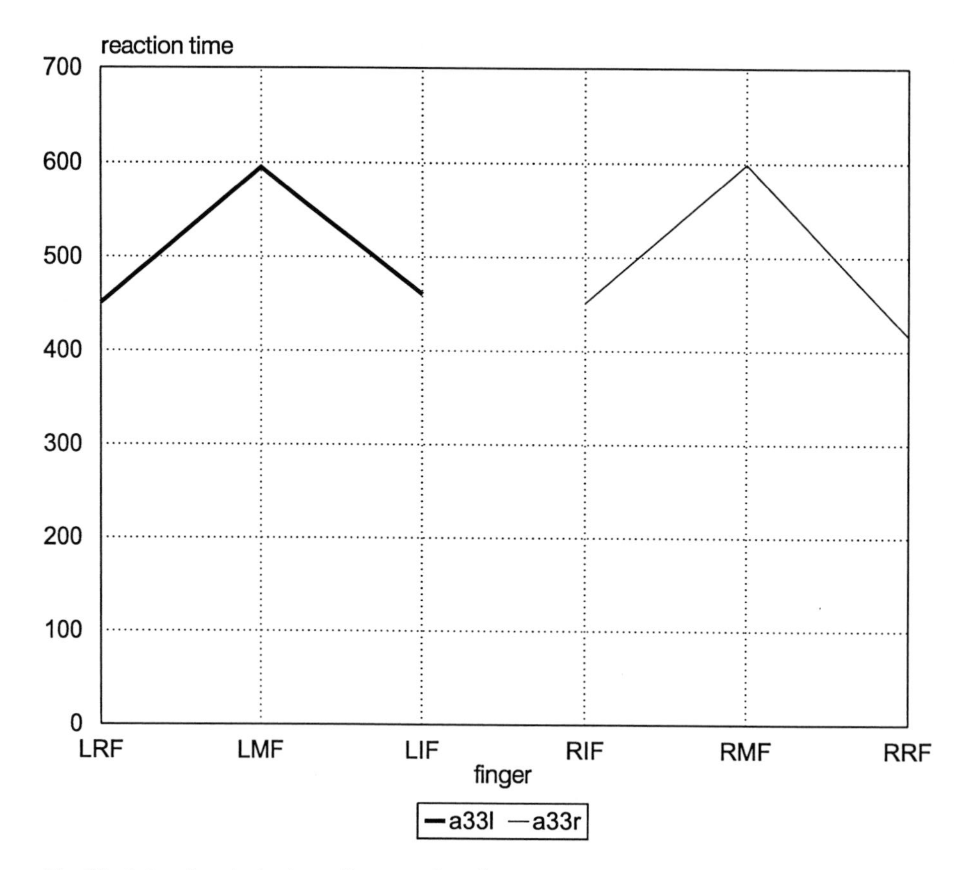

Fig. 31. Monohemispheric auditory median finger reaction times of the tasks a33l and a33r in subject P15. The results show a clear tendency of differences between the three fingers of one hand but no clear side difference between the two hands. P15 is righthanded

Considering the healthy group, one sees *no clear tendency of differentiation* within the finger reaction times of one hand in task a33y: P01, P21, P06, P19, P20, P10, P14 while the following subjects show *a tendency of differentiation* within the finger reaction times of one hand in the task a33y: P03, P07, P10, P12, P15

Summing up, there is no uniform tendency to a differentiation within the fingers of one hand in monohemispheric auditory finger reaction times.

If one compares the finger reaction times of both hands with each other one does not see a side difference in most subjects. Two subjects have a left hand advantage, one subject has a right hand advantage, the others have nearly symmetrical auditory finger reaction times.

The directly observed cycle times in monohemispheric auditory reaction tasks (a22l and a22r)

To illustrate the findings of this section the results of the subject P15 are used as an example once more:

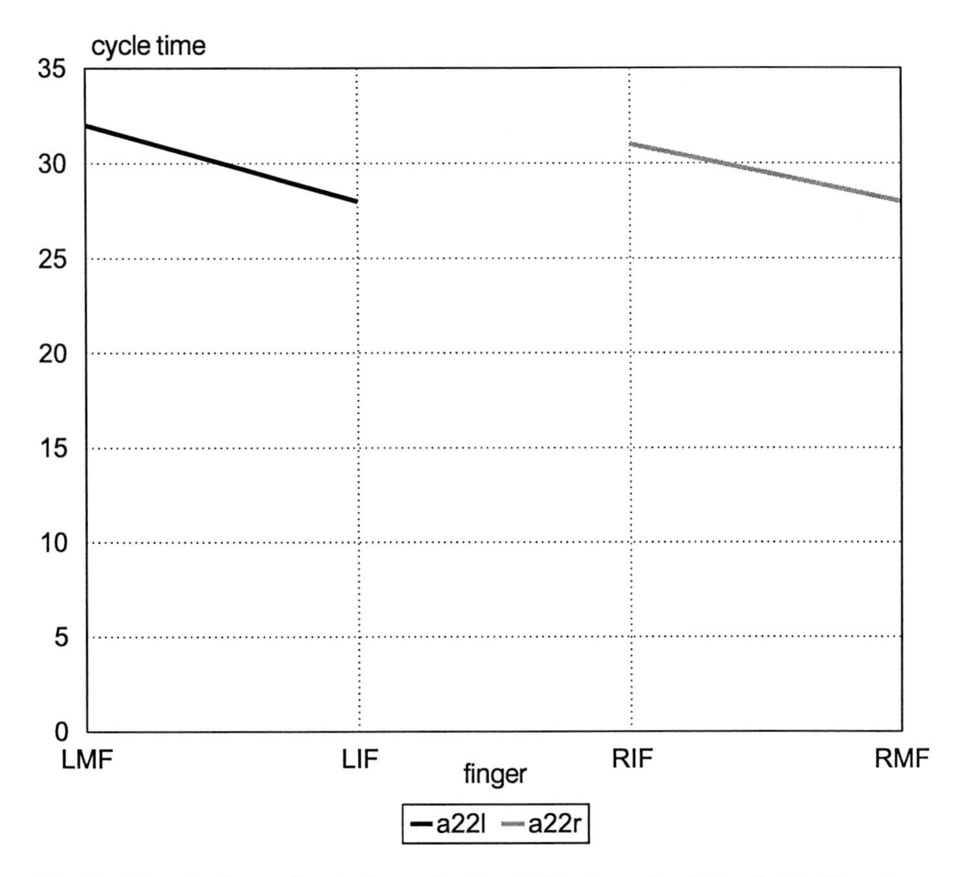

Fig. 32. Directly observed cycle times of subject P15 in the tasks a22l and a22r. There is no significant difference between the left and the right cycle times

The other subjects show the following directly observed cycle times of *single fingers* in the *auditory tasks a22l and a22r*:

Table 14. Comparison of the directly observed cycle times in the tasks a22l and a22r in healthy subjects. vCT is the bihemispheric visual cycle time, i.e. the slope of the linear regression. A small letter behind the initials means repetition of the tasks

	vCT	daCT LMF	daCT LIF	Mean daCTl	daCT RIF	daCT RMF	Mean daCTr	Mean daCT	daCTr-daCTl
P01	19.8	32	30	31	35	25	30	30.5	−1.0
P02b	14.7	30	30	30	37	23	30	30	0
P03	20.8	40	20	30	28	30	29	29.5	−1.0
P04	39.9	26	28	27	30	26	28	27.5	+1.0
P05	28.3	30	33	31.5	37	31	29	30.3	−2.5
P06	32.7	33	27	30	31	38	34.5	32.3	+4.5
P07	24.2	24.2	33	28	30.5	32	26	29	−2.0
P07a	24.5	24.5	36	25	30.5	26	40	33	?????
P08	56.1	33	28	30.5	33	35	34	32.3	+3.5
P09	21.9	26	24	25	24	30	27	26	+2.0
P10	33.6	36	25	30.5	24	32	28	29.3	−2.5
P11	31.7	30	26	28	27	31	29	28.5	+1.0
P12	22.7	25	28	27.5	30	25	27.5	27.5	0
P13	27.3	30	30	30	28	28	28	29	−2.0
P14	24.4	25	27	26	30	32	31	28.5	+5.0
P14a	19.7	30	25	27.5	28	50?	39?	33.3	+11.5
P15	29.5	32	28	30	31	28	29.5	29.8	−0.5
P15a	22.3	26	28	27	28	26	27	27	0
P15 b	18.8	43?	23	33?	25	28	26.5	29.8	−6.5

There are some preliminary observations to be mentioned:

1. vCT is lower than dvCTs in the subjects: P01, P03, P07, P12
   vCT is higher than dvCTs in the subjects: P06
   vCT is about similar to dvCTs in the subjects: P14, P15

2. vCT is lower than daCTs in the subjects: P01, P03, P07, P12
   vCT is higher than daCTs in the subjects:
   vCT is about similar to daCTs in the subjects: P06, P14, P15, P13

3. dvCT is lower than daCT in the subjects: P06
   dvCT is higher than daCT in the subjects:
   dvCT is about similar to daCT in the subjects: P01, P03, P07, P12, P14, P15

Which pattern is generated by these observations?

– *In most of the subjects dvCT is similar to daCT*

– *In a larger group (with vCT <= 24 ms) vCT is lower than dvCT and lower than daCT*

– *In a smaller group (with vCT>=24 ms) vCT is similar to dvCT and similar to daCT*

The most surprising observation is the reduction of the visual cycle time (vCT) relative to the higher directly observed cycle times in the subjects P01, P03, P07 and P12. If the cycle time is not responsible for the the reduced vCT (as the directly observed cycle times propose), then the cycle numbers have to be investigated for being reduced. The next section will show whether these subjects have reduced cycle numbers in the tasks a22l and a22r.

The directly observed auditory cycle times show a slightly higher mean in the absolute values for male subjects but the difference is not as large as the difference in the directly observed visual cycle times. That means that male subjects are more asymmetric than female subjects in monohemispheric auditory tasks but the asymmetry is not as large as in monohemispheric visual tasks.

There are only two subjets who have repeated these tasks. They both show an increase of their asymmetry in the repetition tasks. In P14 dvCTr is longer than dvCTl, in P15 dvCTl longer than dvCTr.

The hypothesis of a correlation between the dominant hemisphere and the auditory side differences has to be looked for.

The directly observed cycle numbers in monohemispheric auditory tasks

Table 15. Comparison of the directly observed cycle numbers in the tasks a22r and a22l in healthy subjects. A small letter behind the initials means repetition of the tasks

	daCN LMF	daCN LIF	Mean daCNl	daCN RIF	daCN RMF	Mean daCNr
P01	3	4	3.5	4	4	4
P02b	3	2	2.5	3	5	4
P03	4	4	4	4	3	3.5
P04	6	6	6	2.5	5	3.8
P05	5	4.5	4.8	5	5	5
P06	4.5	2.5	3.5	4	4	4
P07	3	3	3	3	3	3
P08	6	6	6	3	2.5	2.8
P09	4	6	5	5	2	3.5
P10	4	3	3.5	6	4	5
P11	2	3.5	2.8	4	4	4
P12	4	4	4	3	2	2.5
P13	4	3	3.5	3	5	4
P14	3	3	3	4	3	3.5
P15	5	4	4.5	6	5	5.5
P15a	6	6	6	6	4	5
P15 b	3	3	3	5	4	4.5

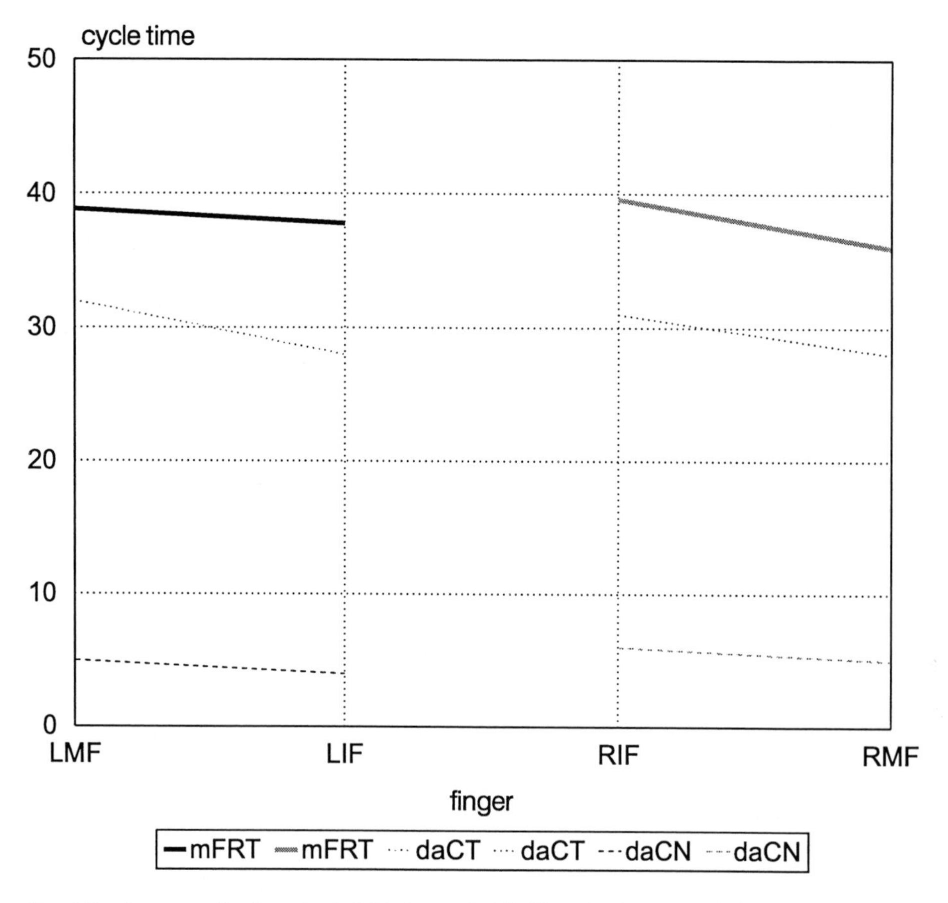

Fig. 33. Common display of mFRT (above, divided by 10), daCT (middle), and daCN (below) in the tasks a22l and a22r (number of trials=200 each) in subject P15. There are no obvious differences between the left and the right mFRT and daCT and a slight difference between the left and the right daCN

The computed number of cycles in monohemispheric auditory reaction tasks

Because the cycle time of these tasks is the same, the extended monohemispheric reaction times must be caused by an increasing number of cycles.

Table 16. In this table the expression mRT(aNN) is abbreviated by aNN. vCT, the visual cycle time (equal to the slope of the linear regression of bihemispheric tasks), is used. The direct observation of CT is taken from a22r –> daCTr and from a22l–>daCTl. daCTr = (daCTri + daCTrm)/2 and daCTl = (daCTli + daCTlm)/2 with daCTr=direct auditory cycle time right, daCTl= direct auditory cycle time left, daCTri = direct auditory cycle time right index finger, daCTrm = direct auditoy cycle time right middle finger and so on

Sub	naive vCT	(a22l-a11l)/ vCT	(a33l-a11l)/ vCT	(a22r-a11r)/ vCT	(a33r-a11r)/ vCT
P01	19.8	6.14	9.68	6.4	10.9
P03	20.8	6.19	8.74	4.76	11.41
P04	39.9	6.2	10.6	6.8	9.4
P05	28.3	5.6	8.6	5.7	8.5
P06	32.7	6.67	10.37	5.89	9.59
P07	24.2	2.23	6.45	3.66	6.04
P08	56.1	6.2	8.1	6.9	8.8
P09	21.9	8.7	11.2	5.4	11.6
P20	19.2	7.98	11.4	6.44	11.7
P10	33.6	4.64	8.18	6.13	8.44
P11	31.7	6.0	11.1	5.7	9.9
P12	22.7	7.52	13	8.42	13.42
P13	27.3	10.6	11.8	9.5	11.9
P14	24.4	3.48	7.42	4.79	7.22
P22	9.65	12	6.88	11.84	
P15	29.5	6.15	9.98	6.34	9.52
P15a	22.3				
P15 b	18.8	7.67	13.44	6.61	11.05

Because of the similar size of visual and auditory cycle times, vCT is used as the representative of cycle time. This may lead to false computed cycle numbers. If one divides the intervals of subject P12 by its directly observed cycle time one gets the following results:

(a22l – a11l)/daCTl = 6.21
(a33l – a11l)/daCTl = 10.74
(a22r – a11r)/daCTr = 6.96
(a33r – a11r)/daCTr = 11.09

Is this valid for the other subjects too?

The subjects who have repeated the tasks need another method to compute vCT than the subjects which are in a naive state. The present method cannot be applied to the repetition. Therefore these lines are empty.

One can make this phenomenon more obvious if one puts (a22l-a11l)/vCT =1 and relates all other expressions to this norm: For some subjects this has be done in the following table.

Table 17. Relations between the medians of different monohemispheric auditory tasks with (a22l-a11l) as unity, that means the numbers tell us how many times a period is longer than (a22l-a11l). x1=(a22l–a11l)/(a22l–a11l), x2 = (a33l–a11l)/(a22l–a11l), x3=(a22r–a11r)/(a22l–a11l), x4=(a33r–a11r)/(a22l–a11l)

Subject	x1	x2	x3	x4
P01	1	1.57	1.0	1.77
P03	1	1.4	0.77	1.84
P06	1	1.6	0.9	1.4
P07	1	2.9	1.64	2.7
P20	1	1.4	0.8	1.46
P10	1	1.76	1.32	1.81
P12	1	1.7	1.12	1.8
P13	1	1.1	0.9	1.1
P14	1	2.13	1.37	2.07
P15	1	1.6	1.0	1.54
P15b	1	1.7	0.9	1.4

The first two subjects show the greatest deviation, the others fit well. The results of P15 and P15b are very instructive. The subject P15 repeats all tasks in P15b: the vCT changes, the daCT remains nearly the same, the quotients of this section change dramatically. If one norms the values, the quotients stay the same. This means the reduction of vCT must have other causes than reduction of cycle time, i.e. the cycle number (if the knowledge which stimulus relates to which key is known at once, no search process has to be done and the number of cycles may be reduced). How many cycles are needed in this case?

This is one of the most fascinating observations of this section. What is the reason of this interindividual constancy of monohemispheric auditory cycle numbers? What do they tell us?

In selecting between competing theories in explaining these quotients, the size of these quotients will be very important. Some theories demand lower quotients, some higher ones.

The dependency of the auditory reaction times from the type of the last target stimuli could give important hints to some kind of backward search. The other possibility is that the scanning mechanism uses the whole set of target stimuli and keys without making a difference. Some considerations will be given in the next section.

The intercepts of monohemispheric auditory reaction tasks

Table 18. The intercepts of the tasks a11y

Subject	a11l	a11r	a11l – a11r
P01	220	221.5	–1.5
P02b	192.5	174.5	+18
P03	288.5	236.5	+52
P04	207	200.5	+6.5
P05	337	322	+15

Table 18. Continued

Subject	a11l	a11r	a11l – a11r
P06	156	151.5	+4.5
P07	250	253.5	–3.5
P08	205	216.5	–11.5
P09	210.5	243	–32.5
P10	192	192	–1.0
P11	217.5	199.5	+18
P12	212.5	195	+17.5
P13	211	200	+11
P14	181	192	–11
P15	204	191.5	+12.5
P15 b	184	181	+3

Here the picture is not as uniform as in the visual case. Again, there is no correlation with the dominant hemisphere, but the number of left-handers is still too small.

Implicit learning in monohemispheric auditory reaction tasks

Table 19. Comparison of the directly observed cycle times in the tasks a22l and a22r in healthy subjects. vCT is the bihemispheric visual cycle time, i.e. the slope of the linear regression. A small letter behind the initials means repetition of the tasks

	vCT	Mean daCTl	Mean daCTr	(a22l-a11l)/ daCTl	(a22r-a11r)/ daCTr	(a33l-a11l)/ daCTl	(a33r-a11r)/ daCTr
P01	19.8	31	30	3.9	4.2	6.2	7.2*
P02b	14.7	30	30	6.1	5.5	10	9.3
P03	20.8	30	29	4.28	3.4	6.1	8.2*
P04	39.9	27	28	9.1	9.7	15.7	13.4
P05	28.3	31.5	29	5.0	5.6	7.7	8.2
P06	32.7	30	34.5	7.3	5.6	11.3	9.1**
P07	24.2	28	26	1.8	3.1	5.1	5*
P07a	24.5	25	40				
P08	56.1	30.5	34	11.3	11.	14.9	14.6
P09	21.9	25	27	7.6	4.4	9.8	9.4
P10	33.6	30.5	28	5.1	7.4	9	10.1**
P11	31.7	28	29	6.9	6.2	12.5	10.8
P12	22.7	27.5	27.5	6.96	6.2	10.7	11.1**
P13	27.3	30	28	9.6	9.3	10.8	11.5
P14	24.4	26	31	3.3	3.8	6.96	5.7*
P14a	19.7	27.5	39?				
P15	29.5	30	29.5	6.06	6.4	9.8	9.5**
P15a	22.3	27	27				
P15b	18.8	33?	26.5	4.4	4.7	7.7	7.8*

To say it exactly, the numbers of the four right columns in the above figure tell us how many cycles the tasks a22y have more than the tasks a11y. It is striking that there are two strategies (*) or (**) which are used. In some sense these are the "directly observed cycle numbers" because they rely only upon observations like decision times and directly observed cycle times.

There are two subjects who show values larger than expected: subject P04 and P08. P04 is 45 years old, P08 is 68 years old. P12 is 48 years without showing these values. Until now the cause of these increased values is unclear.

Discussion of monohemispheric auditory reaction tasks

The hypothetical structure of the task a22y

First hypothesis. The elements of the task a22y are two sounds and two keys. Because the last perceived sound is not present any more, it has to be remembered and compared with the two possible sounds. If the non-target sound is considered first, it is compared with the two possible sounds but without result. Then the target sound is compared to the two possible sounds and is recognized. This comparative process lasts one cycle minimally and four cycles maximally, on average 2.5 cycles. Because the two sounds and the two keys have to be searched for too, one gets an average of 4+2.5=6.5 cycles in a22y

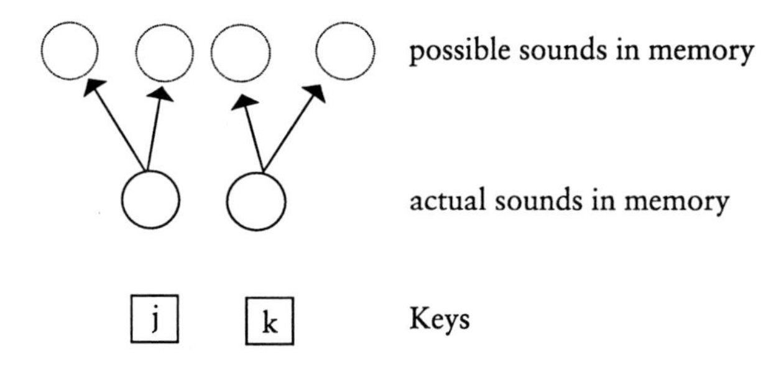

In a33y there are 3 sounds and 3 keys, i.e. 6 elements to consider, and 1 to 9 (3x3) comparing actions, i.e. an average of (1+9)/2=5 cycles. Together one gets on average 6+5 cycles for the tasks a33y.

Second hypothesis. If one does not have all sounds equally represented in ones working memory, one has to use the sequence of the last sounds heard and take them to compare the heard sound with. This lasts longer because a sound can be repeated and therefore the time needed for comparison is extended.

Third hypothesis. Another hypothesis states that each alternative stimulus is compared to the perceived sound. That means, the search process on the possible stimuli has a pairwise appearance. The cycles of the afferent part of the search are doubled. Instead of 2 searching actions on the stimuli, there are four needed and together with the 2 searching cycles on the keys, the task a22y sums up to 2(keys) + 2(stim) + 2(remembering the perceived stimulus) =*6 cycles* and the task a33y 3(keys) + 3(stim) + 3(remembering the perceived stimulus) = *9 cycles* each on the average.

If one looks in Table 16 one sees some evidence for the last cycle numbers. The first two hypotheses would yield higher cycle numbers. If the cycle numbers of 6 and 9 are correct, then they could be used to compute the cycle time in a third manner (first method was the slope of the linear regression in vNN, the second method the directly observed cycle times of the tasks v22y and a22y). The third method would then use the following mathematical expression:

I (a22y – a11y)/CT = 6
II (a33y – a11y)/CT = 9

Because the visual tasks v22y and v33y need not double their cycle number within the afferent part of the search (the alternative stimuli and the target stimuli are within the same perception) these tasks result in the following expressions:

III (v22y – v11y)/CT = 4
IV (v33y – v11y)/CT = 6

Together with the expressions after full implicite learning (Hebb type connections between each stimulus and its coordinated key) one gets another two possibilities:

V (v22y – v11y)/CT = 2.5 (after full implicit learning of the stimulus-key connections)
VI (v33y – v11y)/CT = ?

With a constant cycle time, the above expressions could be used to compute the degree of implicit learning by reduction of the cycle number.

The last two expressions after full implicit learning in the case of the auditory tasks:

VII (a22y – a11y)/CT = ?
VIII (a33y – a11y)/CT = ?

The slope of the linear regression of the tasks v11 to v1010 after full implicit learning has to be computed:

IX slope of v11 to v1010 after full implicit learning = ?

Now the results of P15 and P15b get clearer and the Table 16 results of P15b may be used to compute the CT in P15b.

The slope vCT is dependent on two variables: the difference of medians and the cycle number:

X vCT = (med(task2) – med(task1))/CN

For example, the number of cycles between v11 and v22 is 4 without implicit learning and 2.5 after implicit learning.

Therefore vCT = (med(v22) – med(v11))/4 = (med(v22´) – med(v11´))/2.5.

v22´ and v11´ are repetitions of v22 and v11 after implicit learning.

Fourth hypothesis. What is different in a22y and v22y? In v22y, the frame of possible stimuli is visible, in a22y this is not the case. The auditory competitive stimuli have to be remembered, the frame has to be constructed in the mind. If this is the cause of enhanced cycle numbers in auditory tasks then analogous task lacking any frame should show the same enhancement of cycle number.

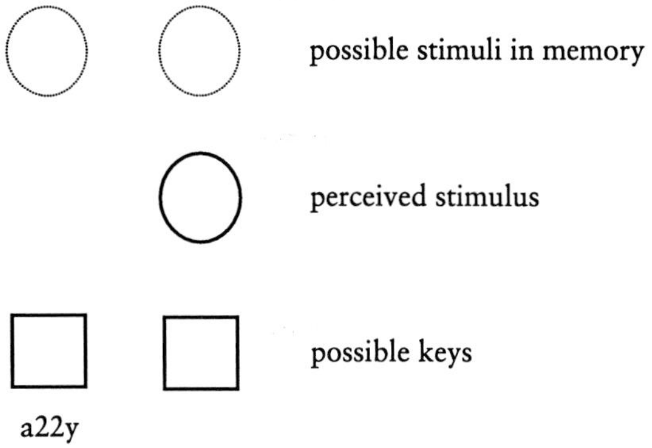

possible stimuli in memory

perceived stimulus

possible keys

a22y

a22y:
first from 5 elements: 3 cycles
then from 4 elements: 2.5 cycles
at last from 3 elements: 2 cycles
sum: 7.5 cycles

1. Search for the perceived stimulus out of 5 elements: 3 cycles

2. Search for the target stimulus (in memory) out of 4 elements (all stimuli minus the perceived stimulus): 2.5 cycles

3. Search for the target key out of 3 elements (all stimuli minus perceived stimulus and minus target stimulus): 2 cycles
   Sum for the task a22y: 7.5 cycles

For the task a33y one gets out of 7 elements: 4 cycles
then out of 6 elements: 3.5 cycles
at last out of 5 elements: 3 cycles
Sum for the task a33y: 10.5 cycles

Fifth hypothesis. A good hypothesis to explain the a22y and a33y reaction times should be simple and symmetrical. The new aspect is the search on elements which are not perceived but are internal representations, eg. of tones. These not present elements have to be remembered. The search on a set with n internally represented elements lasts (n+1)/2 steps on average.

There is no doubt that the task a22y implies four elements, two stimuli and two keys. Other than in the case of v22 the stimuli are not all present but one (the perceived). So we have the follwing searches:

Search 1 on four elements looking for the target stimulus: 2.5 cycles
Search 2 on three elements looking for the non target stimulus: 2 cycles
Search 3 on two elements looking for the target key: 1.5 cycles
Sum: 6 cycles

The objects of search 1 and search 2 can be interchanged by looking for the non target stimulus lasting 2.5 cycles and so on.

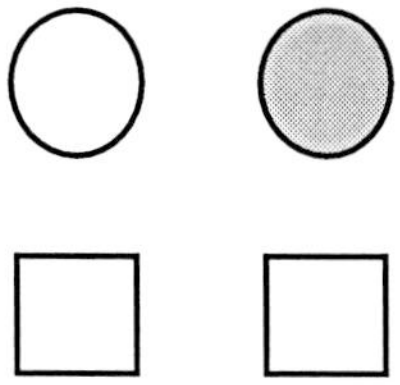

1. Looking for the target stimulus ((4+1)/2=2.5 cycles)

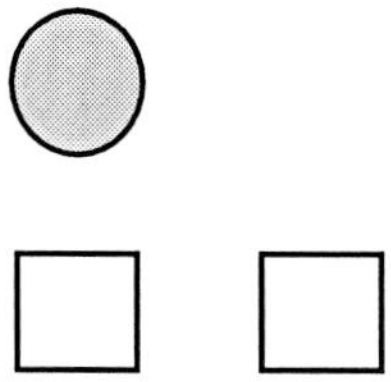

2. Looking for the non target stimulus ((3+1)/2=2 cycles)

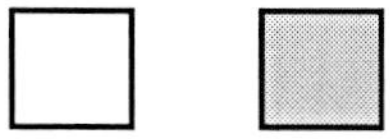

3. Looking for the target key ((2+1)/2=1.5 cycles)

This sums up to 6 cycles for the slow mode of task a22y.

We already know that learning the connection between target stimulus and target key accelerates the decision time by reducing the number of necessary search processes. After implicit learning, we get the following searches in the *visual task v22y:*

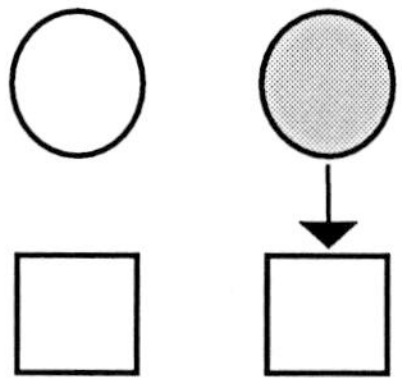

1. Looking for the target stimulus ((4+1)/2=2.5 cycles)

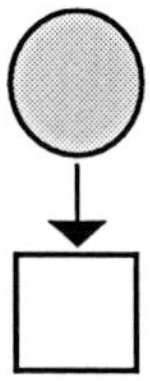

2. Looking directly to the connected target key (1 cycle)

That would attach 3.5 cycles to the fast variant of v22y after implicit learning.

The fast mode of task a22y must be different from the fast mode of task v22y. Here the non target stimuli must be remembered too but these can be connected to each other by Hebbian learning. So we meet the following search processes *in the fast mode of a22y*:

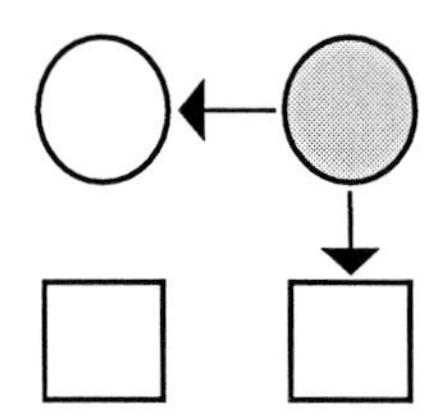

1. Looking for the target stimulus ((4+1)/2=2.5 cycles)

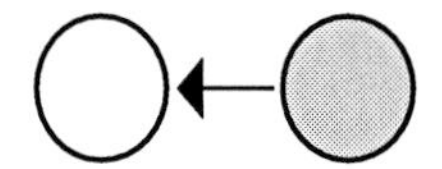

2. Looking for the *connected* non target stimulus (1 cycle)

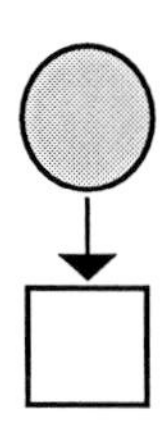

3. Looking for the *connected* target key (1 cycle)

This sums up to 4.5 cycles for the fast mode of a22y.

Now to the tasks a33y. First the slow mode:

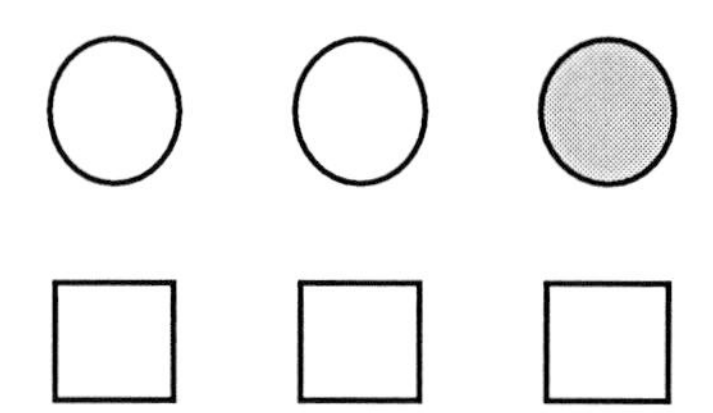

1. Looking for the target stimulus ((6+1)/2=3.5 cycles)

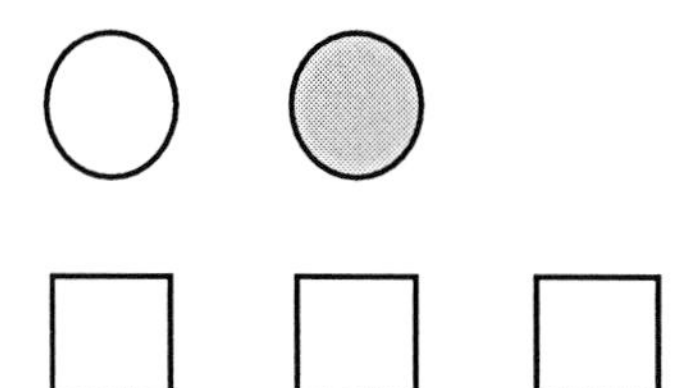

2a. Looking for (remembering) the first non target stimulus ((5+1)/2=3cycles)

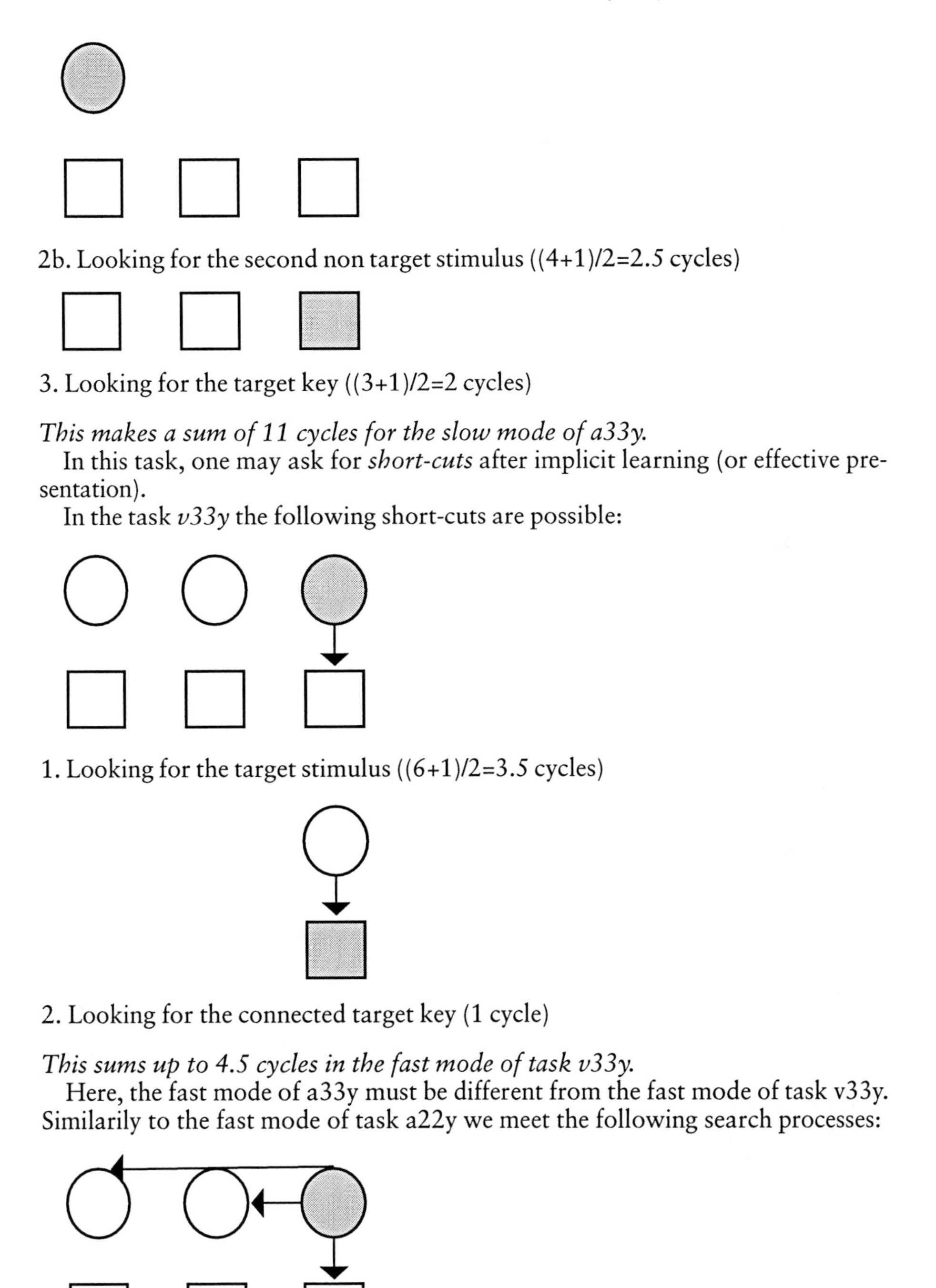

2b. Looking for the second non target stimulus ((4+1)/2=2.5 cycles)

3. Looking for the target key ((3+1)/2=2 cycles)

This makes a sum of 11 cycles for the slow mode of a33y.

In this task, one may ask for *short-cuts* after implicit learning (or effective presentation).

In the task *v33y* the following short-cuts are possible:

1. Looking for the target stimulus ((6+1)/2=3.5 cycles)

2. Looking for the connected target key (1 cycle)

This sums up to 4.5 cycles in the fast mode of task v33y.

Here, the fast mode of a33y must be different from the fast mode of task v33y. Similarily to the fast mode of task a22y we meet the following search processes:

1. Looking for the target stimulus ((6+1)/2=3.5 cycles)

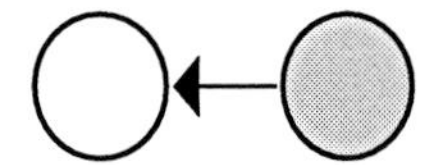

2a. Looking for the first non target stimulus (1 cycle)

2b. Looking for the second non target stimulus (1 cycle)

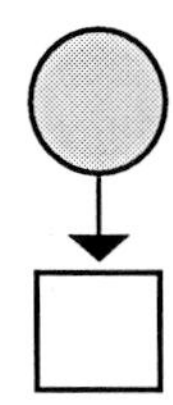

3. Looking for the target key (1 cycle)

Altogether we have 6.5 cycles in the fast mode of task a33y.

Empirical evidence

If one compares the monohemispheric auditory and visual reaction tasks of one subject with another, one sees that the auditory tasks need more time to be done. This is also seen in the computed cycle numbers: in a22y the auditory cycle numbers are 1–2 cycles higher than in v22y and in a33y the cycle numbers are even 3–5 cycles higher than in v33y (see tables above).

An important hint is the recurrence of the number 4 in the directly observed cycle numbers of tasks a22y. This proposes that in these tasks the number of four searching actions plays an essential role. There is an interval between a11y and the first peak in the distribution of a22y (daCN-method). Is there any process in this interval which does not produce a peak? The analogous visual tasks do not show this interval. Here the peaks of v22y start shortly after v11y reaction time.

There could be some other cycles in a22y which are important. These cycles are the minimal necessary cycles to solve the task. There cannot exist trials with reaction times shorter or equal to them.

This minimal number of cycles can be measured empirically and can be predicted by the above hypotheses.

Comparison of the directly observed auditory and visual cycle times

The direct observation of cycle times in the distributions of tasks is difficult. Pöppel has shown a preference of certain reaction times in the auditory trials of a mixed task. If one separates the distribution of single fingers and uses devices with exact timing, similar preferred reaction times can be observed. In these special reaction

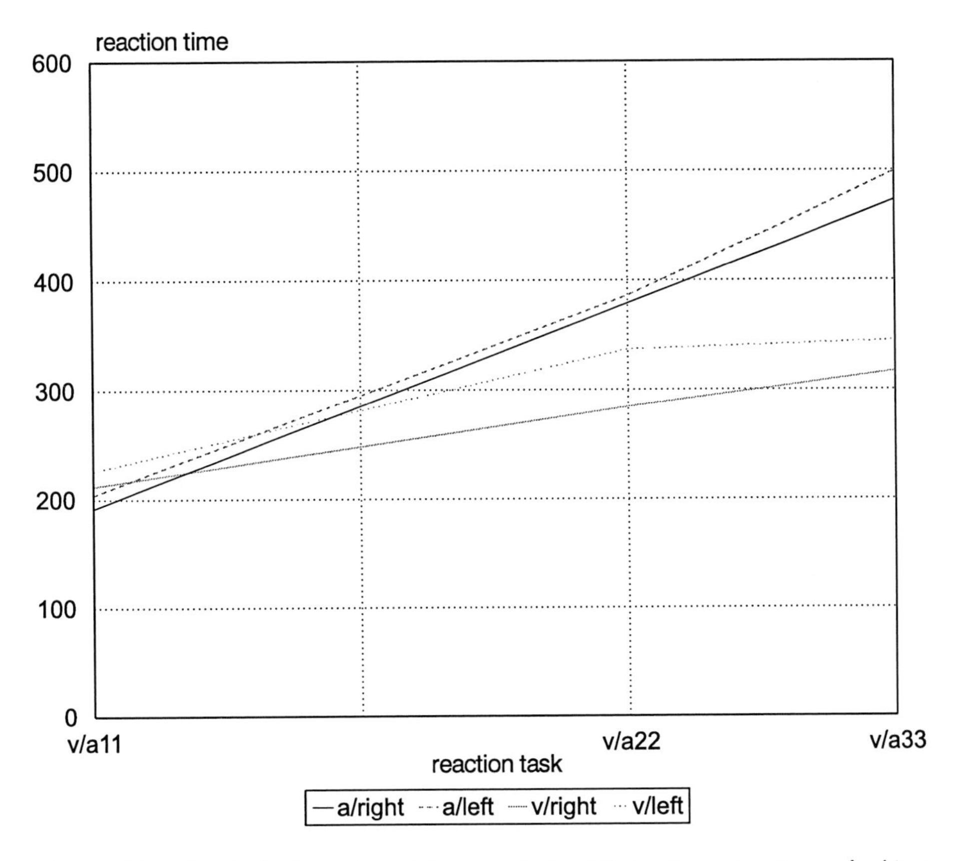

Fig. 34. Monohemispheric auditory (above) and visual (below) reaction times of subject P15

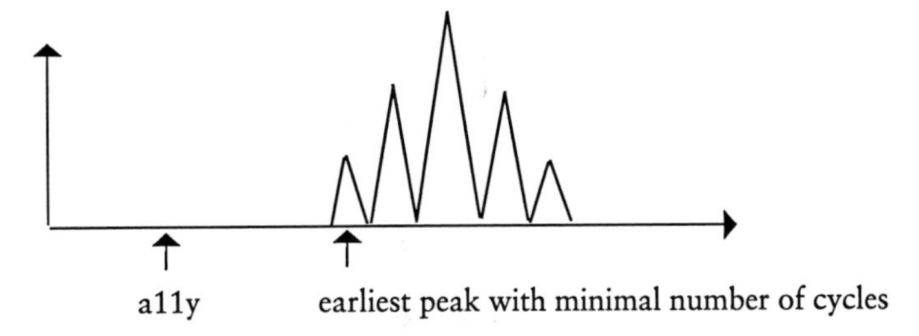

Fig. 35. The location of the earliest peak in a reaction time distribution

times, the frequency increases. The peaks must be relatively high and only the three or four highest are considered. The time differences of these peaks are summed up and the mean value is computed. This method is rather rough, better methods of direct observation have to be developed. Only differences occuring in nearly all subjects may be important.

To see these cycles better one has to use tasks with relatively few cycles like x22y. A different non-decision time causes a phase shift of this finger. The more fingers one uses, the more phase shifted curves sum up, the cycles getting unobservable.

Discussion of auditory and visual cycle times

This work has localized the cycle times between the deciding prefrontal areas of the left and the right hemisphere and their posterior counterparts. The tasks a22r and v22r should use mainly the left hemisphere and the left prefrontal decision area and the tasks a22l and v22l the right hemisphere with the right prefrontal decision area.

If the cycle times of tasks using the same hemisphere are significantly different, then the cycle times are dependent on the magnitude of the activities within the different pathways. It has been proposed that the non-dominant hemisphere (mostly right) is specialized in visual decisions and the dominant hemisphere in auditory decisions.

In right-handed people can be observed:

> In monohemispheric auditory tasks the right hand (i.e. the left hemisphere) is faster than the left hand (i.e. the right hemisphere) and in monohemispheric visual task the left hand (i.e. the right hemisphere) is faster than the right hand (i.e. the left hemisphere).

This specialization of the hemispheres could consist of providing more neurons within the dominant pathway.

Discussion of cycle times of lower areas (sensory or motor)

Is it possible to measure the cycle times of lower areas? In this case the cycle time is the linear time between three successive areas. This question is important because these rhythms may superimpose the cycles of the prefrontal areas and therefore disturb the direct visualisation of prefrontal cycle times.

Discussion of auditory and visual decision times

Pöppel finds the time difference between the two visual and the analog auditory tasks being the same:

mRT(v22) – mRT(v11) = mRT(a22) – mRT(a11) / mRT=mean reaction time

The subject P12 has the following values: mRT(a22rP12)–mRT(a11rP12)=186 ms. Taking the visual cycle time of 22 ms of the subject P12, the difference would imply 8.5 cycles more in the a22rP12 task than in the a11rP12 task. This is much more than the nearly 4 cycles difference in the comparable visual reaction tasks.

If one looks at the task v22r it is obvious, that the scanning mechanism uses the special picture on the screen with the stimuli being in the right half of the visual space to search for. The stimuli do not exist only in the memory but have to be used in their special appearance.

The reason why the number of cycles in monohemispheric auditory reaction tasks is doubled is unknown. One hypothesis is, that the stimuli are not present as in the bihemispheric visual reaction tasks. The layer 5 has to update the elements of the searching set permanently. This can be done either by scanning the elements of the searching set accidentally or in the succession of the last target stimuli and target keys. The updating of the elements is finished when all elements are scanned at least once. This needs 4 cycles only when the memory remembers all scans. An average of 4 cycles is needed, if only the two stimuli are updated, not the keys. This is not necessary because these are present all the time. Therefore it has to be shown, that updating the two stimuli accidentally needs 4 cycles on average.

Does the right hemisphere decide the visual tasks and the left hemisphere the auditory tasks?

Which hemispere does decide if a task is performed in which a sound in the left ear is answered by a key1 and a sound in the right ear is answered by a key2?

The specialization of hemispheres could explain why the reaction time of monohemispheric task v22l is faster than the equivalent bihemispheric time of v22.

If there is only one visual deciding area in the right hemisphere, the pathway of the task vNNr would have to change the hemisphere two times in order to be decided.

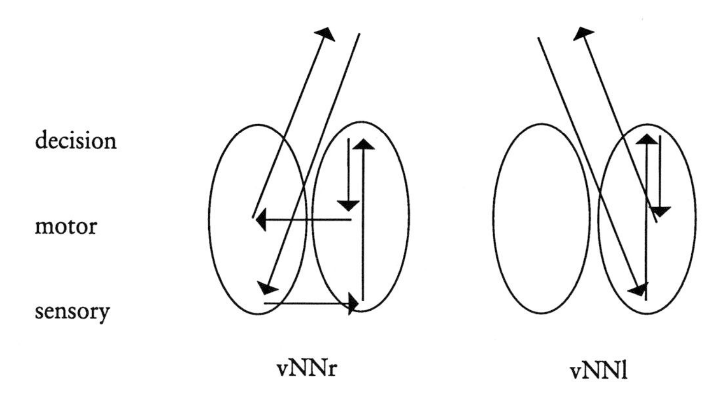

Fig. 36. Hypothetical location of the pathways in the cerebral hemispheres (tasks vNNy)

That would imply the right hemisphere takes part in both tasks, the left hemisphere only in the vNNl task. vNNl would be faster than vNNr. The difference between vNNr and vNNl is the two crossings of the interhemispheric space in vNNr. The pathway of vNNr could also go from left occipital to left frontal and then cross the hemispheres.

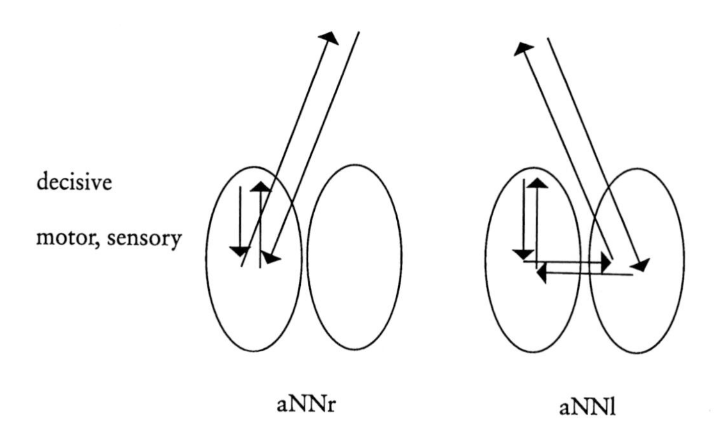

Fig. 37. Hypothetical location of the pathways in the cerebral hemispheres (tasks aNNy)

In the auditory tasks, aNNr is faster because it lacks the two crossings of aNNl.

This different dominance of the hemispheres could explain the different reaction times in monohemispheric tasks. There may be other explanations, e.g. the different number of neurons taking part in the pathways of the hemispheres.

If this theory is correct, the task vNNrl with the stimulus in the right visual field and the left hand pressing the keys must have a reaction time lying between the times of vNNr and vNNl because vNNrl crosses the hemispheres only one time:

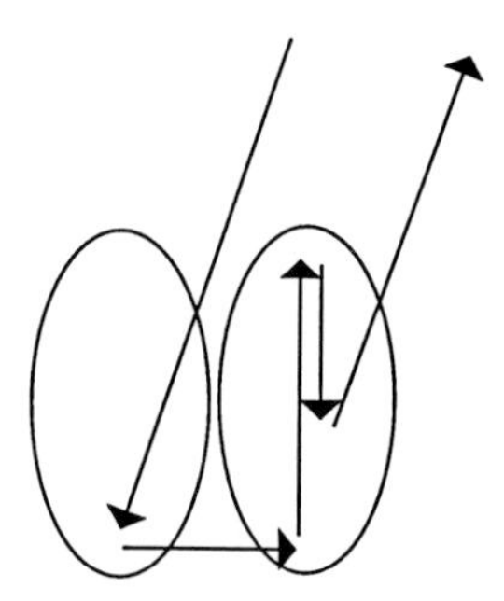

Fig. 38. The location of the pathways in the cerebral hemispheres (task vNNrl)

This has to be investigated.

The monohemispheric cycle times are nearly independent from the target key

The layer5 searches for all stimuli and keys with approximately the same probability in the layer2,3. The target key does occur only once shortly before pressing this key. That means the resulting searching time as the sum of the single searching trials depends mostly on the other elements and only once on the target key. If one separates the distribution of a task by the kind of the pressed target key and directly visualizes the cycle time for each key then these cycle times must be similar because of the previous arguments.

The directly visualized cycle times for every target key show that each finger has its own slightly different cycle time. This, too, is the reason why the slopes of the key reaction times for the left four fingers are different.

The intra-individual variability of reaction time

Introduction

It is known, that the reaction time depends on various factors like smoking, menstrual period, sleep etc. The question is whether the individual cycle time depends on the same factors or whether it shows some stability.

Method

10 subjects repeated the reaction tasks v11 to v1010, v22r, v22l, a22r and a22l. The first ten tasks v11 to v1010 are material for computing the cycle time by linear regression (method 1), the last four tasks are needed for the direct visualisation of cycle time (method 2).

The direct visualized cycle time was obtained by an automatic recognition program which defined a peak if the sum of rise and decline was greater than two in the distribution of reaction times for one finger. The time axis has been divided into 10 ms intervals. The recognition program measured the differences between the peaks, summed them up and computed the mean value. This mean difference between the peaks of the distribution is the direct visualized cycle time of this task. Because it is possible that one cycle does not occur in a distribution, distances between peaks similar to two cycle times obtained by method 1 are not used to compute the mean value. This would give a false mean value if two cycle times are considered as one.

The recognition program computed the slope of the regression line by the same method as SPSS (and was validated by SPSS).

In the repetition, the number of trials within the tasks v11 to v1010 was reduced to 50 trials for each task. The number of trials in the remaining tasks was not reduced.

Results

Reduction of directly observed cycle numbers in subjects

These reductions are imperative for selecting the hypotheses of information processing in reaction tasks. Because the computed cycle numbers have the disadvantage of using vCT until now, the computed cycle numbers (coCN) are not used here but the directly observed cycle numbers (doCNs). In the visual tasks, one gets the following table:

Table 20. Comparison of the directly observed cycle numbers in the tasks *v22r and v22l* in healthy subjects. A small letter behind the initials means repetition of the tasks. Only those subjects with repetitions are considered

	CN LMF	CN LIF	Mean	CN RIF	CN RMF	Mean
P15	5.5	3	4.25	4	2.5	3.25
P15a	4	3	3.5	4	3	3.5
P15b	5	4	4.5	3	3	3
P07	3	3	3	4	3.5	3.75
P07a	2	3	2.5	5	5	5
P14	4	3	3.5	3	3	3
P14a	3	3	3	3	2	2.5
P12	3	4	3.5	3	5	4
P12a	3	4	3.5	2.5	3	2.75
P06	4	4	4	5	5	5
P06a	2	4	3			
P06b	3	4	3.5			

And for the auditory tasks, the table below is valid.

Table 21. Comparison of the directly observed cycle numbers in the tasks *a22r and a22l* in healthy subjects. A small letter behind the initials means repetition of the tasks

	cn LMF	cn LIF	Mean	cn RIF	cn RMF	Mean
P15	5	4	4.5	6	5	5.5
P15a	6	6	6	6	4	5
P15b	3	3	3	5	4	4.5
P07	3	3	3	3	3	3
P14	3	3	3	4	3	3.5
P12	4	4	4	3	2	2.5
P06	4.5	2.5	3.5	4	4	4
P03	4	4	4	4	3	3.5
P01	3	4	3.5	4	4	4
P10	4	3	3.5	6	4	5

Discussion

Preliminary remark

The succession of stimulus and key fits perfectly into Hebb's rule: if two neural events take place together or shortly one after the other, the two events are connected, i.e. their succession is learned.

Reduction of cycle numbers in various tasks after full implicit learning

With the help of Hebb's rule the stimuli are directly connected with their attached keys so that any search for the correct key is superfluous. By this, a number of cycles are saved:

	before implicit learning	after implicit learning
v11y to v22y	4 cycles	2.5 cycles

The number of cycles after full implicit learning has to be *determined experimentally* and *compared to the various predictions* of the competitive hypotheses.

Measurement of reaction times in patients

The reaction times of patients with monohemispheric brain lesions

Investigating patients with a monohemispheric brain damage (CT or MRI) has the great advantage of a well defined reason. Therefore, their reaction time data are well suited for testing the hypotheses of this work. The design of tasks presented to these patients was the following:

v11y, v33y, a11y, a33y (50 trials each),
v22y, a22y (200 trials each)
m22y (100 trials each)
Together 400 + 800 + 200 = 1400 trials.

There are two problems to be mentioned. These subjects often show a right shift of distribution because of their slower reaction times. Therefore the range of interest has to be shifted to the right. The range used in this subjects begins with 300 ms and ends with 680 ms. Within this range the cycle time is observed directly.

The second problem is the interpeak distance in the reaction time distribution. If the cycle time is not prolonged too much, the maximal distance used to compute the mean cycle time is 50 ms (maximum < 2* normal cycle time). With a cycle time of about 30 ms, the maximum of 50 ms is a good upper boundary.

The following items were considered:

- a comparison between a11r and a11l (auditory input/output time prolonged?)
- a comparison between v11r and v11l (visual input/output time prolonged?)
- if the a11y is prolonged on one side and v11y not (or vice versa) then there can be no output damage but the cause of the prolongaton must be an input damage.
- if a11y and v11y are both prolonged on one side then an output damage is possible.
- is (a22y-a11y) and (a33y-a11y) prolonged or (v22y-v11y) and (v33y-v11y)? If the differences are not prolonged and the a11y (v11y) time is prolonged then an input/output damage is prevalent.
- if the difference is prolonged but the quotient (difference/directly observed cycle time) is normal then the cycle time alone is the cause of the prolonged difference.
- if the difference is prolonged and the quotient (difference/directly observed cycle time) is higher than normal then the cycle number is increased.

- is the modality shift prolonged on one side relative to the other, for one modality or for both?

All subjects were naive, that is unpracticed when doing the tasks for the first time. Some subjects repeated the tasks for a second time, then the two investigations are indicated by small letters behind the initials.

The reaction times of patients with schizophrenia

Introduction

The neurobiology of schizophrenia is reviewed by Tamminga (1997). Because neuroleptics reduce cerebral metabolism in the frontal cortex (Holocomb et al. 1996b), it is essential to conduct PET studies on drug-free individuals. Tamminga et al. (1992) found reduced metabolism in anterior cingulate and hippocampal cortices in young, floridly psychotic schizophrenic individuals compared to healthy control subjects. However deficit-type patients showed additional reduced metabolism in the frontal and parietal cortex and thalamus compared with the nondeficit group.

Electrophysiologically schizophrenic patients have an inability to suppress a response to a sensory event. Healthy individuals show a decrease in evoked potential response to a second click (preceded by an identical first click), suggesting that a normal mental refractory period exists to duplicate sensory information. This phenomenon is not apparent in schizophrenic individuals. This loss is also seen in some nonschizophrenic family members who do not display psychotic symptoms (Tamminga 1997, Freedman et al. 1983, Judd et al. 1992, Siegel et al. 1984).

Liddle (1992) found hallucinations/delusions positively associated with rCBF in the left parahippocampal gyrus and the left ventral striatum. Disorganization was associated with flow in the anterior cingulate and mediodorsal thalamus.

Recently in a PET study, patients with hallucinations activated the left and right thalamus, right putamen, left and right parahippocampal area and right anterior cingulate (Silbersweig et al. 1995)

PCP/ketamine is the drug class that most faithfully mimics schizophrenia in healthy persons and most potently and validly exacerbates schizophrenia symptoms in schizophrenic patients. Lahti et al. (1995) showed increased rCBF in the anterior cingulate and decreased rCBF in the hippocampus and lingual gyrus.

Altogether, schizophrenic patients activate brain areas prematurely and not in relationship to the difficulty of the task. Perhaps this could result from abnormal basal or thalamic regulation of some aspects of frontal cortex cognitive function (Tamminga 1997).

The reaction tasks of schizophrenics are altered in some characteristic ways. There is a long known increase of reaction time (eg Maier et al. 1994, Carnahan et al. 1997) and effects like inhibition of return (eg Huey et al. 1994), the modality shift effect (Hanewinkel et al. 1996, Maier et al. 1994, Ferstl et al. 1994) and the crossover effect (Maier et al. 1994). Some authors have examined the relationship between reaction tasks and evoked potentials (Salisbury et al. 1994a, Salisbury et al. 1994b, Strandburg et al. 1994), some the saccadic reaction time (Belin et al. 1995).

The amplitude of P300 is reduced, the latency of P300 prolonged, there are difficulties in screening out distracting stimuli, the P300 abnormalities correlate with left sylvian fissure enlargement (Neylan et al.1997).

Spitzer (1999) says that reducing the contrast impairs the singleness of purpose and the orientation of the procedures to an objective. A reduced contrast can change the formation of the successive sets. The sets may now include extraneous elements and omit task relevant elements. A blurred structure permits certain chances because of ist creativity at best. If unproportional extended, there will be difficulties in achieving its goal because of the interferences which hinder the purposeful advance of the process. The structure would be weakened and its influence on other parts of the brain diminished. These parts would behave on their own.

Methodical adaptions

Because the reaction times of patients with schizophrenia are prolonged, the range of the program by which the cycle time may be directly observed has to be adapted (as it was done in patients with brain damages).

The other constrictions on measured values have to be mentioned: the computer program which recognizes the directly observed cycle times automatically excludes all periods greater than 50 ms when computing the mean value.

Finally, the program which computes the median values excludes all reaction times shorter than 100 ms and longer than 2000 ms. Therefore outstanding reaction times like 17 sec, which were needed once by a patient with schizophrenia, do not contribute to the median value.

Graphical presentation of different influences on reaction time

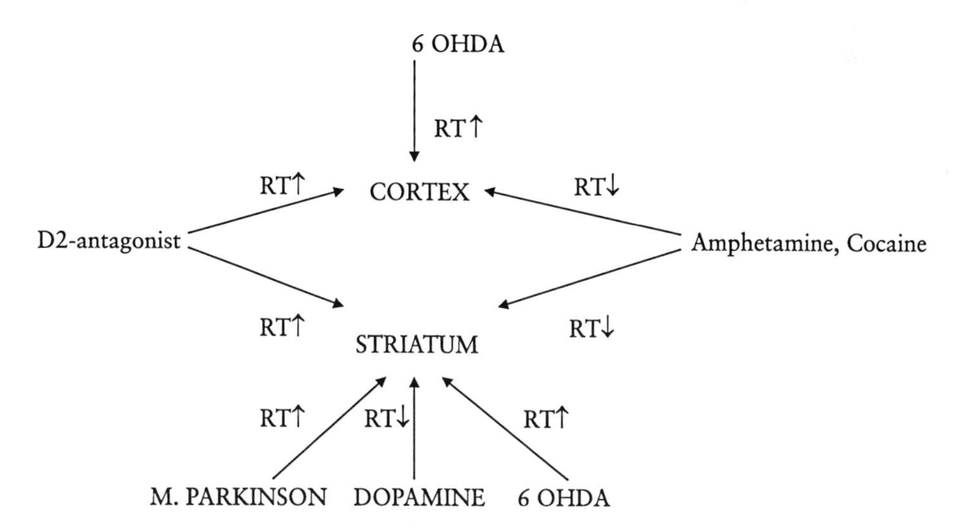

Fig. 39. The different influences on reaction time

The effect of dopamine agonist depends on the dose: high dose agonist lead to an increase of reaction time, low dose agonists to a decrease of reaction time (see Dicussion of references, page 322). In order to decrease the reaction time, dopamine has not only to concentrate the neuronal activity but to increase the activity of the remaining cells.

The event-related potentials of reaction tasks

The ERP of auditory reaction tasks

Introduction

Do the epochs of the reaction time data appear anywhere in the appropriate evoked potentials if a task and its evoked potential are made together? One can try to subtract the evoked potential from the task related potential. Schroeder et al. (1995) have used ERP(a21)-ERP(a11).

Paz-Caballero et al. (1992) have used a task with 9 possible positions in two possible arrangements (square or circle). The subjects were asked to respond to the square containing the spot in the top central position by pressing a key. Therefore their task could be coded as v11,1 (9 locations and two forms). They then applied a subtraction procedure to compute ERP(v11,1)-ERP and obtained very interesting differences between the ERPs in discrimination and passive tasks.

Their difference potential called 2–1 (ERP of discrimination task minus ERP of passive task) lasts about 500 ms.

Paz-Caballero et al.(1992) found three ERP components related to stimulus selection processes: N2, P3 and N3. Especially the N2 component has been associated with stimulus evaluation according to task relevance.

The subtraction potentials

The subtraction potentials of the one stimulus tasks (a11li – a10li) and (a11ri – a10ri) and the two stimuli tasks (a22li – a10li) and (a22ri – a10ri) were computed.

The neural correlates of positive and negative evoked potentials

Stöhr et al. (1989) write about the neural correlates of positive and negative surface potentials.

Because of the complexity of cerebral structures, it is hardly possible to draw conclusions from the polarity of evoked potentials to the basic generators. The excited area is negatively loaded because of the current of positive ions into the cell interior. Depth electrodes in the layer IV of the primary sensible cortex show a negative potential but the simultaneous applied surface electrodes show a positive potential which becomes negative some time later when the dendrites of the layer IV cells become depolarized. A negative surface potential must not be caused by an excitation of the cortical area below the electrode but this can cause a positive surface potential as well, especially if the dipole is tangential to the brain surface.

The knowledge from PET and fMNR about the neural correlates of auditory reaction tasks

Remy et al. (1994) investigated the cerebral blood flow with positron emission tomography induced in 10 healthy subjects by two different tasks: a repeated flexion-extension of all fingers and a repeated flexion-extension of the middle finger. The all-finger movement only activated the primary sensorimotor cortex (SM) and the supplementary motor area (SMA) contralateral to the movement. *However the SMA activation was only observed when the movement was triggered by an auditory cue* but not when it was self-paced.

Holocomb et al. (1996) and Holocomb and Lahti (1997) generated an rCBF comparison between schizophrenic patients and matched healthy control subjects while both groups are performing a practiced auditory recognition task. The task involves discriminating between two auditory tones. The sensorimotor control task involves repetitive tone and motor stimulation, there also is a rest condition task. Image analysis resulted in two subtracted group images: sensorimotor control task minus rest and auditory discrimination task minus sensorimotor control task. *For the decision task, healthy control subjects activated the anterior cingulate, right insula, and right middle frontal cortex.*

Pugh et al. (1996) did a fMRI investigation with 25 adult subjects who discriminated speech tokens ([ba]/[da]) or made pitch judgements on tone stimuli (rising/falling) under both binaural and dichotic listening conditions. Under the dichotic conditions activation within the *posterior (parietal) attention system* and at primary processing sites in the superior temporal and inferior frontal regions was increased. The cingulate gyrus within the *anterior attention system* was not influenced by this manipulation.

The correlation between the latencies of the single potentials and the structure of the task a11

Can the latencies give us some information about the intermediate steps of the task a11?

Stöhr et al. (1989) tell us that in AEP the first cortical acitivity begins at 30ms. Other authors propose times of 10–20 ms for the first cortical response. Therefore Neg1 has something to do with this first cortical acitivity.

On the other hand, it is known that from the motor cortex to the ERP onset about 20 ms are needed and from that to the successful pressing of the key another 50 ms are needed. Because the pyramidal cells of the motor cortex lie in layer5 this potential could be positive. Therefore Pos2 has something to do with this motor potential.

Part II

The spatiotemporal structure of stimulus-response pathways

Measurement of elementary time

The procedure "NESTLE" in a computer program called "FPM31e"

The NESTLE procedure applied to a 5 millisecond time scale of reaction times (program FPM31e, procedure NESTLE)

Within the program FPM the NESTLE procedure computes the difference abs((con+ o*ET)-peak(n)) for each possible elementary time, ET, and for each peak(n) with abs = absolute value, con = constant value (either 70ms in auditory tasks or 120ms in visual tasks), o is that *integer* which minimizes the above difference, o*ET is an integer multiple of the elementary time, the term (con + o*ET) approximates the reaction time of the peak(n) with a minimial remaining difference (it "*NESTLEs to*" the reaction time of peak(n)).

This is done for all peaks of one task with varying integers, o. Then all the minimal differences abs((con+o_n*ET)-peak(n)) are summed up to the value sum(ET). This value is the sum of all minimal differences for one special ET. At last, the ET with the minimal sum is selected.

The NESTLE procedure applied to a 1 millesecond scale of reaction times (program FPM26f58)

For some subjects the distinct peaks fall in a natural distribution of reaction times on a millisecond time scale. For other subjects, this natural distribution is rather equalized having no peaks distinguishing from the background activity. In these cases, the time scale has to be revised using 5 millisecond columns, for example.

The subjects with distinct peaks in the natural distribution, however, offer an additional chance to measure their elementary times. In principle, at least two different methods (Chronophoresis, FPM, natural distribution, repetitions) should show coincident results in one subject. In some subjects, the distribution of reaction times on a 1 millisecond time scale shows distinct peaks which can be used by the NESTLE procedure.

Example: v22rO08A1+2:
ET(v22rO08A)=15

Peak(n)	Peak(n)-con	Quotient	Approximat.	Difference
peak(1)=210	210-120=90	90:15=6		0
peak(2)=225	225-120=105	105:15=7		0
peak(3)=235	235-120=115	115:15=7.7	8*15=120	abs(120-115)=5
peak(4)=245	245-120=125	125:15=8.3	8*15=120	abs(120-125)=5
peak(5)=255	255-120=135	135:15=9		0
peak(6)=270	270-120=150	150:15=10		0
			sum:	10

Fig. 40. Construction of the NESTLE procedure in an example. All values apart from the quotients are in milliseconds

Application of the NESTLE procedure of FPM to all tasks of a subject

Both the 5 millisecond version (FPM31e) and the 1 millisecond version (FPM26f58) are applied to all the tasks of a subject: a11r, a11l, v11r, v11l, a22r, a22l, v22r, and v22l. In some subjects the tasks x33y have been added. In each case, the program (FPM31e or FPM26f58) prints the following figure where the distribution is at the right side and the sums of the NESTLE procedure at the left side:

S16A v11rS16A1 FPM31e 26.05.2000 start: 0

Median: 288.5 Fp: 215 Median-Fp: 73.5 Fp-con: 95
NESTLE index: 11 → 16 14 → 19 16 → 27 20 → 25 23 → 34

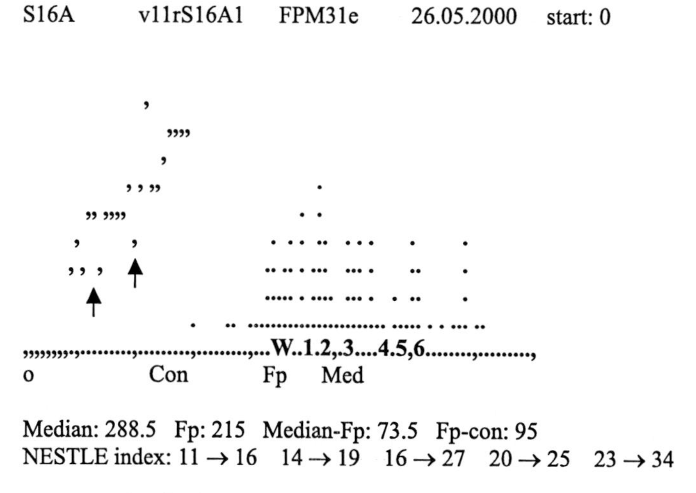

Fig. 41. FPM31e result of subject S16A. There are two relative minima at 14ms and at 20ms (arrows)

The minimal sums of the NESTLE procedure are read from the left part of the above figure (arrows) and registered within the table below. If there is no clear minimum, a question mark is put into the corresponding box of the table.

Table 22. Convergence table of patient S16A

S16A	FPM31e (x11y)	FPM31 (x22y)	FPM26f58 (x11y)	FPM26f58 (x22y)	Results	Chrono-phorese	Median-Fp method
aNNr	15,21	22	15	15,18	15		
aNNl	15,26	15,22,27	(24,35)	14,18,35	15		
vNNr	14,20	10,15,23	?	15,30	15		
vNNl	?	16	?	14,33	15		

Now the agreements are applied and the results are calculated.

In some subjects all the tasks were repeated some time later. Then, the proposed methods were applied to the results of the repeated tasks. In all subjects, a replication of the elementary times (±2ms) was possible. This is the proof that the applied methods measure the elementary times of the subjects with an error maximum of ±2ms. Furthermore the ±2ms constancy of individual elementary times at different testing times has been proven.

In the end, four elementary times have been obtained for each subject. The methods cannot distinguish between elementary times specific for the x11y or the x22y tasks.

Convergence of the results of FPM31e, FPM26f58, and chronophoresis

In most subjects, only the convergence of the four FPM results is investigated. The chronophoresis and the Median – FirstPeakMethod give only further evidence for the correctness of the elementary times. The convergence of the chronophoresis results and the FPM results is striking in some subjects (e.g. H03A and H03B, H05B with SING104r(x11y))

Table 23. Convergence table of subject H03A

H03A	FPM31e (x11y)	FPM31 (x22y)	FPM26f58 (x11y)	FPM26f58 (x22y)	Results	SING100n (x11y)	Median-Fp method
aNNr	14	14	15	13.5	14	13.5	
aNNl	21	15,23	15,19	15	15	15	
vNNr	?	23(2Q)	13	12	12	12	
vNNl	?	?	13,17	12	13	12	

In this case, the *convergence principle* compares all the results (SINGLE100n, SINGLE100r,u, (median – fp)/n method, (fp – con)/m method and NESTLE index and takes that proposal as the elementary time which best fits the most methods.

Problems

If the FPM results of one subject do not converge, then the measurement has to be repeated.

Another problem is that the natural distribution rarely shows half-number multiples of elementary times lying between the whole number multiples. Either these peaks are produced by random or the elementary time is half the value or there is some additional unknown reason for these peaks.

The confrontation of FPM with artificial data

The quality of the NESTLE procedure in FPM26f58 depends on the localisation of the peaks in the 1ms distribution. The version FPM2632a uses artificial latencies to show the results of the NESTLE procedure.

RT1=70+4*17=138
RT2=70+7*17=189
RT3=70+8*17=206
RT4=70+11*17=257

Peaks: 138 189 206 257

Fig. 42. Results of the program FPM2632a with four artificial latencies as input. The coactivation of 23ms with 17ms (arrows) is remarkable. The reason is that both numbers share similar multiples

138/23=6
196/23=8.22
206/23=8.96
257/23=11.17

The less the quotients differ from integers, the more the divisor is proposed as elementary time. If only one reaction time deviates from the construction (integer*elementary time), the NESTLE results deteriorate.

The Chronophoresis of x11y, x22y, and x33y

The NESTLE procedure applied to a set of reaction times (program SINGLE)

The program SINGLE works in the following manner:

1. It takes sequentially all values of possible elementary times (e.g. between 8 ms and 40 ms) and approximates each reaction time of a set by the term RT = con + n*ET + remainder.
2. Then it sums up the remainders of all reaction times of the set. This sum is a measure of the quality of approximation between one possible elementary time and the reaction times of the set. The less this sum is, the better the approximation.
3. The sum of remainders may be figured by a graphical display in which each qustionable elementary time is attached by its sum of remainders (see figure below).
4. The elementary times with the best approximation indicator may be displayed in a so called chronophoresis figure, initially for one set of reaction times.

SYNOPSIS OF RELATIVE MINIMA OF x11y
O13A **singor07** 28.05.1999 variant: 1 threshold: 1

	a11r	a11l	v11r	v11l
6.5	.			
7	.			
7.5	.			
8	.			
8.5	.			
9	.			
9.5	.			
10	.			
10.5	.			
11	.11	22		
11.5	.11	22		44
12	.11	**22**		44
12.5	**.11**	22		**44**
13	.11	22	33	44
13.5	.11	22	33	44
14	.11		**33**	44
14.5	.		33	
15	.		33	
15.5	.			
16	.			

Fig. 43. SINGor07 limits the single reaction times used to compute the elementary time to the upper boundary of (median+50). The subject O13A shows the following plots in x11y (SINGor07): ET(a11r) = P12.5, ET(a11l) = P12, ET(v11r) = P14, ET(v11l) = P12.5

5. This procedure is repeated for many subsets of the 100 trials of a task.
6. At last, a synopsis is displayed in which each questionable elementary time is assigned to the number of sets in which the elementary time showed the minimal sum of remainders. The number of sets is represented as a number of figures.
7. A threshold suppresses step 6 if the number of sets in which the elementary time showed the minimal sum of remainders is below this threshold. Then the program generates the chronophoresis as shown below.

For each task, the best plot has to be found. If the plots possess a maximum, then this is the proposed elementary time, if a plot possesses an uneven number of several adjecent maxima, then the middle value is taken as the elementary time. If a plot possesses an even number of adjecent maxima, the value above the middle is taken as the elementary time. If a plot clearly deviates from the plots of the other tasks, then it may be taken as twice the elementary time if it is twice as much as its neighbors (one searching cycle consists of two elementary times, this 2ET can be detected by chronophoresis, too).

If a task has several plots with the same maximum, then it must be tested whether there are two plots standing in a 1:2 relation (detecting ET and 2ET) or some other relation. If this is not the case, the plot is taken which lies nearer to the plots of the other tasks.

Example: The subject O13A shows the following plots in x11y (SINGor07):

ET(a11r) = P12.5
ET(a11l) = P12
ET(v11r) = P14
ET(v11l) = P12.5

This procedure may be repeated for all the tasks which were performed by the subject. In most cases the x11y and x22y tasks were evaluated by this way, in some subjects the x33y tasks were evaluated as well. The versions SINGLE100n and SINGLE100r use different subsets of reaction times and therefore can be used as controls of each other. The rest of this chapter deals with the limitations of the chronophoresis.

The chronophoresis of x11y gives better results than that of x22y or x33y

The chronophoreses of some subjects (e.g. O07A) have taught me that the chronophoresis of x11y is more precise than that of x22y or x33y. The reason for this observation are the larger reaction times in x22y and x33y. *The larger the reaction times, the more the chronophoresis can find competitive elementary numbers which fit into these reaction times.* (for example 60, 15 is contained four times and 20 three times in 60, the larger the reaction time, the more numbers can be found which are contained in it).

This is true, too, for the nestling procedure in the xNNy distribution where the peaks which lie at the extreme right side of the distribution may be not so reliable as the peaks which lie at the left side of the distribution.

The consequence of this observation may be an improvement of the chronophoresis procedure by

1. limiting the reaction times used to produce the chronophoresis
2. reducing the reaction times (e.g. by linear transformation RT′ = RT – 200).

Difficulties in distinguishing between certain elementary times

The numbers 13 and 15 for example have similar multiples. Therefore the chronophoresis often proposes both numbers as solutions.

13	15
26	30
39	45
52	60
65	75
78	90
91	105
104	120
117	135

The difference between SINGLE104r and SINGLE106n

At the moment, I use two different variations of the SINGLE program: SINGLE106n and SINGLE104r.

Each task consists of 100 trials (reactions). In SINGLE104r, blocks of x11y trials are defined in the following way: 1 to 16, 17 to 32, 33 to 48, 51 to 66, 67 to 82, 83 to 98 and blocks of x22y trials are defined in this way: 1 to 33, 34 to 66, 67 to 99.

In each of these blocks, up to ten trials are taken to compute the elementary time which approximates the reaction times of these trials best.

In the other version, SINGLE106n, the x11y blocks are defined from 5 to 20, 21 to 36, 37 to 52, 53 to 68, 69 to 84, 85 to 100 and the x22y blocks from 5 to 37, 38 to 70, 71 to 100.

The blocks of SINGLE104r and SINGLE106n overlap. Therefore similar results may be expected. It would be better if the blocks of the two versions were totally different.

Artificial data

One can feed SINGLE with artificial data that are numbers which contain a common element: number(n) = con + n*element. If the program is effecient it should find this element. If the artifical reaction times are given by the term art(n)= 120 + n*20.1 (with n=3 to 11), then one gets the chronophoresis below. The program finds the artificial elementary time plus the half of it.

```
artificial data:        singl95f            x11y ( 0 to 0 )
        a11r         a11l          v11r          v11l
6.5  .
7    .
7.5  .
8    .
8.5  .
9    .
9.5  .
10   . 11111        22222         33333          44444
10.5 .
11   .
11.5 .
12   .
12.5 .
13   .
13.5 .
14   .
14.5 .
15   .
15.5 .
16   .
16.5 .
17   .
17.5 .
18   .
18.5 .
19   . 111111111    222222222    333333333     444444444
19.5 .1111111       2222222      3333333       4444444
20   . 111111111    222222222    333333333     444444444
20.5 .11111111      22222222     33333333      44444444
21   . 11111111     22222222     33333333      44444444
21.5 .1111111       2222222      3333333       4444444
22   .
22.5 .
23   .
23.5 .
```

Fig. 44. Artificial data art(n)=120 + 8*17.5 + n*20.1 with n=3 to 11, program SINGL95f

Attributes of elementary times

The intra-individual stability of elementary times

The replication of the tasks by one subject and the independent evaluation results in approximately the same elementary times (±2ms). The reason for this stability is not known.

The most reaction time experiments were conducted between 1 and 6 pm. Is this a prerequisite for the replicability of elementary times? Most of the subjects were young adults between 20 and 30 years old, there were more men than women.

The latencies of event-related potentials prove that the elementary time is no mean value but is used throughout the pathway as a basic component. The elementary time may be explained as the time between two neurons. The elementary time is named in many other ways in the literature. For example, the interhemispheric time is the time which is needed to cross the corpus callosum. Kosslyn et al. (1999) give a value of 15ms.

Because the axons are of different lengths, the time to cross the synapse, to integrate the postsynaptic potentials and to trigger the new action potential may be the steps which determine the elementary time. The most important advantage of a stable, replicable elementary time is its use as the basic component of pathways. So it is of great help in understanding the structure of the pathways of mind.

The symmetry of elementary times

There are the following possible symmetries of elementary times between the different mental pathways.

1. All mental pathways have the same elementary time, ET.
2. There are two different elementary times: ET(xNNr) and ET(xNNl). That means all elementary times at the right side are equal and all elementary times at the left side are equal. This is the case in subject H32A.
3. There are four different elementary times: ET(vNNr), ET(vNNl), ET(aNNr), and ET(aNNl). This is the case in subject H01Z.
4. All mental pathways have different elementary times (ET(a11r), ET(a11l), ET(v11r), ET(v11l), ET(a22r), etc.)

The symmetry decreases from 1 to 4.

Measurement of pathway structure

The linear and cyclical part of the pathway (FPM31e)

The elementary times ET(aNNr), ET(aNNl), ET(vNNr), ET(vNNl)

As has been said before, the elementary time is the fundamental unit in order to understand the structure of mental pathways which are thought as being composed of n elementary times.The number, n, is called the elementary number EN of the pathway. The methods suggested in this book permit the differentiation of four elementary times: auditory or visual, right or left hemisphere. The future must show whether there are less or more elementary times.

The input time and the output time are summed up to the constant time (CON)

The time to reach the first layer in the primary sensory area is called the input time. The time from the last layer of the motor area to the successful press of the key is the output time. The visual input time is approximately 50ms, the auditory input time approximately 0 ms. The visual and the auditory output time is approximately 70 ms. (The input time can be measured by event-related potentials, the output time by single unit recording)

The first peak, FP, of a reaction time distribution is an indicator for the minimal pathway

Minimal pathways have the advantage of being composed of fewer elementary times and therefore showing a simpler structure than longer pathways. Because the computer programs process these data automatically, they have to identify the first peak. For this reason the identification process has to use explicit conditions like:

```
if freq(n)>1 then Fp = n
if freq(n)>0 and (freq(n–2)+freq(n–1)+freq(n)+freq(n+1)+freq(n+2))>=3 then Fp=n
```

The first condition is used to indentify the first peak in the x11y distributions, the second condition in the x22y distributions. The first peak can be detected very well in normal x11y distributions. It is harder to detect in x22y distributions.These distributions are more widespread (complex) than the x11y ones, therefore a more complex condition is needed to avoid incorrect identifications.

(Fp-2ET) divides the linear from the cyclical part in minimal stimulus-response pathways

This is another important point at the x-axis of the reaction time distribution. Its meaning can only be understood on a theoretical basis. The hypothetical neural representation suggests that there are two parts of the pathway within the cortex: a linear part and a cyclical part. According to this theory, the point (Fp-2ET) divides these two parts in the minimal stimulus-response pathway. It has to be shown that this division by (Fp-2ET) is valid in *all* pathways.

The median reaction time MEDIAN

The median reaction time or simply called, MEDIAN, is the reaction time with half of the other reaction times lying below and half of the other reaction times lying above it. The median pathway is used as the standard pathway longer than the minimal pathway, and it is tried to describe its structure mathematically, representing all other pathways of one task. If one imagines the bundle of pathways passable by one task, the median pathway is taken as the representative for all the other pathways.

linEN=(Fp-con)/ET-2 is the number of elementary times in the linear part of the minimal pathway

This statement can be derived from the hypothetical neural representation of the stimulus- response pathways. It is taken as the length (linEN) of the linear part of the minimal pathway. It has to be shown that this is valid for *all* pathways.

cycEN=(Median-Fp)/ET+2 is the number of elementary times between the linear part of the minimal pathway and the median length of the pathway

It is suggested that this number equals the number of elementary times in the cyclical part of the median pathway

This statement, too, can be derived from the hypothetical neural representation. It has to be proven that the median pathway uses the same linear part as the minimal pathway. There are *two evidences* which support this outstanding role of the linear part of the minimal pathway: if the linear part is visible in event-related potentials it must be a component of many different pathways, otherwise it would not emerge in the even-related potenial. The other evidence is the simulation of the reaction time distribution with the linear part of the minimal pathway as an important component of the simulating model.

This evidence does not rule out the possiblilty that some reaction times are composed of some other (longer or shorter) linear pathway. Indeed, the rare reaction times lying left of the first peak must have a shorter linear pathway. More important is the question whether there is a *considerable* number of trials (reaction times) which use a longer linear pathway.

Altogether, the length of the linear pathway is very important because it determines the length of the cyclical pathway and therefore the strategy which is used during this period.

Example of the NESTLE results and the reaction times distribution of one task

In both the chronophoresis and the FPM, only the responses of the index finger are evaluated.

FPM31e uses the distribution of reaction times of one task xNNy with 5 ms intervals. FPM computes the median of the xNNy distribution (only for the index finger), the first peak, the epochs (median – fp), and (fp – con).

It takes the peaks of the xNNy distribution and computes an elementary time from these peaks in the same way as the chronophoresis works. But this time, the input are the peaks of the xNNy distribution ("NESTLE index").

These Figures can be produced for every task x11y, x22y, and x33y.

H23 a11rH23 FPM31e 28.05.2000 start: 0

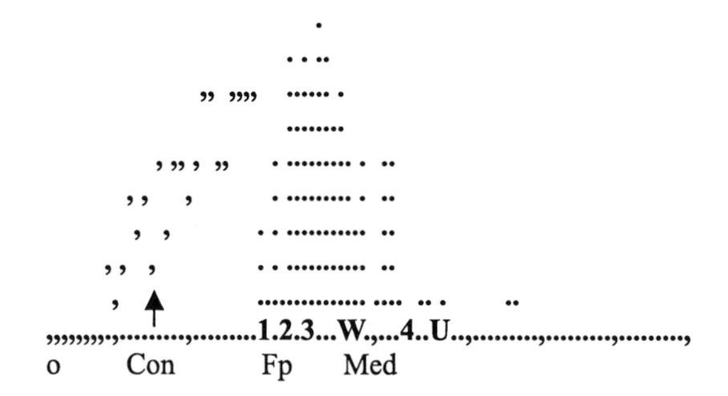

Median: 195 Fp: 150 Median-Fp: 45 Fp-con: 80
NESTLE index: 10 → 5 13 → 20 *15 → 10* 17 → 21 20 → 25 24 → 27

Fig. 45. The NESTLE results (left) and the reaction time distribution (right) of the task a11rH23. The length of the linear pathway is $linEN = (Fp-con)/ET - 2 = (150-70)/15 - 2 = 80/15 - 2 = 5.3-2 = 3.3$, and the length of the cyclical pathway $cycEN = (median - Fp)/ET + 2 = (195-150)/15 + 2 = 45/15 + 2 = 5$ The constant time, the first peak and the median reaction time are indicated by the program

Pathway information from the 1ms-distribution (FPM26f58)

Is it possible to measure the first peak in the 1ms-distribution of FPM26f58 more precisely than in the 5ms- distribution of FPM31e?

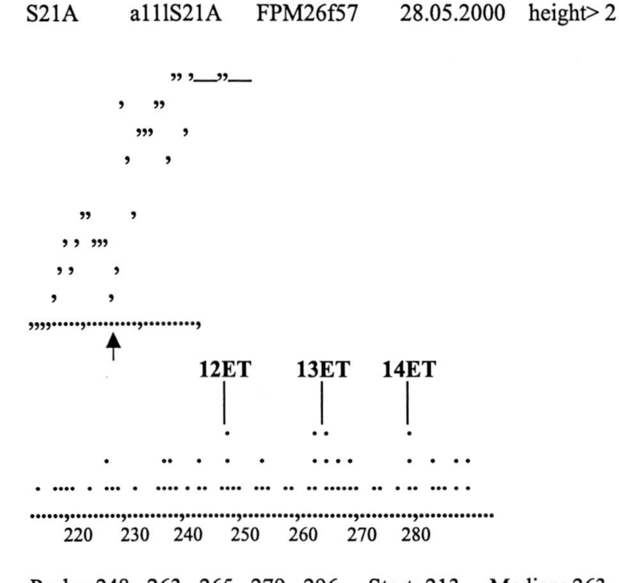

Fig. 46. The fine structure of FPM26f58(a11lS21A) contains peaks at 12ET, 13ET, 14ET, and 15ET. That means, a constant linEN plus multiples of 2ET cannot explain these peaks

Therefore, different linEN must exist, for example:

6 linET + 6 cycET = 12ET
5 linET + 8 cycET = 13ET
6 linET + 8 cycET = 14ET
5 linET + 10cycET = 15ET

Other structures may give the same results:

9 linET + 4 cycET = 13ET
etc.

It is safe to say that different linEN must exist to explain the observed peaks.

Simulation of a reaction time distribution using the program SIMx11y

The observed distribution of a11lS21A is different from the simulated one because the former starts with few fast reaction times and more intermediate reaction times:

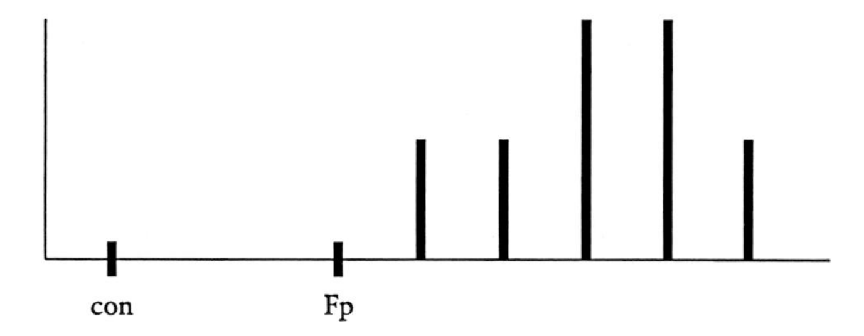

Fig. 47. Type of observed distribution of a11lS21A

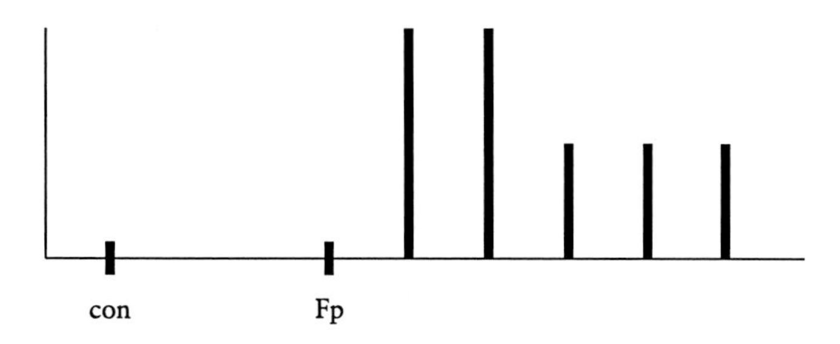

Fig. 48. Type of simulated distribution of a11lS21A

If different pathways use different linEN, the median cannot be used to determine the strategy of the searching set. The median can only be used for this purpose if linEN is (relatively) constant.

Hypothetical neural representation

The cortical structure of the sensory portion of the visual pathway

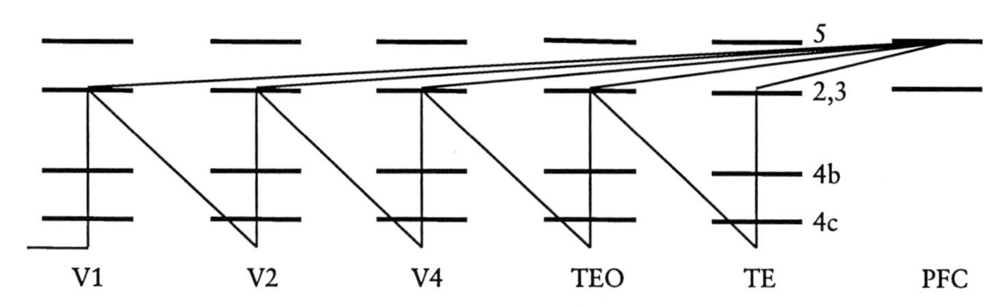

Fig. 49. The hypothetical cortical structure of the temporal visual pathway. The sequence of sensory areas has been taken from Reid (1999)

The prefrontal cortex may insert at each area in the layer 2,3. Therefore the visual pathway may vary between one visual area (-> 3linET), two visual areas (6linET), three visual areas (9 linET), four visual areas (12 linET), and five visual areas (15 linET). A linear portion of the visual pathway which is longer than 15 linET may not be explained by this model. The linear pathway contains (n–1) ET in the visual areas and 1 ET in the motor area.

RT = con + (linEN–1)*ET + cycEN + 1*ET

The existence of the connections between each visual area and the prefrontal cortex has to be established.

Hypothetical division of the stimulus-response pathway into a linear and a cyclical part

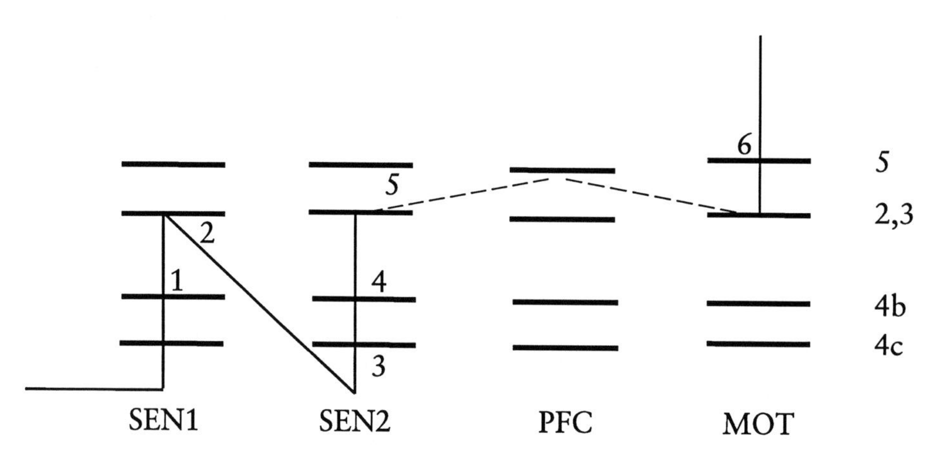

Fig. 50. Hypothetical structure of a stimulus-response pathway. The linear part of this stimulus-response pathway comprises 6 elementary times (bold lines), the minimal cylical Part 2 elementary times (broken lines)

The structure (equation) of mental pathways

Minimal pathway

The minimal pathway of a stimulus-response task has the structure: (con + minlinEN*ET + 2*ET).

Median pathway

It is the question whether the median pathway and with it all the other pathways also use the minimal linear pathway: (con + minlinEN*ET + n*ET).

The variability of the linear part of the pathway

Because of its all-importance, it shall be repeated once more: the length of the linear portion of the pathway is essential for the determination of the length of the cyclical portion of the pathway. That length in turn determines the strategy of the searching (scanning) memory set.

The variability of the linear pathway within one task (with 100 trials)

The fact that there are reaction times left of the first peak proves that the length of the linear pathway may be shorter in some trials. In some subjects these fast trials show a very interesting distance of 3ET from each other. This is the time that is needed to cross one area. Therefore one has to accept, that in some trials one or two sensory areas are saved. If this is true for faster trials one has to expect that there are some slower trials too which need more areas than the dominant linear pathway possesses.

S13A a22lS13A1+a22lS13A2 FPM31e 28.05.2000 start: 100

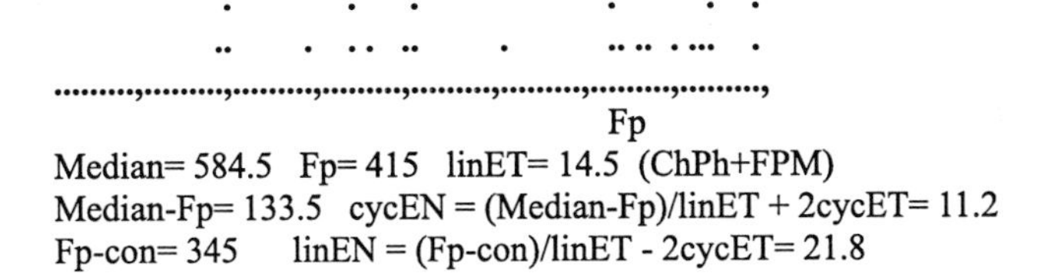

Fig. 51. Part of the reaction time distribution of the task a22lS13A

The existence of a dominant linear pathway is derived from event-related potentials and simulation programs (SIM43,44) which generate the real distribution inclusive the real median if fed with the linear pathway and the elementary time (e.g. SIM43(a11rS16A))

The variability of the linear pathway in a rapid succession of tasks

In this case, the design of the study remains the same but one task is performed repeatedly.

Subject H23E

v11rH23A = 120 + (3 + 2(2 + §2))14   3+5.1
v11rH23B = 120 + (3 + 2(2 + §2))12   2.6+5.5
v11rH23C = 120 + (5 + 2(2 + §2))13   4.9+5.7

v11rH23E1= 120 + (3 + 2(2 + §2))13   3.4+5.4
v11rH23E2= 120 + (4 + 2(2 + §2))13   4.2+4.7
v11rH23E3 = 120 + (5 + 2(2 + §2))13   5.3+4.9
v11rH23E4 = 120 + (5 + 2(2 + §2))13   4.9+5
v11rH23E5 = 120 + (5 + 2(2 + §2))13   4.5+6.2
v11rH23E6 = 120 + (5 + 2(2 + §2))13   4.9+5.5

The H23E tasks were performed shortly each after the other. The mean of the previous tasks v11rH23A – v11rH23C was taken as the elementary time ET(v11rH23) = 13. The variation of ET(v11rH23) alone could explain the variation of the linear pathway. It is remarkable that the linEN rises from 3.4 to 4.9 in the repetitions (decreasing attention?). Therefore the hypothesis can be made that the linEN of single trials may vary within the same range. This has to be shown by single trial evoked potentials, measuring the time of activation of the prefrontal cortex as the beginning of the cyclical part of the pathway.

Subject H12C

a11rH12A = 70 + (2 + 2(2 + §2))16.5   1.6+4.5
a11rH12B = 70 + (3 + 2(2 + §2))16.5   2.8+4.4*

a11rH12C0 = 70 + (3 + 2(2 + 1))16.5   2.5+3.6
a11rH12C1 = 70 + (3 + 2(2 + 1))16.5   2.8+4
a11rH12C2 = 70 + (3 + 2(2 + 1))16.5   2.8+4.5*
a11rH12C3 = 70 + (3 + 2(2 + 1))16.5   2.8+4.1

Remarks:

1. The implicit learning axiom demands slow mode in a11rH12B and fast mode in a11rH12C3
2. The linEN are remarkable stable with the same tendency to increase at the beginning.

The variability of the linear pathway due to experimental distractions

Empirical investigations how certain distractions influence the task v11rH23

1. Simultaneous counting (repeatedly from 1 to 50):
   Median of v11rH23D1: 270.5 ms (counting with moving the mouth but not aloud) A short time later the control task was performed: v11rH23D2: 255 ms

2. Simultaneous tapping with fingers of the left hand. For example, five finger tapping (5FT): little finger, ring finger, middle finger, index finger, thumb, little finger, and so on. Median of v11rH23D3: 372 ms with 5FT.
   I notice that pressing the key with the right index finger interrupts the motor sequence of the left hand. After a short pause there is a new start of the little finger (reset).

3. Simultaneous tapping with two fingers (left index finger and middle finger): Median of v11rH23D4: 300 ms with 2FT.

4. Simultaneous tapping with three fingers (left ring finger, middle finger, index finger)
   Median of v11rH23D5: 306 ms with 3FT. Median of v11rH23D6: 240.5 ms (this control task was performed shortly afterwards)

Hypothetical interactions between two tasks

What do these observations say about the pathway of v11rH23D if simultaneous actions of the left hand influence the reaction time of v11rH23D? The two pathways must interact anywhere on their route, for example in the prefrontal cortex.
The sensorimotor structure of the 2 finger tapping looks like:

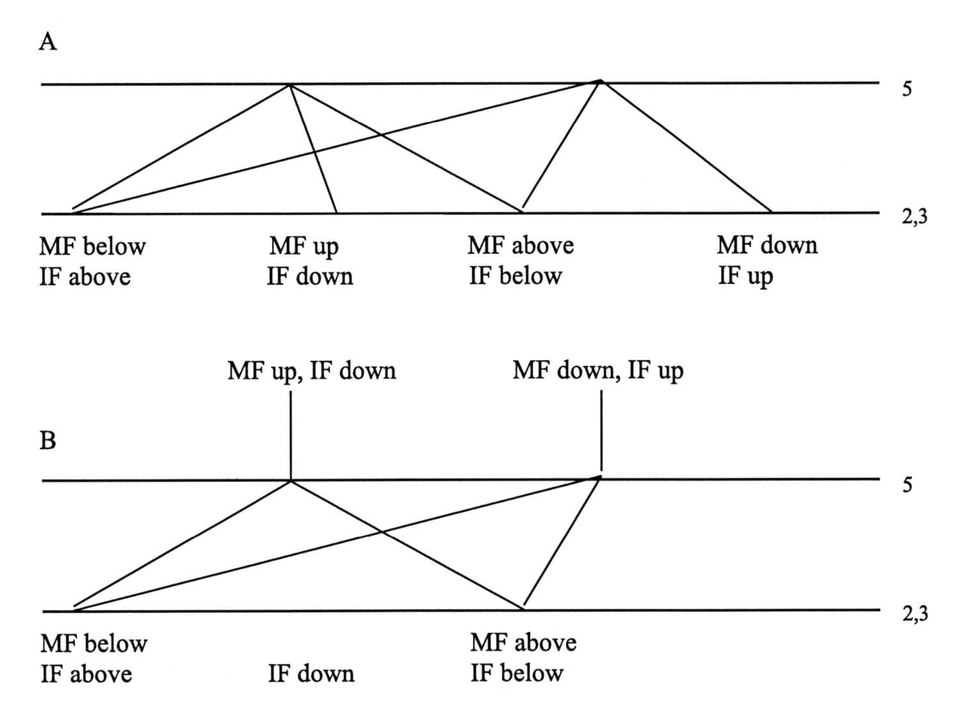

Fig. 52. The sensorimotor structure of the 2 finger tapping (*MF* middle finger, *IF* index finger) B shows another way to represent the motor outputs of Fig. A

This set system may work as a motor generator but it must influence the pathway of v11rH23D. The PFC contains two sensorimotor sets (SMS). These two SMS interact with each other in an unknown way. At the present time, only hypotheses can be made how the two task sets are coordinated with each other:

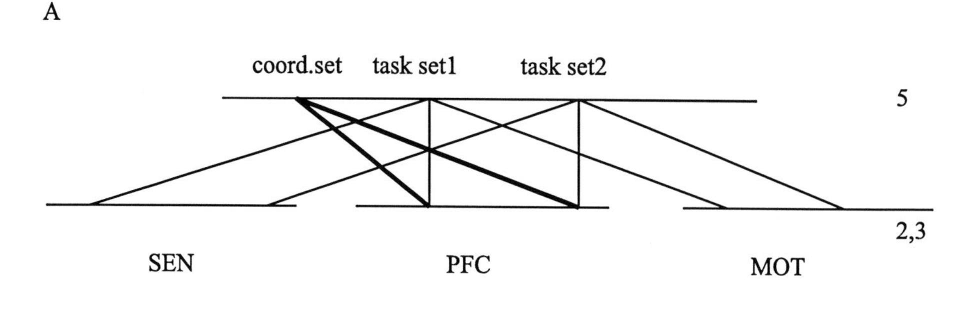

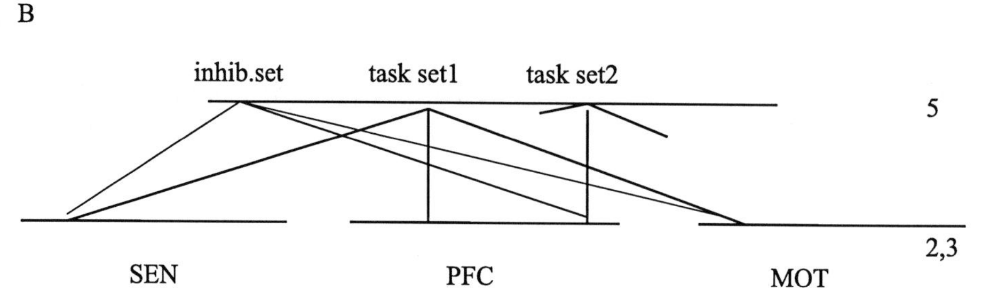

Fig. 53. The hypothetical coordination of two different task sets. (A) A coordination set could decide which of the two task sets may be active at a time. (B) An inhibition set may help to suppress one of the task sets instantly if the other task set is needed

The delay of the task v11r caused by the nFT task of the left hand correlates with the number of tapping fingers

The nFT task is running when the target stimulus of v11r appears on the screen. My observation, that the finger tapping halts when the target stimulus is responded by a key press, indicates that the n finger tapping is inhibited by v11r for a short time. It takes 10 linET = 5 cycles to inhibit the 5 finger task. The task set of 5 finger tapping is located in the PFC together with the inhibition set of this task (not painted). The peripheral structure is located in the motor area. There are 5 basic elements and 5 top elements.

The 5FT task needs 10 linET=5 cycles on average to be inhibited. The inhibition of one basic element costs one cycle.

The relations which are confined to PFC represent the motivational factors dealing with the sensorimotor task. The output of these relations goes to emotionally active areas and the inputs to these relations come from these areas.

In doing the combined task (5FT plus v11r), the appearance of the target stimulus of v11r interrupts the task 5FT. The action structure gives priority to v11r, the

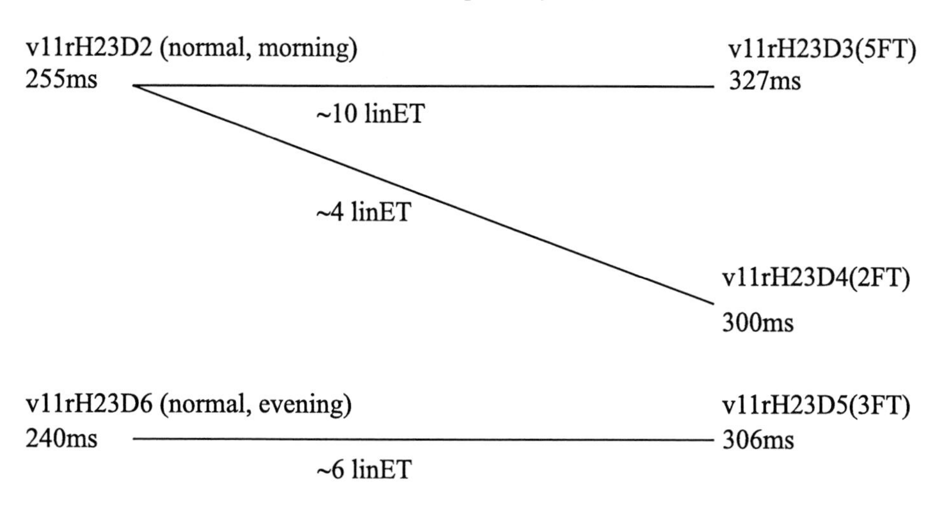

Fig. 54. The different median reaction times for the tasks v11r while different nFT tasks are performed with the left hand

5FT is inhibited for a short time, v11r is carried out, then 5FT begins anew often after a reset starting with the left little finger. The inhibition of 5FT lasts some time (10 linET on average). During this time v11r has to wait until the system is free to do v11r.

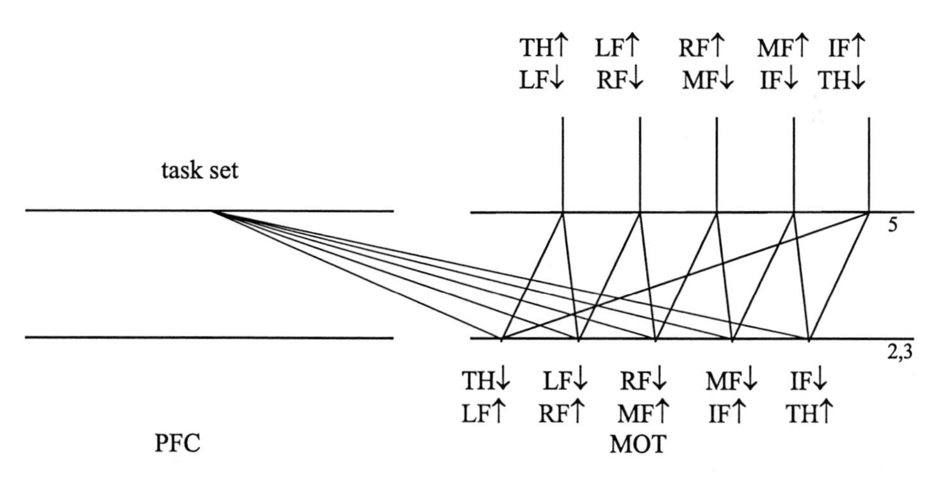

Fig. 55. Hypothetical structure of the task 5FT (five finger tapping). (*LF* Little finger, *RF* ring finger, *MF* middle finger, *IF* index finger, *TH* thumb)

The variability of the linear pathway in repetitions of tasks after days to months

This is investigated in later chapters of Part III and IV in subjects which have repeated the tasks. The figure below shows the relation between the linear pathway of the first and the second series.

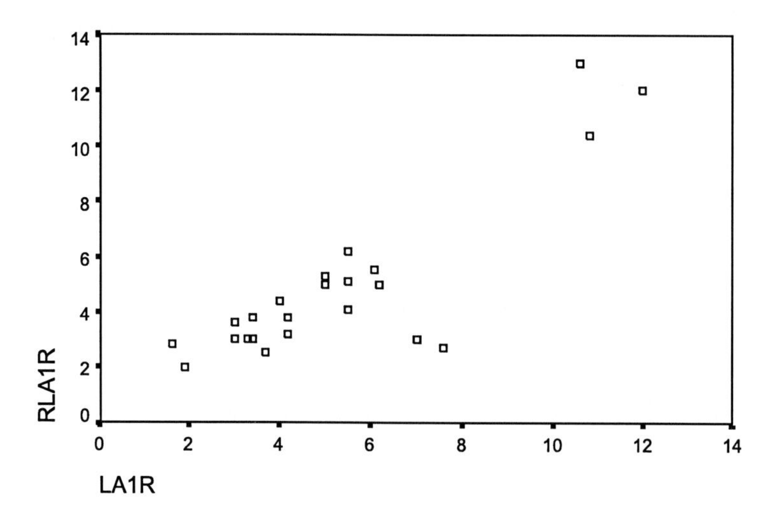

Fig. 56. The length of the linear pathway in 23 healthy subjects in series A (LA1R) and series B (RLA1R). The Pearson correlation coefficient is r=0.861, the significance level is p< 0.01

The variability of the linear pathway in single trials of event-related potentials (ERP)

This theme is treated in Part IV.

The neural representation of variations of the linear pathway

Examples of linear pathways of the task x11y

The subsequent figures show variations with increasing length of the linear pathway, starting with linEN=2.

The linear portion of the pathway (version B) comprises 3 steps. The sensory area needs two steps and the motor area one step, this sums up to three steps for the linear portion of this x11y pathway. The linear pathways often vary with multiples of 3. So the length of the linear portion fits very well to the structure of the cortex. The linear portion of this x11y pathway (version C) needs 6 steps.

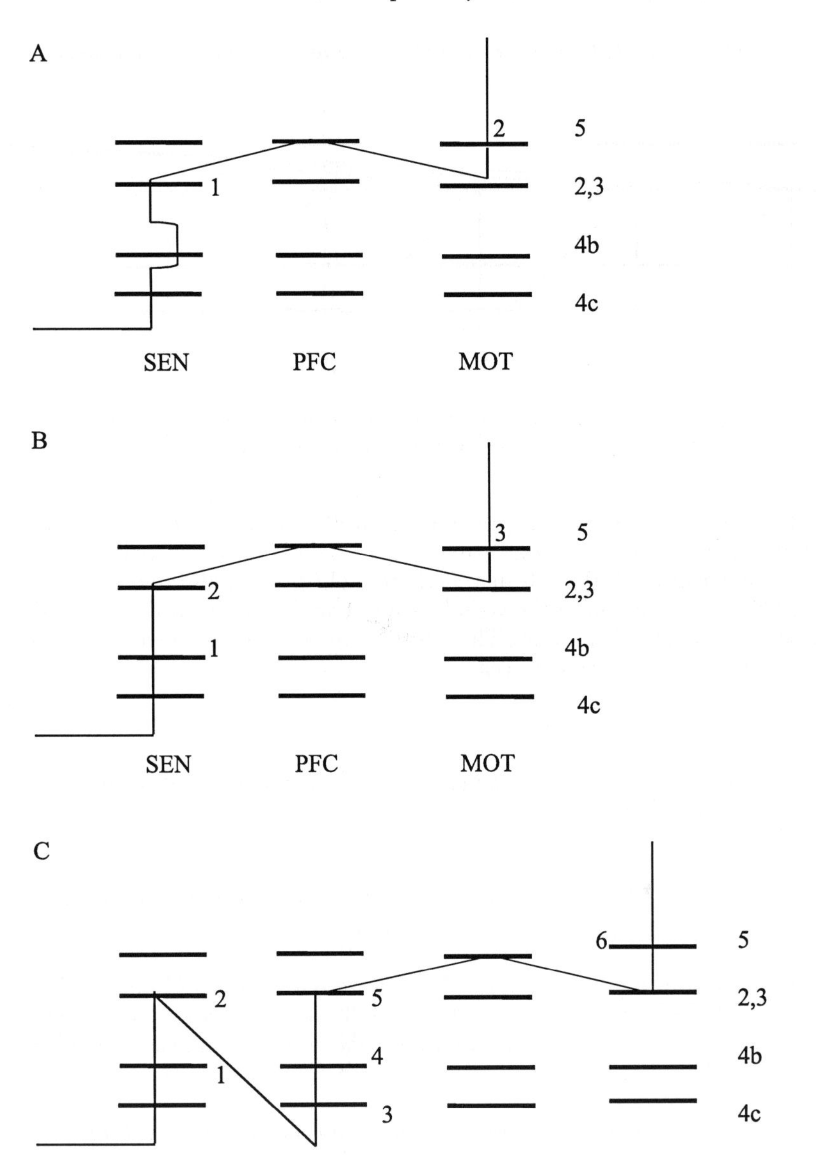

Fig. 57. Examples of linear pathways of the task x11y. (A) linEN=2, (B) linEN=3, (C) linEN=6

The cortical structure of the sensory part of the visual pathway

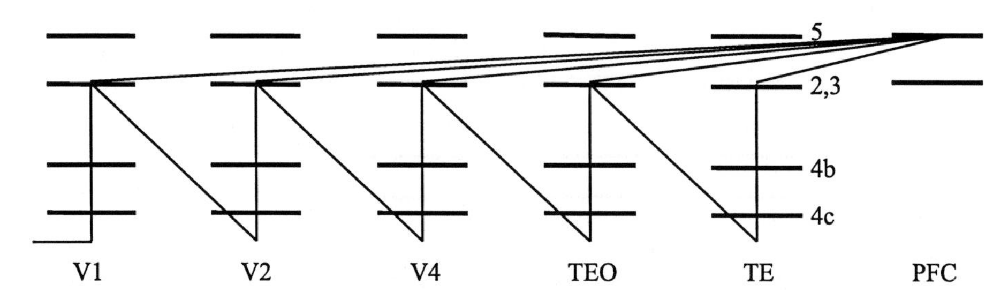

Fig. 58. The hypothetical cortical structure of the temporal visual pathway. The sequence of sensory areas has been taken from Reid (1999)

The frontal cortex may insert at each area in the layer 2,3. Therefore, the visual pathway may vary between one visual area (–> 3linET), two visual areas (6linET), three visual areas (9 linET), four visual areas (12 linET), and five visual areas (15 linET). A linear portion of the visual pathway which is longer than 15 linET may not be explained by this model. The linear pathway contains (n–1) ET in the visual areas and 1 ET in the motor area.

RT = con + (linEN–1)*ET + cycEN + 1*ET

The existence of the connections between each visual area and the prefrontal cortex has to be established.

The cortical structure of the sensory portion of the auditory stimulus-response pathway

The primary auditory cortex is area A1 or Brodmann's areas 41 and 42. It is highly probable that the human auditory cortex is subdivided into numerous functional areas, but the positions and roles of such areas remain to be determined (Hudspeth 1999).

The length of the linear portion and the attentional state

In the relaxed state, a subject can use more pathways than in stress and anxiety.This is due to the influence of neuromodulators (norepinephrine, dopamine) from subcortical structures. They provide a highly focused activation of the really necessary pathways by increasing both the glutamate and the GABA effects, the norepinephrine in the sensory areas, the dopamine in the frontal areas.The reduction of the linear part of the stimulus-response pathway by norepinephrine would mean faster reaction times, without "thinking too much".The high attention would fade out the extra pathways and concentrate on short "emergency pathways".

The most effective attentional state is accompanied by intermediate levels of norepinephrine, higher levels are used in dangerous or unknown situations. They cause a permanent change of the attention. In the relaxed state, however, there is a low focused activation with the chance of musing and creativity (Dehaene et al. 1999).

The influence of attention on the length of the linear pathway has to be proven by measuring both the attentional state and the length of the linear part. The attentional state of a subject performing these tasks can be measured for example by self-rating scale with 10 items or by objective methods measuring the function of the autonomous nerve system.

Prediction: healthy subjects with an increased linEN can reduce it by stress, those with a reduced linEN can increase it by relaxing. In patients with schizophrenia, this is not possible any more.

Focusing works properly provided that there are some attributes distinguishing the "necessary" pathways from the "luxury" ones. This may be an increased spontaneous activity and/or a lower threshold to activate them.

In schizophrenia, this advantage is lost either by insufficiency or by decay of this prefered pathways. Therefore the unusual "luxury" pathways have to substitute this (functional or structural) loss. An additional focusing on these unusual pathways by increased neuro-modulators produces the symptoms of schizophrenia.

Important note: The standard pathways and the non-standard pathways of patients with schizophrenia do not have the same probability of being activated. The non-standard pathways have a better chance than the standard pathways. *Perhaps the support of the neuro-modulatory system is necessary in order to activate these non-standard pathways.*

The distinction between standard and non-standard pathways is caused by the prefrontal cortex with its different pre-activation of memory sets. On the other hand, the early sensory areas are prefered to recruit the basic elements of memory sets because of their earlier activation. Perhaps memory sets which are born first in ontogenesis have to age and die first, memory sets which are born later, may live longer. Perhaps this process is accelerated in patients with schizophrenia. Healthy subjects who grow old and loose their memory sets do not become paranoid because their neuromodulatory system has aged too, only people with an asynchronous decay of the memory sets and the neuronmodulatory system can experience these symptoms.

Why should the linear pathway be constant in a task with 100 trials?

The formation of differentiated, mature memory sets from undifferentiated, immature ones results in the preferred linear pathways. This formation during ontogenesis results from an interaction between the genetically based starting structure and the influences coming from the outside. The product is a number of differently active mature memory sets which make up the individual person (together with their interrelations in the internal prefrontal structure). This internal structure of the PFC can influence the frequency of activation of the memory sets too.

Why is the linEN not as variable as the cycEN?

There are one or more immature, undifferentiated memory sets which search in all the sensory areas. That memory set which first finds an active stimulus-image (in the multitude of sensory areas) may learn this relation and has an advantage in the next trial. The early sensory areas have a time advantage for delivering this stimulus-image. If a mature memory set is shaped with a clear preference of a stimulus element and and a response element this memory set tends to be dominant in all the subsequent tasks (then it is named task set). It is an interesting question how the task sets of different tasks are interrelated in some way. That means, is the task set of say x22y completely independent from the task set of x11y or not, is the linEN(x22y) interrelated with the linEN(x11y) or not?

There may be some variability of linEN but not very much and with a dominant, most frequent length. There is evidence (subject S02B for example) that in patients with schizophrenia the linEN does vary more than in healthy subjects.

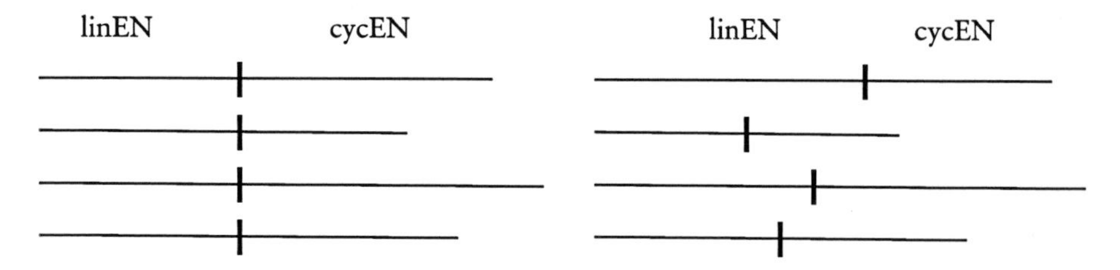

Fig. 59. Is the transition point (break) between the linear and the cyclical part of the pathway constant or variable?

The linEN would vary with 3ET in contrast to cycEN which varies with 2ET. Is there a linEN which can be summed up by a multiple of 3ET and a multiple of 2ET?

3ET + 2ET	6ET + 2ET	9ET + 2ET	12ET + 2ET
3ET + 4ET	6ET + 4ET	9ET + 4ET	12ET + 4ET
3ET + 6ET	6ET + 6ET	9ET + 6ET	12ET + 6ET
etc.			

A peak with the structure (con + 11ET) would be suspicious of containing linEN=9.

If the corresponding distribution contains a minlinEN=3 or minlinEN=6, this would be the proof that *in the same distribution different linEN are present.*

A second test for the occurrence of different linEN for one task is the shape of the distribution. The observed distribution must be compared to the simulation with one linEN, then to the simulation with several linEN with a maximum, then to the simulation with several linEN without a maximum.

A third test for the occurrence of different linEN for one task is the shape of the evoked potential. If there is a distinct latency or attribute of the ERP at (minlinEN*ET) this would support the notion of *one dominant* linEN (either constant or maximum).

Is there a frontal potential which always occurs at the same time after the stimulus, when the PFC scans the sensory areas? Or is there some indication for several potentials occurring at 3ET, 6ET, 9ET etc.?

In the case of constant (maximal) linEN, the border between linEN and cycEN should be visible in event-related potentials. One should find a correlation: First peak in the distribution <–> distinct latency in ERP

If the linear portion were not constant then there would be trials with longer linear pathways. This would mean a shorter cyclical pathway for these trials because the median would be the same. A shorter cyclical pathway would imply a shorter search caused by a more effective search strategy (e.g. by using a working memory for non-target task elements).

The strategy of the searching set depends, therefore, on the length of the linear pathway. The equations are hypotheses with the constancy of the linear portion and the strategy of the cyclical portion determining each other. These hypothetical equations have to be confirmed by the validity of the "implicit learning axiom", the shape of the simulated distribution, the differences between various tasks, homogeneous changes of the linear pathway in replications, and the moment of PFC activation in ERP and in single trials. The linEN could have a distribution of its own:

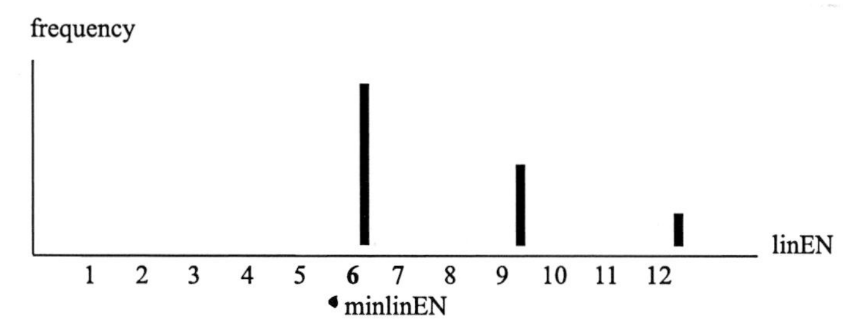

Fig. 60. Distribution of the length of the linear pathway (linEN)

Do the lengths of the linear pathways in one series of a subject change homogeneously (with the same difference) compared to a subsequent series?

This would be an indication for a *general* influence (like attention or motivation etc.)

The variability of the cyclical part of the pathway

Requirements

There exists a mature memory set (=task set). This task set is spontaneously active. A stimulus element of the memory set becomes activated by sensory afferents This is the end of the linear part. Now the cyclical part begins.

The minimal cyclical pathway

The task set performs a random search on its basic elements in order to find the target stimulus element. Because the task set is spontaneously active, it may scan the stimulus element when this is activated by the sensory pathway. The activation goes back to the layer 5 of the PFC and then by chance immediately activates the response elements. This is the minimal cyclical pathway ever possible. It lasts two elementary times.

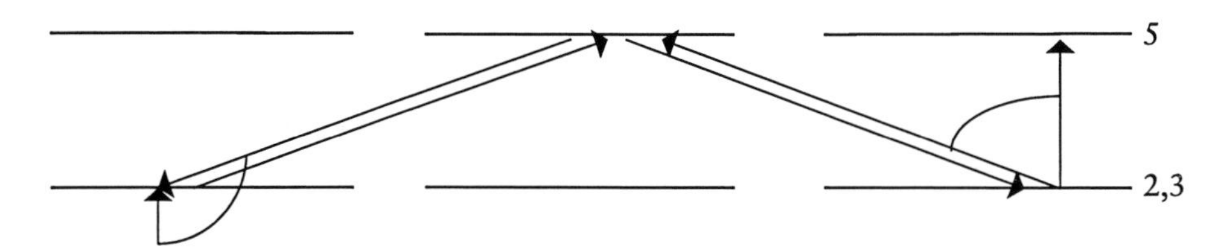

Fig. 61. The minimal cyclical pathway. Simultaneous relations are connected by arcs

The median cyclical pathway depends on the mode and the number of searching sets

In the non-minimal pathway, the search of the task set for the stimulus element is succeeded by the activation of the corresponding sensorimotor set and its search for the response element. *This* search may use slow mode or fast mode, one searching set (the SMS) or two searching sets (the SMS plus the task set). The first case is called double search, the last triple search. The difference between the two search and the three search pathways is the fact that in the first pathway, the tSMS activates the response element without the help of the task set and in the second path-

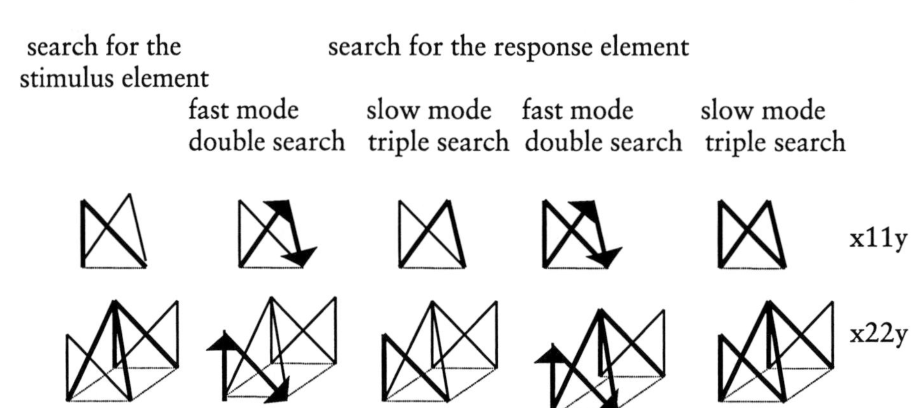

Fig. 62. The searching sets of the tasks x11y and x22y are shown with the search for the stimulus element and the alternative searches for the response element

way, the tSMS stimulates the response element. Its activation needs the help of the task set. The number of searches is caused by the synaptic strength and direction of the relations of the sensorimotor set (SMS). Both can be changed by implicit learning.

The difference between the internal mean, the external mean, and the external observable median number of searching cycles

The first equations which described the structure of the pathways were wrong. One has to distinguish between the mean number of searching cycles performed within the brain and the number of ETs observable in the reaction times. The observable number of searching cycles is 2 ET smaller than within the brain because some processes run simultaneously (the first descending relation and the arrival of the stimulus, the last ascending relation of the searching mechanism and the relation from layer2,3 to the motor area.

The second discrepancy is the difference between the median and the mean number of searching cycles. The mean is the arithmetic mean of all searching cycles. But the short searching cycles are more freqent than the long searching cycles, therefore the median is shifted towards the short searching cycles.

If the response element is not found within two elementary times, the task set has to continue the search for the stimulus element. Each searching cycle lasts two more elementary times. The *mean* number of searching cycles in order to find the stimulus element is n (with n being the number of basic elements). This number cannot be seen externally because there are always two relations of the searching cycles which are simultaneous to other relations of the pathway: the first relation of the searching cycle is simultaneous to the last relation of the linear pathway and the last relation of the searching cycle is simultaneous to the last relation of the linear pathway. Therefore the observable mean number = (internal mean number – 2). The mean number is not used but the median value. Because the distribution of reaction times is shifted to the left, the median is less than the mean value. Therefore the observable median = (internal mean – 2)*factor with factor < 1.

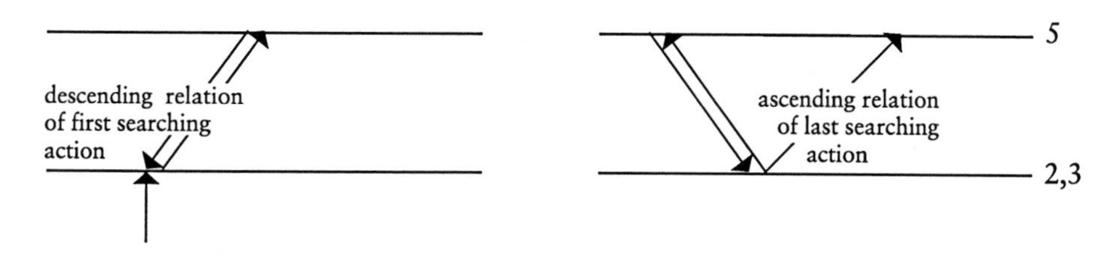

Fig. 63. Partial structure of the searching set (= task set which searches for the stimulus element)

The descending relation of the first searching action occurs before the arrival of the target stimulus because of the spontaneity of the task set. The ascending relation of the last searching action occurs simultaneously to the first motor relation.

The internal mean number of searching cycles is 2cycCT + 2cycCT = 4cycCT = 8cycET but externally observable is only 6 cycET.

Is this valid for two subsequent searches too?

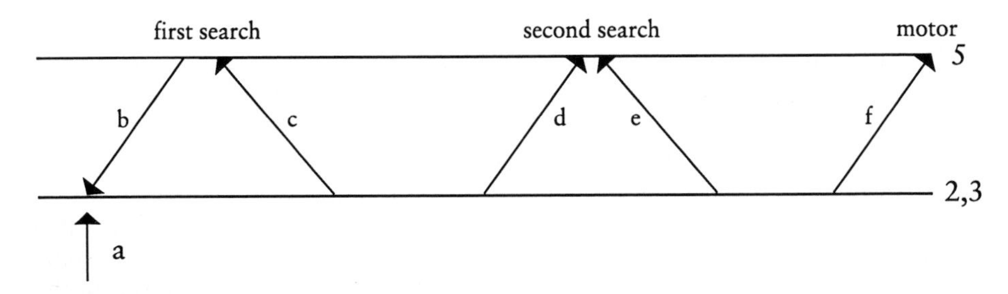

Fig. 64. Partial structure of two subsequent searching sets with (= first the task set which searches for the stimulus element and secondly the sensorimotor set which searches for the response element)

The first descending relation(b) of the first search is simultaneous to the arrival of stimulus(a). The last ascending relation(c) of the first search is simultaneous to the first ascending relation (d) of the second search. The last ascending relation (e) of the second search is simultaneous to the ascending relation to the motor area (f).

The mathematical structure of mental pathways

There are the following pathways for x11y. As the cycle number, the *internal mean number* of searching cycles is given.

x11y = con + (n + 2*(2+1))* ET = con + (n + 6)ET /random search for the target stimulus, then immediate activation of the response element (=fast mode)

x11y = con + (n + 2*(2+2))*ET = con + (n + 8)ET /random search for the target stimulus, then random search for the target response.
Observable mean = 8 cycET – 2 cycET= 6cycET
Observable median = 5 to 6 cycET

x11y = con + (n + 2*(2+1*2))*ET = con + (n+8)ET /random search for the target stimulus, then *simultaneous search* of the tSMS (fast mode) and the task set for the target response
Observable mean=8 cycET–2 cycET=6cycET
Observable median = 5 to 6 cycET

x11y = con + (n + 2*(2+2*2))*ET = con + (n+12)ET /random search for the target stimulus, then *simultaneous search* of the tSMS and the

task set for the target response
Observable mean=12 cycET–2 cycET=10cycET
Observable median = 8 to 9 cycET

There are the following pathways for x22y:
In the next two cases, the target response element is activated by the tSMS alone without the help of the task set. These processes occur preferentially *in visual tasks.* All cycET are internal and mean values.

fast mode

v22y = con + (n + 2*(4 + 1))*ET
= con + (n + 10)*ET

1. random search of the task set for the target stimulus,
2. search of the target sensorimotor set for the target response element (fast mode of the tSMS search)

slow mode

v22y = con + (n + 2*(4 + 2))*ET
= con + (n + 12)*ET

1. random search of the task set for the target stimulus,
2. search of the target sensorimotor set for the target response element (slow mode of the tSMS search)

In the next two cases, the sensorimotor set needs the help of the task set to activate the target response element (= "third search"). These processes prefer to occur *in auditory tasks.* All cycET are internal and mean values. The mean observable cycET are two cycET smaller than the internal ones because the first and the last cycET run simultaneously to an afferent, respectively efferent, relation. The median number of cycET is still smaller.

fast mode

a22y = con + (n+ 2*(4 + 1 * 4))*ET
= con + (n + 16)*ET

1. random search of the task set for the target stimulus,
2. simultaneous search of the task set and the target sensorimotor set for the target response element. (fast mode of the tSMS search)

slow mode

a22y = con + (n + 2*(4 + 2 * 4))*ET
= con + (n + 24)*ET

1. random search of the task set for the target stimulus,
2. simultaneous search of the task set and the target sensorimotor set for the target response element. (slow mode of the tSMS search)

Remark: the simultaneous search means that the two searching sets (tSMS and task set) have to activate the target response element at the same time. If tSMS has two elements and the task set (TS) has four elements, the probability for the tSMS is ½ and that of the task set ¼ to stimulate the target response element (tRE). From this follows the probability to stimulate the tRE at the same time of ½ * ¼ = 1/8.

That means the set system needs 8 searching cycles (2*4cycCT) on average to achieve this event. If the tSMS works in the fast mode, the tSMS stimulates the target response element every time the task set stimulates any of its elements. That means the probability of the tSMS to stimulate the tRE is 1.

Table 24. Presented is the mean number of internal cycles (internal mean). The external mean = (internal mean – 2). The internal, respectively external, median is computed by an multiplicative factor. In the brackets is the externally observable median number of searching cycles

		2 searches	3 searches
x11y	fast	2+1 (3–4 cycET)	2+1*2 (6 cycET)
	slow	2+2 (5–6 cycET)	2+2*2 (8–9 cycET)
x22y	fast	4+1 (7–8 cycET)	4+1*4 (12–14 cycET)
	slow	4+2 (9–10 cycET)	4+2*4 (20 cycET)
x33y	fast	6+1 (10–12 cycET)	6+1*6 (20 cycET)
	slow	6+2 (13,14 cycET)	6+2*6 (30 cycET)

A (2+1*2) search can only be assumed in a repeated task if the preceding task had used a (2+2*2) search.

The irreversibility of the searching mode (the implicit learning axiom)

If a task x11y uses a fast mode search, with the *immediate activation* of the target response after the activation of the tSMS, the task x22y (with identical x and y) cannot use the slow mode. This is caused by the fact that the task x22y uses the sensorimotor set of the task x11y.

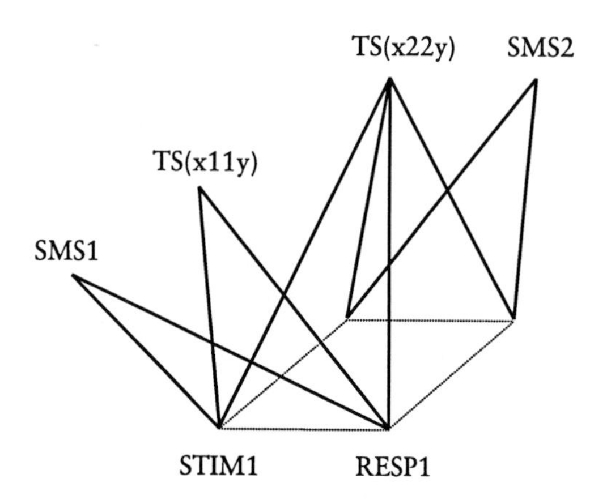

Fig. 65. The task x22y uses the sensorimotor set SMS1 of x11y again. TS = task set

The transition from fast mode to slow mode between x11y and x22y or x22y and x33y is forbidden because the effect of implicit learning cannot be annulled within minutes.

The reversibility of the number of searches in x11y and x22y

There are two adaptive changes of a set system:

1. Transition from slow mode to fast mode of the sensorimotor set (SMS).
2. Transition from double search to triple search and the reverse.
   Both transitions are independent from each other.The question is, whether the number of searching sets is reversible? In the double search, the sensorimotor set (tSMS) can activate the response element (tRESP) without the help of the task set (TS). This may be an effect of implicit learning, too. In this case it would be irreversible in the succeeding task (x11y –> x22y).

Another interpretation of the triple search is the participation of awareness. This would imply that double searches are performed without awareness of the response element, triple searches with awareness of the response element. In this case the existence of double search would be reversible. The stimulus elelment could be aware in every trial. Are there times when triple searches are abundant?

Examples of pathways

The subject O13A

Though SMS1 can activate RESP1 alone in a11rO13A, that means without help of TS(x11y) this is not possible in a22rO13A. Here it needs the help of TS(x22y). Why? Hypotheses:

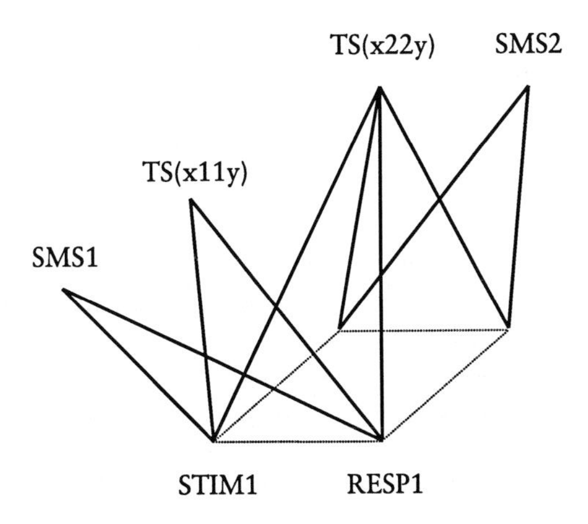

Fig. 66. The hypothetical structure of the set systems of the tasks x11y and x22y

If TS(x22y) searches for the stimulus element, it scans the basic elements less frequently than in the task set TS(x11y), the basic elements are less reagible und need therefore the support of awareness (=triple search).

The reverse case, that x11y uses triple search and x22y double search, is present in v11rO05A and v22rO05A. The awareness interpretation would explain this by the reduction of awareness from x11y to x22y.

The evaluation of v11rH01C

If one takes ET(v11rH01C)=17ms, (Fp–con)=50ms contains only 3 linET and (Med–Fp)=35ms contains only 2 cycET.

v11rH01C
has a linear pathway of length 2 linET. The minimal cyclical pathway has the length of 2cycET.

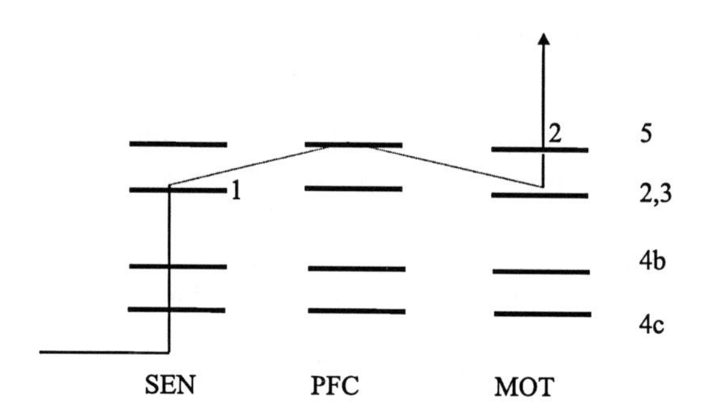

Fig. 67. The pathway of x11y with linEN=2 (shortcut of layer 4b in SEN) and mincycEN=2

The evaluation of v11yH01A

Here (Fp-con) contains 4 linET and (Med-Fp) contains 3 cycET, that means the linear part is 2 ET long and the cyclical part 5 ET.

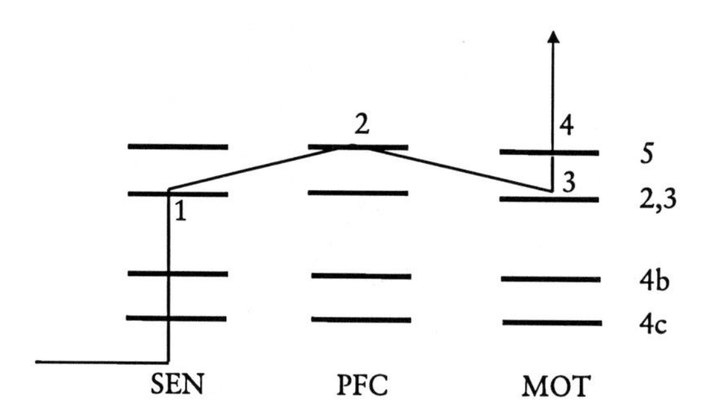

Fig. 68. Hypothetical pathway of v11yH01A with linEN=2 and cycEN=5 (not shown)

A linear pathway of 2 linET is shown below (it can be constructed by skipping the layer 4c of SEN). The 5 cycET are needed to find the stimulus element in layer 2,3 and the response element in the same layer.

v11lH01A has a linear pathway of length 3 linET and a cyclical pathway of length 5 cycET. That means v11lH01A does not skip the layer 4c of SEN. It searches for the target stimulus with (2+2) = 4 cycles or (2+1*2) = 4 cycles (internal mean).

This corresponds to an observable median of 5–6cycET.

Discussion

Critical considerations

General principles for evaluating the data

One must take extra caution in evaluating the data. In order to minimize errors, several measures have been undertaken.

1. Convergence principle
   The different methods (chronophoresis, first peak method, equations, ERP) should give the same results in one subject (± 2 ms).
2. Replication principle
   All results have to be replicated intra-individually, i.e. subjects who repeat the tasks sometime later (days to months) should nearly show the same results (± 2 ms).
3. Correspondence principle
   The ultimate security is obtained if the reaction time data and the ERP data give the same results, eg. the same elementary times or the same length of the linear pathway.

Does the model explain the observed differences between the mean reaction times (x22y – x11y) and (x33y – x22y)?

Table 25. Presented is the mean number of internal cycles (internal mean). The external mean = (internal mean – 2). The internal respectively external median is computed by an multiplicative factor. In the brackets is the externally observable median number of searching cycles

		8ET 2 searches	12ET 3 searches
x11y	fast	2+1 (3–4 cycET)	2+1*2 (6–7 cycET)
	slow	2+2 (5–6 cycET)	2+2*2 (8–9 cycET)
x22y	fast	4+1 (7–8 cycET)	4+1*4 (13–14 cycET)
	slow	4+2 (9–10 cycET)	4+2*4 (20 cycET)
x33y	fast	6+1	6+1*6
	slow	6+2	6+2*6

A (2+1*2) search can only be assumed in a repeated task if the preceding task had used a (2+2*2) search.

If the visual tasks may use triple searches too, the difference of 4 cycles between v11y and v22y can be explained. Provided that the linear portion of both pathways is of equal length, a (2+2) system in v11rO15A and a (4+1*4)system in v22rO15A differ by (4+1*4) – (2+2) = 4 cycles.

It remains to be demonstrated that the difference between x33y and x22y may also be explained by the possible structures of the pathways. It is striking that the difference in one theory is (a33y – a22y) = (3.5 + 3 + 2.5) – (2.5 + 2 + 1.5) = 3/see below.

Table 26. Presented is the mean number of internal cycles (internal mean). The external mean = (internal mean – 2). The internal respectively external median is computed by an multiplicative factor. In the brackets is the externally observable median number of searching cycles

		4ET 2 searches	6ET 3 searches
x11y	fast	2+1 (3–4 cycET)	2+1*2 (6–7 cycET)
	slow	2+2 (5–6 cycET)	2+2*2 (8–9 cycET)
x22y	fast	4+1 (7–8 cycET)	4+1*4 (13–14 cycET)
	slow	4+2 (9–10 cycET)	4+2*4 (20 cycET)
x33y	fast	6+1 (10–12 cycET)	6+1*6 (20 cycET)
	slow	6+2 (13–14 cycET)	6+2*6 (30 cycET)
		4ET	

A (2+1*2) search can only be assumed in a repeated task if the preceding task had used a (2+2*2) search.

Unexplained observations

Gesine's effect

Gesine's effect: the single reaction times at the end of a block of 10 trials are faster than at the beginning. Alexander's idea: is there any correlation between the time between two blocks of ten trials and the Gesine effect.

Multiple results in chronophoresis and NESTLE procedure

Some plots are visible in the chronophoresis and the NESTLE procedure (e.g. v11lO07A). What does that mean? Is there some mathematical reason for two numbers appearing together?

For example *15**4 = *20**3

Methodical shortcomings

Collateral conditions

Gesine's observation: patients tend to put 3 fingers at the keyboard. The mono-hemispheric performance in visual tasks is not controlled. The sound volume in auditory tasks is not controlled. The influences on the elementary time are not investigated sufficiently.

Ambiguity of data evaluation: elementary times

Is it possible that chronophoresis and NESTLE procedure give the double elementary time?

This would mean a doubled elementary number. Are such doubled EN compatible with the prevailing model? Both procedures cannot exlude that the elementary time is half the value.

Multiple plots

The various chronophoreses of S21A get the same motive but are chronologically shifted. Which is the reason for this shift? There are three arcs in the chronophoreses of S21A: an upper arc, an intermediate arc, and a lower arc. The lower arc is twice as much the upper arc. To decide which of the arcs describes the elementary times one needs additional methods like FPM.

Alternative linear pathways

The equation: medianRT = con + minlinEN*ET + cycEN*ET does not mean, that *all* reaction times are represented by this equation. There will be reaction times that will use a longer linear pathway linEN > minlinEN. But the equation states that a prominent subset of reaction times uses minlinEN. How can this be proven?

If a larger subset uses minlinEN then this value should occur in the *event-related potentials*.

How would a *simulated* distribution of reaction times look like if all obey the above equation.

Altogether the relative constancy of linEN is still insufficiently secured. This can imply the existence of other cycEN and other search strategies of memory sets.

The tasks xNNy use artificial task sets

The task sets of xNNy are artificial products, learned to perform the experimental tasks. Are there natural task sets? Do they behave in the same manner as the artificial ones? How can the natural task sets be investigated?

The neural representation of elements and memory sets

Each element (basic or top element) is represented by a set of neurons. The number of the participating neurons and their delineation (structural or functional) is not known. The location of memory sets is not known precisely. Especially the fast sensorimotor sets may lie in the prefrontal cortex or the basal ganglia.

There is no explanation for the size of the elementary time nor for its relative stability over time.

Insufficiency of the used hardware and software

The possible error of 2ms in measuring the elementary time is relative large: 2 ms of 12 ms are 16.7%. The frequencies of the screen and the keyboard are not known precisely, therefore their influence on the measured values cannot be excluded. The only argument against any significant influence is the replicability of elementary times. The precision used by the software in evaluating the raw data (1ms) may be increased if the hardware conditions are proven to give precise data.

The possibility of other theoretical explanations for the measured data

It is still possible that there are other neural mechanismus which can be used to explain these data. Perhaps the variability of the cyclical pathway is due to irrelevant memory sets having to be finished before the reaction task can be processed.

Part III

Applications of stimulus-response pathways in neurology and psychiatry

The pathways of healthy subjects

Convergence tables

The first subject whose pathway structure was investigated was a healthy student, H01. For that reason, a detailed enumeration of the considerations for understanding the pathways for H01 is presented here.

Winning the equations of H01 step by step

The chronophoretic result ET=15 fits best into the distribution of a11lH01D

Experiment D
only auditory, the chronophoresis shows a very clear result of ET=13 to 15 ms. The 15 ms fit best into the periods of the distribution. Therefore, the structure of a11lH01D is as follows.

a11lH01D = $(con + 6ET + 2CT + 2^{*} 2CT)_{ET=15}$ 2 sensory areas, slow mode

search for the stimulus

tSMS and task set have to scan the response element tRESP simultaneously (probability: ½ * ½ = 1/4). This means an average of 4 searching cycles to find the target response.

because two relations are parallel and therefore do not appear in the reaction time

observable mean:
a11lH01D = $(con + 6linET + 2cycCT + 2^{*} 2cycCT - 2cycET)_{ET=15}$

median (lies left to the mean):
a11lH01D = $(con + 6linET + 8\ cycET)_{ET=15}$

The calculation of the median number of searching cycles has to be achieved.

Applying the ET=15 of a11lH01D to the task a11lH01A

If I apply the structure of a11lH01D to a11lH01A, which elementary time does fit best? The median structure of a11lH01A = (con + 9 linET + 8 cycET) is possible, if one takes ET=15.

median:
$a11lH01A = (con + 9ET + 8\ ET)_{ET=15}$

That means, a11lH01A has the same cyclical structure as a11lH01D but its linear structure is one area longer than in H01D.

9 linET which is equivalent to 3 sensory areas. The first trials of a11lH01A lie at (con + 9 linET + 2 cycET)

*Applying the structure of a11lH01A = (con + 9 linET + 2 cycCT + 2*2cycCT) to a11rH01A*

If I take ET = 12, I get the structure of a11rH01A = $(con + 12\ linET + 8\ cycET)_{ET=12}$

internal mean $a11rH01A = (con + 12linET + 2cycCT + 2^*\ 2cycCT)_{ET=x}$
observable mean $a11rH01A = (con + 12linET + 2cycCT + 2^*\ 2cycCT - 2cycET)_{ET=x}$
observable median $a11rH01A = (con + 12linET + 8cycET)_{ET=x}$

This means, a11rH01A uses four sensory areas. It is striking that until now all linEN are multiples of 3 which is compatible to the idea that there the pathway uses three layers of a sensory area. The last pages show how beginning with a11lH01D the structure of the tasks a11lH01A and a11rH01A was obtained.

Can the elementary times of aETrH01A = 12 and aETlH01A=15 contribute to unterstand the structure of a22yH01A?

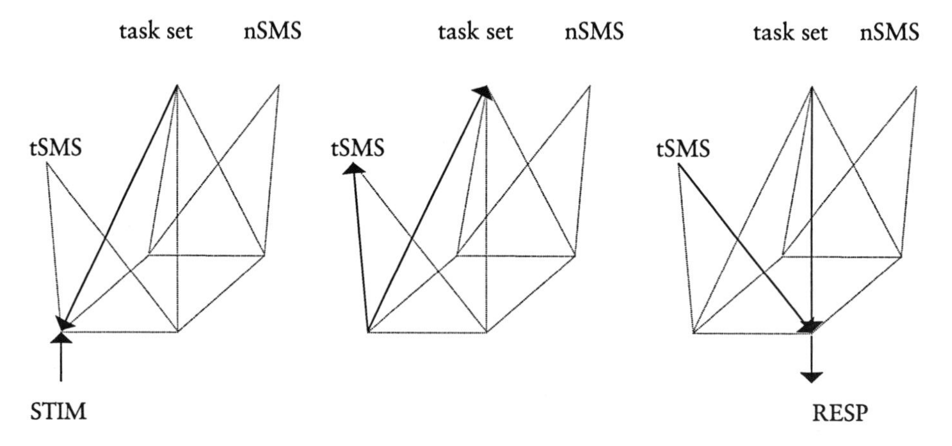

Fig. 69. In this example of x22y, the simultaneous search of the task set and the tSMS activate the response element. In a double search, the help of the task set would not be necessary

The fastest x22y pathway needs only 2 cycET for the search. This is equivalent to an intern mean value of 4 cycET. Another x22y pathway needs 4CT to find the target stimulus and 2CT to find the target response. This is equivalent to an intern mean value of 12 cycET. The longest x22y pathway needs 4 CT to find the target stimulus and 2 * 4 CT to find the target response (by coincidence of the task set and the target sensorymotor set in the target response element). This is equivalent to an internal mean value of 24 cycET. In the last two cases one has to subtract 2 cycET from the internal mean values. The observable *means* are therefore 10 cycET and 22 cycET. The *medians* lie to the left of the means because of the lopsided distribution of x22y.

Because the structure of a22rH01A = $(\text{con} + 15\ \text{linET} + 20\ \text{cycET})_{ET=12}$ one has to favor the third variant: after the search for the target stimulus, the sensorimotor set and the task set have to activate the target response element simultaneously in order to generate a motoric output.

a22rH01A = $(\text{con} + 15\ \text{linET} + 4\ \text{cycCT} + 2 * 4\ \text{cycCT})_{ET=15}$

random search of the task set for the target stimulus (→ 4 cycCT)

coincident search of the task set and the target sensorimotor set for the target response element. (→ 2 * 4 cycCT)

Can the elementary time aETlH01A=15 and the above structure of a22rH01A be used to unterstand the structure of a22lH01A?

A possible partition of the a22lH01A reaction time is: Median - first peak = 9 cycET, left of the first peak lie 2 or 4 additional cycET (dependent from the first trials). If one takes 4 additional cycET one has the following structure:

a22lH01A = con + 15 linET + 13 cycET

Which set system produces such a time structure? If one looks at Table 26 one sees the fast mode of a22y with a mean internal number of 16 cycET, which can correspond to the external median number of 13 cycET. The first trials lie at (con + 15 linET + 2 cycET), the first peak at (con + 15 linET + 4 cycET). The internal mean structure is a22l = $(\text{con} + 15^{*}\ \text{linET} + 4\ \text{cycCT} + 1 * 4\ \text{cycCT})_{ET=15}$

Remark: an alternative structure has been a22lH01A = (con + 15 linET + 2.5 cycET + 1.5 cycET + 2.5 cycET). These were the numbers of searching cycles for the normal mode with a working memory. Now I think, one has to use the simplest structure which is able to explain the pathway. If this is possible without working memory, the better.

Which are the pathway structures of a11yH01B and a11yH01C?

a11rH01B

If I take aETrH01 = 12ms, I get the structure of a11rH01B = $(\text{con} + 12\ \text{linET} + 9\ \text{cycET})_{ET=12}$

This means, a11rH01B uses four sensory areas. A median number of 9 cycET is unusual. Nevertheless I would attach it to the structure 2s3. Altogether it has the same structure as a11rH01A.

$x11y = (con + n^*linET + 2cycCT + 2^* 2cycCT)_{ET=x}$

observable mean a11rH01B= $(con + 12linET + 2cycCT + 2^* 2cycCT - 2cycET)_{ET=12}$
observable median a11rH01B= $(con + 12linET + 8cycCT)_{ET=12}$

a11lH01B
If I take aETlH01 = 15ms, I get the structure of a11lH01B = $(con + 9.6\ linET + 6\ cycET)_{ET=12}$

This means, a11lH01B uses three sensory areas. I would attach it to the structure 2s2 or 2f3.

Because the median structure of a11lH01A is a11lH01A= (con + 9 linET + *8 cycET*) which is 2s3, the task a11lH01B uses a faster set system than the task a11lH01A. Usually this is due to a fast mode of the sensorimotor set. Hence 2s3 to 2f3 would describe this transition best.

$x11y = (con + n^*ET + 2CT + 2CT)_{ET=x}$

This structure is indistinguishable from the structure

$x11y = (con + n^*ET + 2CT + 1^*2CT)_{ET=x}$

In the first structure the task set searches first, then the sensorimotor set. In the second structure the task set searches first too, then the sensorimotor set stimulates the response element each time, the task set searches for the response element.

observable mean a11lH01B= $(con + 9linET + 2cycCT + 1^* 2cycCT - 2cycET)_{ET=12}$
observable median a11lH01B= $(con + 9linET + 6cycCT)_{ET=12}$

a11rH01C
If I take aETrH01 = 12ms, I get the structure of a11rH01C= $(con + 6\ linET + 6.5\ cycET)_{ET=12}$

This means, a11rH01C uses two sensory areas. I would attach it to the structure 2s2 or 2f3.

This means the linear pathway is as shortened as the cyclical pathway. Because a11rH01A has the pathway $x11y = (con + n^*linET + 2cycCT + 2^* 2cycCT)_{ET=x}$ I would prefer the fast mode hypothesis of a11rH01C. *But I should consider, that the structure 2s2 is simpler than the structure 2f3.*

$x11y = (con + n^*ET + 2CT + 2CT)_{ET=x}/2s2$

This structure is indistinguishable from the structure

$x11y = (con + n^*ET + 2CT + 1^*2CT)_{ET=x}/2f3$

In the first structure the task set searches first then the sensorimotor set. In the second structure the task set searches first too, then the sensorimotor set stimulates the response element each time, the task set searches for the response element.

observable mean a11rH01C= $(con + 6linET + 2cycCT + 1^* 2cycCT - 2cycET)_{ET=12}$
observable median a11rH01C= $(con + 6linET + 6cycCT)_{ET=12}$

a11lH01C
If I take aETlH01 = 15ms, I get the structure of a11lH01C = (con + 7.3linET + 6.7 *cycET*)$_{ET=12}$

This means, a11lH01B uses three sensory areas with two shortcuts. I would attach it to the structure 2s2 or 2f3. Because a11lH01B has the structure a11lH01B = (con + 9.6 linET + 6 *cycET*)$_{ET=12}$, the linear portion is shortened but the cyclical portion remains unchanged.

Here too two structures are possible:

x11y = (con + n*ET + 2CT + 2CT)$_{ET=x}$

This structure is indistinguishable from the structure

*x11y = (con + n*ET + 2CT + 1*2CT)*$_{ET=x}$

In the first structure the task set searches first, then the sensorimotor set. In the second structure the task set searches first too, then the sensorimotor set stimulates the response element each time, the task set searches for the response element.

observable mean a11lH01B= (con + 9linET + 2cycCT + 1* 2cycCT - 2cycET)$_{ET=12}$
observable median a11lH01B= (con + 9linET + 6cycCT)$_{ET=12}$

I should decide which structure is better suited to represent the 6cycET: 2s2 or 2f3? This procedure is of historical value only. Now the way to get the elementary times and the pathway structure has changed.

The measurement of elementary times and the evaluation of the pathway structure

Presently the following methods are used to measure the elementary times:

First the elementary times are read from the nestle procedure of the programs FPM31e(x11y), FPM31e(x22y), FPM26f58(x11y), and FPM26f58(x22y). Then the results are filled into the first four columns of a convergence table like the following:

Table 27. Convergence table of subject H02A. The results are filled in by help of the subsequent agreements

H02A	FPM31e (x11y)	FPM31 (x22y)	FPM26f (x11y)	FPM26f (x22y)	Chronophoresis	Agreement	Result
aNNr	?	13,25	13,26	13			13
aNNl	16,20	13,19	35	?			14.5
vNNr	13	15	15	22,33			15
vNNl	?	17	?	34			17

Agreements:
1. I take the even number *occuring in both* FPM26g(x11y) and FPM26g(x22y).
2. If they have no common number, I take the *mean* of both methods.
3. If there is no number in one method, I take the common number of xNNr and xNNl.
4. If there is no common number of xNNr and xNNl, I take the mean of both.

Agreements:
1. I take the even number *occuring in both* FPM27d(x11y) and FPM27d(x22y).
2. If they have no common number, I take the *mean* of both methods (if two numbers are proposed) or the mean of the two adjacent numbers (if more than two numbers are proposed).
3. If there is no number in one method, I take the common number of xNNr and xNNl.
4. If there is no common number of xNNr and xNNl, I take the mean of both.

Now, the final elementary times are used to calculate the linEN and cycEN in the reaction time distribution. The table below is used to get the mode and the number of search processes from the cycEN.

Table 28. Presented is the mean number of internal cycles (internal mean). The external mean = (internal mean - 2). The internal respectively external median is computed by an multiplicative factor. In the brackets the externally observable median number of searching cycles

		2 searches	3 searches
x11y	fast	2+1 (3-4 cycET)	2+1*2 (6 cycET)
	slow	2+2 (5-6 cycET)	2+2*2 (8-9 cycET)
x22y	fast	4+1 (7-8 cycET)	4+1*4 (12-14 cycET)
	slow	4+2 (9-10 cycET)	4+2*4 (20 cycET)
x33y	fast	6+1 (10-12 cycET)	6+1*6 (20 cycET)
	slow	6+2 (13,14 cycET)	6+2*6 (30 cycET)

A (2+1*2) search can only be assumed in a repeated task if the preceding task had used a (2+2*2) search. At last the equations are written down.

Design: 100 trials of the tasks x11y, 200 trials of the tasks x22y, and 300 trials of the tasks x33y

In all subjects, only the index finger has been evaluated. In order to get 100 trials performed by the index finger, one has to do 200 trials of the tasks x22y and 300 trials of the task x33y The 100 trials of the index finger allow to compare the results of the different tasks adequately. Only some healthy subjects performed the tasks x33y.

Survey over the healthy subjects

Table 29. Survey over all healthy subjects. *Series*: Number of series the subject has performed. *Differ*: the results were evaluated independent from each other. *Conver*: there exists a convergence table in 16.5, *x33y*: the subject has performed 300 trials of x33y

Number	Initials	Repetit.	Differ.	Conver.	x33y	Sex	Age	Hand
1	H01A	4			+	m	30	r
2	H02A	2	+	+		f	29	r
3	H03A	2		+		m	24	r
4	H04	2	+	+		m	25	
5	H05A	2	+	+		m	24	r
6	H06A	2				m	22	r
7	H07A	2	+	+		m	25	r
8	H08	1		+		f	24	r
9	H09A	3	+	+		m	28	l
10	H10	3	+	+		m	26	r
11	H11A	2		+		f	45	r
12	H12A	2				f	23	r
13	H13A	3		+		m	22	r
14	H13	1		+		m	30	r
15	H15	1		+		m	68	r
16	H16A	2		+		m	23	r
17	H17A	1		+	+	f	22	r
18	H18A	1		+	+	m	25	r
19	H19A	2	+	+		f	25	r
20	H20A	3		+		m	25	r
21	H21A	2		+		m	22	r
22	H22A	2		+		m	36	r
23	H23	3	+			m	48	r
24	H24	1		+		m	24	r
25	H25	1		+		m	22	r
26	H26A	2	+	+		f	23	
27	H27	1		+		f	23	l
28	H28B	2	+	+		f	34	
29	H29A	2				m	25	r
30	H30A	2	+			m	26	r
31	H31A	2	+	+		m	34	r

Convergence tables

H02

H02A	FPM31e (x11y)	FPM31 (x22y)	FPM26f (x11y)	FPM26f (x22y)	Chronophoresis	Agreement	Result
aNNr	?	13,25	13,26	13			13
aNNl	16,20	13,19	35	?			14.5
vNNr	13	15	15	22,33			15
vNNl	?	17	?	34			17

H02B	FPM31e (x11y)	FPM31 (x22y)	FPM26f (x11y)	FPM26f (x22y)	Chrono-phoresis	Agree-ment	Result
aNNr	?	11,22	?	15			13
aNNl	?	14	13	16			14
vNNr	13	15	13	15			14
vNNl	16	16,18,25	14	?			16

H03

H03A	FPM26f 47(x11y)	FPM27d (x11y)	FPM27d (x22y)	SING100n (x11y)	SING100r (x11y)	Agree-ment	Result
aNNr	14	14	15	13.5		1	14
aNNl	21	15,23	15,19	15		1	15
vNNr	?	23(2Q)	13	12		2a	12
vNNl	?	?	13,17	12		3	13

The SING100n results are not used to get the elementary times.

H03B	FPM26f 47(x11y)	FPM27d (x11y)	FPM27d (x22y)	SING100n (x11y)	SING100r (x11y)	Agree-ment	Result
aNNr	22,26,30	14,18	22	16.5			14
aNNl	15	14,20	?	14.5		2b	14.5
vNNr	14	?	17,23	12.5		2b	13
vNNl	?	?	13,16,20	14			?

H03	H03A	H03B	Max.diff.	Agree	Results
aNNr	14	14			14
aNNl	15	14.5			15
vNNr	12	13			12.5
vNNl	13	?			13

H04

H04A	FPM31e (x11y)	FPM31 (x22y)	FPM26f (x11y)	FPM26f (x22y)	Chrono-phoresis	Agree-ment	Result
aNNr	15,25	11,19,27	14	?			14
aNNl	?	28	12,14	10,11			11.5,14
vNNr	17	20	14,18	14			18
vNNl	?	20,22	19,21	23,31			21

H04B	FPM31e (x11y)	FPM31 (x22y)	FPM26f (x11y)	FPM26f (x22y)	Chrono-phoresis	Agree-ment	Result
aNNr	20	25,28	22,33	?			15,21
aNNl	19	26	16,18,20	16			16,18.5
vNNr	15	20	18	16			15.5,19
vNNl	?	21,25	28	?			21,25,28

Remarks: 1. If the results are ambiguous, these results are preferred, which have a counterpart in H04A.

H05

H05A	FPM31e (x11y)	FPM31 (x22y)	FPM26f (x11y)	FPM26f (x22y)	Chronophoresis	Agreement	Result
aNNr	14,17,23	14,19	?	22			4,18,22
aNNl	14,17	11,20	17	?			12.5,17
vNNr	15	15	16	16			15.5
vNNl	19	25	19,24	?			19,24.5

H05B	FPM31e (x11y)	FPM31 (x22y)	FPM26f (x11y)	FPM26f (x22y)	Chronophoresis	Agreement	Result
aNNr	14,16	21	14	21			14,21
aNNl	14,17	26	12,14	12			13,17
vNNr	?	23,26	22	26			13,22.5
vNNl	19	18,21,25	?	?			18.5

H07

H07A	FPM31e (x11y)	FPM31 (x22y)	FPM26f (x11y)	FPM26f (x22y)	Chronophoresis	Agreement	Result
aNNr	25	18	21	?			18
aNNl	25	16,20	20	24			16,20
vNNr	17	27	18	24,27,31			17
vNNl	?	17,25	34	?			17

H07B	FPM31e (x11y)	FPM31 (x22y)	FPM26f (x11y)	FPM26f (x22y)	Chronophoresis	Agreement	Result
aNNr	16	29	28	?			16
aNNl	15	26,23	22,25	23			15,23
vNNr	14,17,25	?	24,25	18			17.5,25
vNNl	15,27	27,28	?	21,33			16,27

H08

H08A	FPM31e (x11y)	FPM31 (x22y)	FPM26f (x11y)	FPM26f (x22y)	Chronophoresis	Agreement	Result
aNNr	21	15,18	27	?			15
aNNl	15	21	?	14			14.5
vNNr	15,23	15,25	?	26			14
vNNl	21,27	15,22	15	34			15

H09

H09A	FPM26f 20(x11y)	FPM26g (x11y)	FPM26g (x22y)	SING10 On(x11y)	SING10 Or(x11y)	Agree-ment	Result
aNNr	14	15	14	13.5	12, 17	2	14.5
aNNl	14,15	15	15	20	12.5, 24	1	15
vNNr	14?	15	17	21.5	21.5	2	16
vNNl	15	15	15	11,16.5 24.5	12.5, 18	1	15

To describe the plots of the chronophoresis, the *center* of the plot is used. One has to consider that the FPM programs use the same method: the same dependence of the constant con (70ms or 120ms) for example. The chronophoresis is independent from this constant and the (Fp-con) and (median-Fp) methods use different procedures to estimate the elementary times.

H09B	FPM26f 20(x11y)	FPM26g (x11y)	FPM26g (x22y)	SING10 On(x11y)	SING10 Or(x11y)	Agree-ment	Result
aNNr	17	15	15	11,17,25	21	1	15
aNNl	16	13	17	12,15,20	?	2	15
vNNr	13,15	15	16	11	17	2	15.5
vNNl	12,14,16	13,16	13	16,18,23	17,22	1	13

The results of the three series H09A, H09B, and H09C in an overview:

The subject H09C has the subsequent measures:

H09C	FPM26f 20(x11y)	FPM26g (x11y)	FPM26g (x22y)	SING10 On(x11y)	SING10 Or(x11y)	Agree-ment	Result
aNNr	16	14,16	16	19	18.5	1	16
vNNl	17	17	15	13	13.5	2	16
vNNr	?	15,16	15	13	17.5	1	15
vNNl	15	15,17	?	17	15.5	3	15

Agreement:
1. I take the even number *occuring in both* FPM26g(x11y) and FPM26g(x22y).
2. If they have no common number, I take the *mean* of both methods.
3. If there is no number in one method, I take the common number of xNNr and xNNl.
4. If there is no common number of xNNr and xNNl, I take the mean of both.

H09	H09A	H09B	H09C	Max.diff.	Agree	Results
aNNr	14.5	15	16	1.5	2	15
aNNl	15	15	16	1	1	15
vNNr	16	15.5	15	1	2	15.5
vNNl	15	13	15	2	1	15

That means the elementary times are reproducible with an error of ±2ms.
The chronophoresis has to be improved. There are two ways to get the equations of the pathways in subjects with multiple series: the equations could be computed from elementary times of this *special* series or from elementary times of *all* series.

H10

H10B	FPM26f44 (x11y)	FPM27d (x11y)	FPM27d (x22y)	SING100n (x11y)	SING100r (x11y)	Agreement	Result
aNNr	14	12,13	13,15				13
aNNl	13?	12,13,17	12,17		15.5		12,17
vNNr	15	11,15	15				15
vNNl	13	13,16,17	13,15,17		12		13,17

H10C	FPM26f44 (x11y)	FPM27d (x11y)	FPM27d (x22y)	SING100n (x11y)	SING100r (x11y)	Agreement	Result
aNNr	?	11,13,15	21,25				11,13
aNNl	?	14,16	12,19				15
vNNr	?	13	15				14
vNNl	?	?	12,15,24				?

H103 comes after H10B (therefore the 3).
ET(aNNl) is taken as the median between 12,14,16,19.

H10D	FPM26f44 (x11y)	FPM27d (x11y)	FPM27d (x22y)	SING100n (x11y)	SING100r (x11y)	Agreement	Result
aNNr	14	12	13	13,17.5		2a	12.5
aNNl	16	17	15	15		2a	16
vNNr	14	16,19	11-16?	15?		1	16
vNNl	13	15	15	12		1	15

The result columns computes its values from FPM27d(xNNy) by the agreements stated above.

The subject H10D shows distinct peaks in the natural distribution, therefore FPM26f44 can be applied. These results differ from the above results by up to 2 milliseconds. The subject H10X shows an additional correspondence between different chronophoresis results.

H10	H10B	H10C	H10D	Max.diff.	Agree	Results
aNNr	13	11,13	12.5			13
aNNl	12,17	15	16			16
vNNr	15	14	16			15
vNNl	13,17	?	15		2a	14

I have to decide whether to use the results of *one* task within the equation systems or the results of *all* task.

H11

H11A	FPM26f47 (x11y)	FPM27d (x11y)	FPM27d (x22y)	SING100n (x11y)	SING100r (x11y)	Agree-ment	Result
aNNr	17	16	?			2a	**16.5**
aNNl	?	14	13,16			2b	**13.5**
vNNr	13,15	11,15	12,13,18			2b	**11.5**
vNNl	19	15,17	17,18			1	**17**

Agreement:
1. I take the even number *occuring in both* FPM26g(x11y) and FPM26g(x22y).
2. If they have no common number, I take the *mean* of both methods.
3. If there is no number in one method, I take the common number of xNNr and xNNl.
4. If there is no common number of xNNr and xNNl, I take the mean of both.

H11B	FPM26f47 (x11y)	FPM27d (x11y)	FPM27d (x22y)	SING100n (x11y)	SING100r (x11y)	Agree-ment	Result
aNNr	17	16	16			1	**16**
aNNl	14	14	20,25?			1	**14**
vNNr	14	?	15			2a	**14.5**
vNNl	15	15	16,17			2b	**15.5**

Remarks: In H11A and GH10 FPM26f47 is used if FPM27d(x11y) or FPM27d(x22y) is invalid.

H11	H11A	H11B	Max.diff.	Agree	Results
aNNr	16.5	16	0.5		16
aNNl	13.5	14	0.5		14
vNNr	11.5	14.5	3		13
vNNl	17	15.5	1.5		16

Integer numbers have been preferred.

H13

H13A	FPM26f20 (x11y)	FPM27d (x11y)	FPM27d (x22y)	SING100n (x11y)	SING100r (x11y)	Agree-ment	Result
aNNr		14,17	15	14		2b	**14.5**
aNNl		11,15,17	17	15		1	**15**
vNNr		12	14	13.5		2a	**13**
vNNl		16	15	14		2a	**15.5**

Agreement:
1. I take the even number *occuring in both* FPM27d(x11y) and FPM27d(x22y).
2. If they have no common number, I take the *mean* of both methods (if two numbers are proposed) or the mean of the two adjacent numbers (if more than two numbers are proposed).
3. If there is no number in one method, I take the common number of xNNr and xNNl.
4. If there is no common number of xNNr and xNNl, I take the mean of both.

H13B	FPM26f20 (x11y)	FPM27d (x11y)	FPM27d (x22y)	SING100n (x11y)	SING100r (x11y)	Agree-ment	Result
aNNr		15,17	11			2b	13
aNNl		16,19	15			2b	15.5
vNNr		13,18	15,21			2b	14
vNNl		11,13	?				

H13C	FPM26f20 (x11y)	FPM27d (x11y)	FPM27d (x22y)	SING100n (x11y)	SING100r (x11y)	Agree-ment	Result
aNNr		23(->12)	24(->12)			1	12
aNNl		15,18	13			2b	14
vNNr		15,19	12			2b	13.5
vNNl		29(->14)	17			2a	15.5

H13	H13A	H13B	H13C	Max.diff.	Agree	Results
aNNr	14.5	13	12	2.5	2c	14
aNNl	15	15.5	14	1.5	2b	15
vNNr	13	14	13.5	1	2b	13.5
vNNl	15.5		15.5	0	1	15.5

Rule 2c: If the numbers have the same distance then the mean of all numbers is taken.

H14

H14A	FPM31e (x11y)	FPM31 (x22y)	FPM26f (x11y)	FPM26f (x22y)	Chrono-phoresis	Agree-ment	Result
aNNr	16	18,29	?	?			15
aNNl	15,24,28	17,20,27	15,18	13,18			14,18
vNNr	15	17,25	?	17			12,16
vNNl	26	12,24	12	13,29,35			12.5

H15

H15A	FPM31e (x11y)	FPM31 (x22y)	FPM26f (x11y)	FPM26f (x22y)	Chrono-phoresis	Agree-ment	Result
aNNr	20,23	24.5	?	11			22
aNNl	22	17,23	34	11			17,22
vNNr	24	15	17	?			16,24
vNNl	19	21,28	15,19,22	24			19,21.5

H16

H16A	FPM26f 47(x11y)	FPM27d (x11y)	FPM27d (x22y)	SING100n (x11y)	SING100r (x11y)	Agree-ment	Result
aNNr	15	?	29				15
aNNl	14	11	13,17			2b	13.5
vNNr	8,17	17	12,17			1	17
vNNl	14,21	13,27	21				13.5

Remark: The occurrence of an elementary time and its double value is especially important and increases its probability to be true.

H16B	FPM26f 47(x11y)	FPM27d (x11y)	FPM27d (x22y)	SING100n (x11y)	SING100r (x11y)	Agreement	Result
aNNr	**13,15**	?	14,20				**14**
aNNl	13	**14**	12,25				**13**
vNNr	8	**17**	15,17				**17**
vNNl	14,15,17	?	15,29				**15**

H16	H16A	H16B	Max.diff.	Agree	Results
aNNr	15	14	1	2a	14.5
aNNl	13.5	13	0.5	2a	13
vNNr	17	17	0	1	17
vNNl	13.5	15	1.5	2a	14

(13.5+13)/2 = 13.25 is rounded to 13
(13.5+15)/2 = 14.25 is rounded to 14

H17

H17A	FPM26f 53	FPM31a	SING106 n	SING106 r	Agreement	Result	ET(x33y)-ET(x11y)
a11r	14,20,28	**21**			2b	20.5	
a11l		**25**			2a	23.5	
v11r						22	
v11l	20	**17**			2b	**17**	
a22r			14,23,31				
a22l	**22**		16,31				
v22r			20				
v22l			11,17,25				
a33r	24		23,33				
a33l			13,23				
v33r	**22**		12,18,27				
v33l	**15,19**		11,17,23				

Remarks:
1. The results of H17A are not as frequent as to differentiate between ET(x22y) and ET(x33y). The bold numbers were used to compute the elementary times.
2. There are plots in the chronophoresis of H17A at the locations suggested by the convergence of FPM-results.

H18

H18A	FPM31e (x11y)	FPM31 (x22y)	FPM26f (x11y)	FPM26f (x22y)	Chronophoresis	Agreement	Result
aNNr	10	13,20	22,28	14,19	12		**10,13**
aNNl	13	12,25	23	12,19	14		**12**
vNNr	15,23	12,21,24	15	12,15,17	14		**12,15**
vNNl	14,17	11,19	14	?			**14,18**

H19

H19A	FPM31e (x11y)	FPM31 (x22y)	FPM26f (x11y)	FPM26f (x22y)	Chrono-phoresis	Agree-ment	Result
aNNr	15	15	26	17,19			15
aNNl	16,18	15,28	?	15			15
vNNr	15	18	14	14.5			14.5
vNNl	15,28	19,28	?	?			14.5

H19B	FPM31e (x11y)	FPM31 (x22y)	FPM26f (x11y)	FPM26f (x22y)	Chrono-phoresis	Agree-ment	Result
aNNr	15	15,21	12	?			15
aNNl	15	18	13,15	17			15
vNNr	15	17	?	24,25			15
vNNl	14,16,18	26	14,20,33	17,23			15

H20

H20A	FPM26f 47(x11y)	FPM27d (x11y)	FPM27d (x22y)	SING100n (x11y)	SING100r (x11y)	Agree-ment	Result
aNNr		15	15,20				15
aNNl		26	13,16,21				13
vNNr		15					15
vNNl							

Remarks: In H20A only the auditory tasks and v11rH20A were performed.

H20B	FPM26f 47(x11y)	FPM27d (x11y)	FPM27d (x22y)	SING100n (x11y)	SING100r (x11y)	Agree-ment	Result
aNNr	28	13	13				13
aNNl	27	13,26	?				13
vNNr	11,13	13	?				13
vNNl	16,22	21	?				16?

H20C	FPM26f 47(x11y)	FPM27d (x11y)	FPM27d (x22y)	SING100n (x11y)	SING100r (x11y)	Agree-ment	Result
aNNr	10,15	15	?				15
aNNl	?	15,22	23				13
vNNr	15	15	15				15
vNNl	?	15	17				16

H20	H20A	H20B	H20C	Max.diff.	Agree	Results
aNNr	15	13	15			15
aNNl	13	13	13			13
vNNr	15	13	15			15
vNNl		16?	16			16

H21

H21A	FPM31e (x11y)	FPM31 (x22y)	FPM26f (x11y)	FPM26f (x22y)	Chronophoresis	Agreement	Result
aNNr	17	14,26,29	14	14			14
aNNl	25	12,16,21	?	19			12, 16
vNNr	?	15,26	13	14,20			14
vNNl	15	15,19	?	16,19			15

H21B	FPM31e (x11y)	FPM31 (x22y)	FPM26f (x11y)	FPM26f (x22y)	Chronophoresis	Agreement	Result
aNNr	14,23	28	18	?			14
aNNl	23,27	15	21	15			15
vNNr	13	15	23.5, 33	23,29			14
vNNl	15	?	?	13			14

H22

H22A	FPM26f 20(x11y)	FPM27d (x11y)	FPM27d (x22y)	SING100n (x11y)	SING100r (x11y)	Agreement	Result
aNNr		15	15		18	1	15
aNNl		15	18		17	2a	16.5
vNNr		15,17	15?		16	2b	15
vNNl		14	13,19		13.5, 20	2b	13.5

H22B	FPM26f 20(x11y)	FPM27d (x11y)	FPM27d (x22y)	SING100n (x11y)	SING100r (x11y)	Agreement	Result
aNNr		14,15	14			1	14
aNNl		16	12			2a	14
vNNr		16	13,21			2b	14.5
vNNl		14?	15			2a	14.5

H22	H22A	H22B	Max.diff.	Agree	Results
aNNr	15	14	1	2a	14.5
aNNl	16.5	14	2.5	2a	15
vNNr	15	14.5	0.5	2a	15
vNNl	13.5	14.5	1	2a	14

ET(aNNl) and ET(vNNr) are rounded.

H24

H24A	FPM31e (x11y)	FPM31 (x22y)	FPM26f (x11y)	FPM26f (x22y)	Chronophoresis	Agreement	Result
aNNr	16,26	14,24,26	16	?			12.5,16
aNNl	16,18	19	21,26	28,30			16.5
vNNr	24	16,23	?	19			12
vNNl	15	20,28	?	25			14.5

H25

H25A	FPM31e (x11y)	FPM31 (x22y)	FPM26f (x11y)	FPM26f (x22y)	Chrono-phoresis	Agree-ment	Result
aNNr	21,28	19,22	13	14			14
aNNl	14	24	28	14			14
vNNr	12	15	12,15,24	16			12,15
vNNl	18	15,28	?	30			15

H26

H26A	FPM31e (x11y)	FPM31 (x22y)	FPM26f (x11y)	FPM26f (x22y)	Chrono-phoresis	Agree-ment	Result
aNNr	20	11,20	20	11			10.5
aNNl	13,26	13,24	12	?			12.5
vNNr	22,27	26	22	19,27			13.5
vNNl	13,26	?	27	15,28			13.5

H26B	FPM31e (x11y)	FPM31 (x22y)	FPM26f (x11y)	FPM26f (x22y)	Chrono-phoresis	Agree-ment	Result
aNNr	?	19,21	12,21	12			11
aNNl	11	19	21,27	?			11
vNNr	10,20	12,15,17	14,21	12,19,31			13
vNNl	23	19	30	29			13

H27

H27A	FPM31e (x11y)	FPM31 (x22y)	FPM26f (x11y)	FPM26f (x22y)	Chrono-phoresis	Agree-ment	Result
aNNr	29	15,23	29	?			14.5
aNNl	28	11	28	15			14
vNNr	12,17,21	17,22	?	27,32			17
vNNl	17,23	16,26	15,17	18,33			16.5

H28

H28B	FPM31e (x11y)	FPM31 (x22y)	FPM26f (x11y)	FPM26f (x22y)	Chrono-phoresis	Agree-ment	Result
aNNr	13,27	13,19,24	?	21,24			13
aNNl	16,22	12	13	17			12.5
vNNr	17,21	9,16	17,21	16			16.5
vNNl	16	9,21	?	?			16

H28C	FPM31e (x11y)	FPM31 (x22y)	FPM26f (x11y)	FPM26f (x22y)	Chrono-phoresis	Agree-ment	Result
aNNr	23,29	9,18	?	14			13
aNNl	24	15,25	20,25	15,30			13
vNNr	16,19,24	15,18	14,34	12			16.5
vNNl	25	11,21,29	15	28,32			15

H31

H31A	FPM31e (x11y)	FPM31 (x22y)	FPM26f (x11y)	FPM26f (x22y)	Chrono-phoresis	Agree-ment	Result
aNNr	18	25	?	16.5			17
aNNl	16	29	31,34	?			15.5
vNNr	?	25	21	16,32			16
vNNl	21	13	30	17			15,16

H31B	FPM31e (x11y)	FPM31 (x22y)	FPM26f (x11y)	FPM26f (x22y)	Chrono-phoresis	Agree-ment	Result
aNNr	15.5,12	13	17	?			12,16
aNNl	15	16	?	19,27			15
vNNr	17	27	?	29			16
vNNl	?	14	?	16			15

Collection of Equations

Design: 100 trials of x11y and 200 trials of x22y

The prerequisite for the determination of the neural structure of mental pathways is the correct measurement of the elementary times. Because the pathways (2 + 1*2) and (2 + 2) are undistinguishable. They are written as (2 + §2).

H01A

auditory

a11rH01A = 70 + (12 + 2(2 + 2*2))12	12 + 8
a11lH01A = 70 + (9 + 2(2 + 2*2))13	9 + 8.6
a22rH01A = 70 + (15 + 2(4 + 2*4))12	14.6 + 19.3
a22lH01A = 70 + (15 + 2(4 + 1*4))13	15 + 13
a33rH01A = 70 + (24 + 2(6 + 2))12	24.7 + 12.7
a33lH01A = 70 + (27 + 2(6 + 1))13	26.5 + 11.3

visual

v11rH01A = 120 + (2 + 2(2 + §2))17	2 + 5
v11lH01A = 120 + (3 + 2(2 + §2))16	3 + 5
v22rH01A = 120 + (3 + 2(4 + 1))17	3 + 7
v22lH01A = 120 + (6 + 2(4 + 1))16	6 + 7
v33rH01A = 120 + (6 + 2(6 + 1))17	6.5 + 8.4
v33lH01A = 120 + (9 + 2(6 + 1))16	9.9 + 7.3

Remarks:

1. a33rH01A uses two searches. That means, tSMS does not need anymore the help of the task set in order to activate the response element. But tSMS works with slow mode. Is this sequence: first the transition from three searches to two searches then the transition from slow mode to fast mode frequent? Or are both processes independent from each other? The last argument is supported by the transition of a (2 + 2) system into a (4 + 1*4) system.

H01B

auditory

a11rH01B = 70 + (12 + 2(2 + 2*2))12	12 + 9
a11lH01B = 70 + (12 + 2(2 + §2))13	11.5 + 6
a22rH01B = 70 + (15 + 2(4 + 1*4))12	16 + 14.6
a22lH01B = 70 + (21 + 2(4 + 1*4))13	21.5 + 15.8

visual

v11rH01B = 120 + (3 + 2(2 + §2))17	3.8 + 5
v11lH01B = 120 + (8 + 2(2 + §2))16	7.5 + 6
v22rH01B = 120 + (9 + 2(4 + 1))17	9 + 7
v22lH01B = 120 + (9 + 2(4 + 2))16	9 + 9.4

Remarks:

1. The linear period is lengthened relativ to x11yH01A. Perhaps there are some inhibiting mechanisms because the subject had to instruct another subject during his doing the task.
2. v22lH01B must be fast mode because v22lH01A uses fast mode.

H01C

auditory

a11rH01C = 70 + (6 + 2(2 + §2))12	6 + 6.5
a11lH01C = 70 + (9 + 2(2 + §2))13	8.7 + 7.5
a22rH01C = 70 + (16 + 2(4 + 1*4))12	16 + 13
a22lH01C = 70 + (15 + 2(4 + 1*4))13	15 + 13

visual

v11rH01C = 120 + (2 + 2(2 + 1))17	2 + 3
v11lH01C = 120 + (4 + 2(2 + §2))16	4 + 5.4
v22rH01C = 120 + (6 + 2(4 + 1))17	5.4 + 5.6
v22lH01C = 120 + (6 + 2(4 + 1))16	6 + 7.2

Remarks:

1. The extremely low periods of v11rH01C can *only* be explained by the current model with the internal-external and mean-median dichotomy.

H01D

auditory

a11lH01C = 70 + (6 + 2(2 + 2*2))13	6 + 8

Remarks:

1. Is is forbidden (implicit learning axiom) that a fast mode of a previous task becomes a slow mode in the subsequent task (in the sequence x11y, x22y, x33y). The reverse case is possible.

H02A

auditory

a11rH02A = 70 + (3 + 2(2 + §2))13	3.4 + 4.8
a11lH02A = 70 + (3 + 2(2 + §2))14.5	2.5 + 4.7
a22rH02A = 70 + (6 + 2(4 + 1*4))13	6.8 + 14.5
a22lH02A = 70 + (7 + 2(4 + 2*4))14.5	7 + 15.2

visual

v11rH02A = 120 + (2 + 2(2 + 1))15	2 + 3.2
v11lH02A = 120 + (2 + 2(2 + 1))17	2.1 + 4.1
v22rH02A = 120 + (2 + 2(4 + 1))15	2.3 + 7.1
v22lH02A = 120 + (3 + 2(4 + 1))17	3.6 + 7.4

H02B

auditory

a11rH02B = 70 + (3 + 2(2 + 1))13	3.8 + 4.5
a11lH02B = 70 + (5 + 2(2 + 1))14	4.4 + 4
a22rH02B = 70 + (12 + 2(4 + 1*4))13	11.5 + 12.7
a22lH02B = 70 + (12 + 2(4 + 1*4))14	11.9 + 12.8

visual

v11rH02B = 120 + (2 + 2(2 + 1))14	2 + 3.1
v11lH02B = 120 + (2 + 2(2 + 1))16	2 + 3.8
v22rH02B = 120 + (2 + 2(4 + 1))14	1.9 + 6.7
v22lH02B = 120 + (4 + 2(4 + 1))16	3.6 + 7.4

H03A

auditory

a11rH03A = 70 + (5 + 2(2 + 2*2))14	5.5 + 7.2
a11lH03A = 70 + (6 + 2(2 + 2*2))15	6.7 + 7.8
a22rH03A = 70 + (15 + 2(4 + 2*4))14	15.5 + 18.4
a22lH03A = 70 + (18 + 2(4 + 2*4))15	17.3 + 19.1

visual

v11rH03A = 120 + (4 + 2(2 + 2*2))12.5	4 + 9
v11lH03A = 120 + (3 + 2(2 + 2*2))13	3.8 + 7.6
v22rH03A = 120 + (8 + 2(4 + 1*4))12.5	8 + 12.7
v22lH03A = 120 + (9 + 2(4 + 1*4))13	8.8 + 11.4

H03B

auditory

a11rH03B = 70 + (5 + 2(2 + 2*2))14	5.1 + 7.5
a11lH03B = 70 + (6 + 2(2 + §2))15	5.7 + 5.4
a22rH03B = 70 + (14 + 2(4 + 2*4))14	14.1 + 17
a22lH03B = 70 + (11 + 2(4 + 2*4))15	10.7 + 18.4

visual

v11rH03B = 120 + (4 + 2(2 + 2*2))12.5	4.4 + 7.9
v11lH03B = 120 + (6 + 2(2 + 2*2))13	6.1 + 6.7
v22rH03B = 120 + (6 + 2(4 + 1*4))12.5	6 + 13.3
v22lH03B = 120 + (9 + 2(4 + 2))13	8.8 + 10.7

Remarks:

1. All auditory tasks use slow mode!
2. v11rH03A(2 + 2*2), v22rH03A(4 + 1*4), v11rH03B(2 + 2*2), v22rH03B(4 + 1*4) is not possible
3. v11lH03A(2 + 2*2), v22lH03A(4 + 1*4), v11lH03B(2 + 2*2), v22lH03B(4 + 2) is not possible

The structures in italic script are incompatible with the implict learning axiom which states the irreversibility of implicit learning in short periods.

But H03A was performed in November/December 1998 and H03B was performed in September 1999, which was 9 months later. The subject could have learned in the first series and forget it during the nine months, using slow mode again. During the second series only v22rH03B learns implicitely.

H04A

auditory

a11rH04A = 70 + (6 + 2(2 + 2*2))14	6.2 + 7.9
a11lH04A = 70 + (9 + 2(2 + 2*2))14	9.1 + 7.8
a22rH04A = 70 + (15 + 2(4 + 2*4))14	15.5 + 16.8
a22lH04A = 70 + (18 + 2(4 + 2*4))14	18 + 19.1

visual

v11rH04A = 120 + (2 + 2(2 + §2))18	2.4 + 5.9
v11lH04A = 120 + (3 + 2(2 + 2*2))21	3 + 7.8
v22rH04A = 120 + (6 + 2(4 + 2))18	6.6 + 9.8
v22lH04A = 120 + (6 + 2(4 + 1*4))21	5.9 + 12.1

H04B

auditory

a11rH04B = 70 + (5 + 2(2 + §2))15	5 + 6.3
a11lH04B = 70 + (5 + 2(2 + 2*2))16	5.2 + 6.8
a22rH04B = 70 + (15 + 2(4 + 2))15	14.7 + 10.2
a22lH04B = 70 + (11 + 2(4 + 1*4))16	10.8 + 12

visual

v11rH04B = 120 + (2 + 2(2 + 1))19	2 + 4.2
v11lH04B = 120 + (3 + 2(2 + 1))21	2 + 4.9
v22rH04B = 120 + (4 + 2(4 + 1))19	4.1 + 6.8
v22lH04B = 120 + (4 + 2(4 + 1))21	4 + 7.3

Remarks:

1. I take cycEN(v11lH04B) = 4.9 for fast mode because v22lH04A and v22llH04B use fast mode.

H05A

auditory

a11rH05A = 70 + (2 + 2(2 + §2))14	1.9 + 4.4
a11lH05A = 70 + (3 + 2(2 + §2))12.5	2.8 + 4.5
a22rH05A = 70 + (9 + 2(4 + 1*4))14	9.8 + 13.1
a22lH05A = 70 + (15 + 2(4 + 1*4))12.5	15.6 + 11.4

visual

v11rH05A = 120 + (2 + 2(2 + 1))15.5	2 + 2
v11lH05A = 120 + (2 + 2(2 + 1))19	2 + 3.3
v22rH05A = 120 + (3 + 2(4 + 2))15.5	3.5 + 9.2
v22lH05A = 120 + (5 + 2(4 + 1))19	5.1 + 7

Remarks:
1. v22rH05A violates the "implicit learning axiom".

H05B

auditory

a11rH05B = 70 + (2 + 2(2 + 1))14 2 + 4.1
a11lH05B = 70 + (3 + 2(2 + 1))13 3.4 + 3.8
a22rH05B = 70 + (9 + 2(4 + 1*4))14 9.4 + 10.6
a22lH05B = 70 + (9 + 2(4 + 1*4))13 9.9 + 12.5

visual

v11rH05B = 120 + (2 + 2(2 + 1))13 2 + 3.5
v11lH05B = 120 + (2 + 2(2 + 1))18.5 2 + 3.8
v22rH05B = 120 + (3 + 2(4 + 2))13 3.4 + 10.5
v22lH05B = 120 + (5 + 2(4 + 1))18.5 4.8 + 6.1

Remarks:
1. ET(vNNrH05B) = 13 and ET(vNNrH05A) = 15.5
2. cycEN = 10.6 must be fast mode in order to fulfil the implicit learning axiom
3. v22rH05B is slow mode again. Either cycEN = 10.5 implies fast mode (4 + 1*4) or v11rH05B is slow mode too.

H06A

auditory

a11rH06A = 70 + (5 + 2(2 + §2))15 5 + 4.5
a11lH06A = 70 + (5 + 2(2 + §2))12.5 4.4 + 6.5
a22rH06A = 70 + (9 + 2(4 + 2))15 8.7 + 8.8
a22lH06A = 70 + (12 + 2(4 + 1*4))12.5 11.2 + 13.4

visual

v11rH06A = 120 + (2 + 2(2 + 1))12 2 + 3.5
v11lH06A = 120 + (2 + 2(2 + §2))14.5 2.1 + 4.8
v22rH06A = 120 + (3 + 2(4 + 1))12 2.6 + 7.2
v22lH06A = 120 + (5 + 2(4 + 1))14.5 4.9 + 8.3

Remarks:
1. The elementary times are read from the chronophoresis by comparison of the various tasks (various subsets).
2. In v11r the difference (Fp-con) is only 2ET. That means the first peaks of the distribution cannot be produced by search but by linear variation. Because minimal search takes 4ET, the trials at Fp must be produced by linear pathway of a complete length of 2 ET.

H06B

auditory

a11rH06B = 70 + (5 + 2(2 + 1))15 5:3 + 3.8
a11lH06B = 70 + (6 + 2(2 + §2))12.5 6 + 4.8
a22rH06B = 70 + (11 + 2(4 + 1))15 11 + 7.6
a22lH06B = 70 + (15 + 2(4 + 1))12.5 14.4 + 8.4

visual

v11rH06B = 120 + (2 + 2(2 + §2))12   2 + 4.6
v11lH06B = 120 + (2 + 2(2 + §2))14.5   2.1 + 5.1
v22rH06B = 120 + (5 + 2(4 + 1))12   5.1 + 8.7
v22lH06B = 120 + (5 + 2(4 + 1))14.5   5.2 + 6.8

Remarks:

1. Very interesting: a11lH06B must be (2 + 1*2) because a22lH06A used fast mode.
2. The same is true for v11rH06B and v11lH06B.
3. v22rH06B must be fast mode because of the fast mode of the preceding tasks v11rH06B, v22rH06A and v11rH06A. That means 8.7 which lies between 8 and 9 (see table) must be fast mode.

H07A

auditory

a11rH07A = 70 + (6 + 2(2 + 2*2))18   6.1 + 8.9
a11lH07A = 70 + (6 + 2(2 + §2))16   6.8 + 5.8
a22rH07A = 70 + (12 + 2(4 + 2))18   11.9 + 9.1
a22lH07A = 70 + (13 + 2(4 + 2))16   13 + 9.3

visual

v11rH07A = 120 + (2 + 2(2 + 1))17   2 + 3.4
v11lH07A = 120 + (2 + 2(2 + 1))17   2 + 4.1
v22rH07A = 120 + (5 + 2(4 + 1))17   5.1 + 7.7
v22lH07A = 120 + (6 + 2(4 + 1))17   6.3 + 8

H07B

auditory

a11rH07B = 70 + (6 + 2(2 + §2))16   5.5 + 6.1
a11lH07B = 70 + (6 + 2(2 + §2))15   6.7 + 5.5
a22rH07B = 70 + (12 + 2(4 + 2))16   12.4 + 8.6
a22lH07B = 70 + (12 + 2(4 + 2))15   12 + 10.3

visual

v11rH07B = 120 + (2 + 2(2 + 1))17.5   2 + 3.8
v11lH07B = 120 + (3 + 2(2 + 1))16   2 + 3.8
v22rH07B = 120 + (5 + 2(4 + 1))17.5   4.9 + 5.7
v22lH07B = 120 + (5 + 2(4 + 1))16   4.9 + 8.1

H08A

auditory

a11rH08A = 70 + (5 + 2(2 + §2))14   4.8 + 6
a11l H08A = 70 + (5 + 2(2 + §2))14.5   4.6 + 5.8
a22r H08A = 70 + (12 + 2(4 + 2))14   11.9 + 8.4
a22l H08A = 70 + (10 + 2(4 + 2))14.5   10.1 + 9

visual

v11rH08A = 120 + (2 + 2(2 + §2))14   2.1 + 4.7
v11lH08A = 120 + (2 + 2(2 + §2))15   2.3 + 4.5
v22rH08A = 120 + (6 + 2(4 + 1))14   6.6 + 6.6
v22lH08A = 120 + (4 + 2(4 + 1))15   4.3 + 7.2

H09A *(with ET = 14,14,15,15)*
auditory

a11rH09A = 70 + (8 + 2(2 + §2))14	7.6 + 5
a11lH09A = 70 + (6 + 2(2 + §2))14	6.2 + 5
a22rH09A = 70 + (9 + 2(4 + 1*4))14	9.8 + 14.5
a22lH09A = 70 + (12 + 2(4 + 2*4))14	11.6 + 16.8

visual

v11rH09A = 120 + (2 + 2(2 + §2))15	2.3 + 5
v11lH09A = 120 + (3 + 2(2 + 2*2))15	2.7 + 7
v22rH09A = 120 + (8 + 2(4 + 2))15	8 + 9
v22lH09A = 120 + (11 + 2(4 + 1*4))15	10.7 + 11.4

Remarks:
1. The cycEN of v11lH09A lies between (2 + §2) and (2 + 2*2).

H09A *(with ET = 15,15,15.5,15)*
auditory

a11rH09A = 70 + (7 + 2(2 + §2))15	7 + 5.1
a11lH09A = 70 + (6 + 2(2 + §2))15	5.7 + 5.2
a22rH09A = 70 + (9 + 2(4 + 1*4))15	9 + 13.7
a22lH09A = 70 + (11 + 2(4 + 2*4))15	10.7 + 15.8

visual

v11rH09A = 120 + (2 + 2(2 + §2))15.5	2.3 + 5
v11lH09A = 120 + (3 + 2(2 + 2*2))15	2.7 + 7
v22rH09A = 120 + (8 + 2(4 + 2))15.5	8 + 9
v22lH09A = 120 + (11 + 2(4 + 1*4))15	10.7 + 11.4

H09B
auditory

a11rH09B = 70 + (3 + 2(2 + *1*))15	2.7 + 4.5
a11lH09B = 70 + (3 + 2(2 + §2))15	2.7 + 5.6
a22rH09B = 70 + (8 + 2(4 + 1*4))15	7.7 + 12.5
a22lH09B = 70 + (8 + 2(4 + 1*4))15	8 + 13.8

visual

v11rH09B = 120 + (2 + 2(2 + 1))15.5	2 + 3.1
v11lH09B = 120 + (2 + 2(2 + §2))15	2 + 5.5
v22rH09B = 120 + (4 + 2(4 + 1))15.5	4.1 + 8.2
v22lH09B = 120 + (7 + 2(4 + 1))15	7 + 8.1

Remarks:
1. a11rH09B rather fast mode with cycEN = 4.5 because a22rH09A uses fast mode too.

H09C
auditory

a11rH09C = 70 + (5 + 2(2 + 1))15	4.7 + 4.6
a11lH09C = 70 + (3 + 2(2 + §2))15	3.7 + 5.1
a22rH09C = 70 + (6 + 2(4 + 1*4))15	6.3 + 12.1
a22lH09C = 70 + (8 + 2(4 + 1*4))15	7.7 + 12.5

visual

v11rH09C = 120 + (2 + 2(2 + 1))15.5	2 + 2.9
v11lH09C = 120 + (2 + 2(2 + 1))15	2 + 4.5
v22rH09C = 120 + (2 + 2(4 + 2?))15.5	2.2 + 9
v22lH09C = 120 + (8 + 2(4 + 2?))15	7.7 + 10.3

Remarks:

1. a11rH09C rather fast mode with cycEN = 4.6 because a22rH09B uses fast mode too.
2. a11lH09C must be (2 + §2) = (2 + 1*2) because a22lH09B uses fast mode.
3. v11lH09C must use fast mode because v22lH09B uses fast mode.
4. v22yH09C cannot use slow mode, that means the cycEN = 9 must be fast mode! *These are serious flaws, especially v22lH09C with cycEN = 10.3! What ist wrong with these two tasks? The series were performed on 22.10.(H09A), 26.10.(H09B), and 28.10.(H09C).*

It is the median which is responsible for these flaws. It is computed by 50 trials but the distribution comprises 100 trials and sometimes the median(50 trials) ≠ median(100 trials).

There are multiple oppurtinities to fail: the first peaks, the elementary times. v11lH09C may use slow mode too.

v11lH09A	v22lH09A	v11lH09B	v22lH09B	v1lH09C	v22lH09C
(2 + 2*2)	(4 + 1*4)	(2 + §2)	(4 + 1)	(2 + 1)	(4 + 2)

Attention: 4.5 and 4.6 may imply (2 + §2) = (2 + 1*2) !!

H10B

auditory

a11rH10B = 70 + (4 + 2(2 + 1))13	4.2 + 3.9
a11lH10B = 70 + (3 + 2(2 + 1))16	3.3 + 4.3
a22rH10B = 70 + (11 + 2(4 + 2))13	11.5 + 9.3
a22lH10B = 70 + (9 + 2(4 + 2))16	9.6 + 9.5

visual

v11rH10B = 120 + (3)15	3.2
v11lH10B = 120 + (2 + 2(2 + 1))14	1.6 + 4
v22rH10B = 120 + (1 + 2(4 + 1))15	1 + 7.1*
v22lH10B = 120 + (4 + 2(4 + 1))14	4.4 + 5.3

Remarks:

1. The tasks of this subject start with H10B because the subject has performed a task H10A with another design (bihemispheric tasks).
2. In v11rH10B the (Median-constant) = 48ms, that means 48/15 = 3.2ET. This implies the median pathway of v11rH10B is a pure linear pathway without any search. The reason may be the training this subject had had by similar bihemispheric tasks before.
3. In v22rH10B the first peak is supposed to be changed about 15 ms to the right. This would cause an equation of v22r = con + (2 + 2(4 + 1))15 with 2 + 6 ETs.
4. In v22lH10B the cycET = 5.3 is very low. One has to consider the possiblity that the real elementary times are half of the proposed ones.

H10C

auditory

a11rH10C = 70 + (3 + 2(2 + 1))13	3.8 + 4.6
a11lH10C = 70 + (3 + 2(2 + 1))16	2.7 + 3.7
a22rH10C = 70 + (9 + 2(4 + 2))13	8.4 + 10.5
a22lH10C = 70 + (9 + 2(4 + 1))16	9.6 + 8.3

visual

v11rH10C = 120 + (2 + 2(2 + 1))15	2 + 2.7
v11lH10C = 120 + (2 + 2(2 + 1))14	1.9 + 4
v22rH10C3 = 120 + (2 + 2(4 + 1))15	2.3 + 5.3
v22lH10C3 = 120 + (3 + 2(4 + 1))14	3.7 + 6.6

Remarks:

1. a11rH10C could use slow mode with cycEN = 4.6
2. In v11lH10C the first peak is false. I have to reconsider the definition of the first peak in tasks with 100 trials.
3. a22rH10C with cycEN = 10.5 could be fast mode.
4. The series a11rH10B(2 + 1), a22rH10B(4 + 2), a11rH10C(2 + 1), and a22rH10C(4 + 2) is not possible. Alternative series: a11rH10B(2 + 2), a22rH10B(4 + 2), a11rH10C (2 + 2), and a22rH10C(4 + 2)

In order to get fast mode, the first peak of a11rH10B has to be shifted 10 ms to the left. Then the equation would be: a11rH10B = con + (3 + 2(2 + 2))13 with linEN = 3.4 and cycEN = 4.7.

H10D

auditory

a11rH10D = 70 + (2 + 2(2 + 2))13	2.2 + 5.4
a11lH10D = 70 + (3 + 2(2 + 1))16	3 + 4.3
a22rH10D = 70 + (12 + 2(4 + 2))13	11.5 + 10.2
a22lH10D = 70 + (11 + 2(4 + 1))16	10.8 + 8.5

visual

v11rH10D = 120 + (2 + 2(2 + 1))15	2 + 2.7
v11lH10D = 120 + (2 + 2(2 + 1))14	2 + 4.3
v22rH10D = 120 + (2 + 2(4 + 1))15	2 + 5
v22lH10D = 120 + (3 + 2(4 + 1))14	3.7 + 5.4

Remarks:

1. The considerations about the aNNr series are supported by the result of a11rH10D.

H11A

auditory

a11rH11A = 70 + (3 + 2(2 + §2))16	3 + 5.2
a11lH11A = 70 + (5 + 2(2 + §2))14	4.8 + 5
a22rH11A = 70 + (9 + 2(4 + 2*4))16	8.9 + 18.1
a22lH11A = 70 + (9 + 2(4 + 2*4))14	9.4 + 19.5

visual

v11rH11A = 120 + (2 + 2(2 + §2))13	2.2 + 5.6
v11lH11A = 120 + (2 + 2(2 + §2))16	2.1 + 5.5
v22rH11A = 120 + (9 + 2(4 + 1*4))13	9.5 + 11.7
v22lH11A = 120 + (6 + 2(4 + 1*4))16	6.4 + 12.5

H11B

auditory

a11r H11B = 70 + (3 + 2(2 + §2))16	3.6 + 5.5
a11l H11B = 70 + (5 + 2(2 + §2))14	5.5 + 5.5
a22r H11B = 70 + (14 + 2(4 + 1*4))16	13.6 + 13.1
a22l H11B = 70 + (14 + 2(4 + 2*4))14	13.4 + 20.3

visual

v11r H11B = 120 + (2 + 2(2 + §2))13	1.8 + 5.8
v11l H11B = 120 + (2 + 2(2 + §2))16	2.1 + 4.6
v22r H11B = 120 + (6 + 2(4 + 1*4))13	5.7 + 14.9
v22l H11B = 120 + (5 + 2(4 + 1*4))16	5.2 + 11.1

Remarks:

1. Is the border between linEN and cycEN visible in ERPs?
2. The cycEN of v11r H11B lies at the limit between fast mode (4 + 1*4) and slow mode (4 + 2*4).
3. v11y H11B must be (2 + §2) = (2 + 1*2) because v22yH11A use fast mode.

H12A

auditory

a11rH12A = 70 + (2 + 2(2 + §2))16.5	1.6 + 4.5
a11lH12A = 70 + (2 + 2(2 + 1))18	2.2 + 3.5
a22rH12A = 70 + (6 + 2(4 + 2))16.5	5.6 + 10.5
a22lH12A = 70 + (6 + 2(4 + 2))18	6.3 + 10.4

visual

v11rH12A = 120 + (3 + 2(2 + §2))15.5	3.5 + 5.2
v11lH12A = 120 + (3 + 2(2 + §2))18	2.7 + 4.6
v22rH12A = 120 + (4 + 2(4 + 1))15.5	4.1 + 7.9
v22lH12A = 120 + (5 + 2(4 + 1))18	4.7 + 8.4

Remarks:

1. There is one inconsistency between the fast mode of a11lH12A and the slow mode of a22lH12A. a11lH12B and a22lH12B work in the fast mode. That means, either a11lH12A is wrong or a22lH12A is wrong! The distribution of a11lH12A is unusually distorted to the left.
2. Is there any event within the history of subject H12 which can explain this assymetry?

This assymetry may be due to the right prefrontal cortex because both sonsory modalities are affected.

H12B

auditory

a11rH12B = 70 + (3 + 2(2 + §2))16.5   2.8 + 4.4*
a11lH12B = 70 + (3 + 2(2 + 1))18   2.4 + 4.2
a22rH12B = 70 + (6 + 2(4 + 2))16.5   6.5 + 10.4
a22lH12B = 70 + (11 + 2(4 + 1))18   10.5 + 6.3

visual

v11rH12B = 120 + (2 + 2(2 + 1))15.5   1.5 + 5*
v11lH12B = 120 + (3 + 2(2 + 1))18   3 + 4.5
v22rH12B = 120 + (6 + 2(4 + 1))15.5   5.4 + 6.6
v22lH12B = 120 + (6 + 2(4 + 1))18   6.1 + 6.1

Remarks:

1. The nestle results of a22r and a22l differ from the chronophoresis results, the nestle results of v22r and v22l confirm the chronophorese results
2. In a11r *slow mode* and 2 searches have to be taken because of the slow mode of a22r in spite of cycET = 4.4 (lying between fast mode and slow mode).
3. v11rH12B must be *fast mode* because v22rH12A is fast and v22rH12B is fast.
4. The behavior of the sensorimotor set during the tasks (x11yNNA -> x22yNNA > x11yNNB -> x22NNB etc. is an important test for the correctness of the equations.

H13A

auditory

a11rH13A = 70 + (6 + 2(2 + 1))14   5.5 + 4.4
a11lH13A = 70 + (3 + 2(2 + §2))15   3.7 + 5.5
a22rH13A = 70 + (8 + 2(4 + 1*4))14   8 + 15.8
a22lH13A = 70 + (9 + 2(4 + 1*4))15   9.3 + 13.4

visual

v11rH13A = 120 + (9 + 2(2 + 2*2))13.5   9.1 + 9.6
v11lH13A = 120 + (7 + 2(2 + 2*2))15.5   7.4 + 6.8
v22rH13A = 120 + (12 + 2(4 + 1*4))13.5   12.1 + 14.5
v22lH13A = 120 + (12 + 2(4 + 1))15.5   11.9 + 7.1

Remarks:

1. In a22rH13A, the cycEN = 15.8 is too high for (4 + 1*4). Either the real elementary time is higher in this series or the median number of cycET is aberrant.
2. It is striking that the subject H13 uses fast mode in the auditory task of the first series (The program FPM26f34 proposes *half the elementary time*, then there would be no fast mode). But half of the elementary times would cause an explosion of cycle numbers e.g. in v11rH13A, the cycling would last 19cycET, the linear portion would be 18 linET.
3. The evaluation of the pathway structure uses elementary times due to *all* three series of this subject.
4. In v11lH13A the cycEN = 6.8 is slightly too small for (2 + 2*2) mode.

H13B

auditory

a11rH13B = 70 + (4 + 2(2 + §2))14*   4.1 + 4.9
a11lH13B = 70 + (3 + 2(2 + §2))15   2.7 + 4.7

a22rH13B = 70 + (9 + 2(4 + 1*4))14 9.8 + 12.9
a22lH13B = 70 + (6 + 2(4 + 1*4))15 6.3 + 14.5

visual
v11rH13B = 120 + (2 + 2(2 + §2))13.5 1.7 + 6.3
v11lH13B = 120 + (2 + 2(2 + §2))15.5 2.2 + 5.5
v22rH13B = 120 + (5 + 2(4 + 1))13.5 4.7 + 7.3
v22lH13B = 120 + (3 + 2(4 + 1))15.5 3.2 + 8.2

Remarks:
1. * The task a11rH13B is compatible with the tasks a11rH13A and a22rH13A because (2 + §2) may mean (2 + 1*2). This must not be forgotten and is valid for all task x11yH13B because all x22yH13A use fast mode.

H13C

auditory
a11rH13C = 70 + (4 + 2(2 + §2))14 4.4 + 5.5
a11lH13C = 70 + (2 + 2(2 + §2))15 2.3 + 5.6
a22rH13C = 70 + (9 + 2(4 + 1*4))14 8.4 + 13
a22lH13C = 70 + (8 + 2(4 + 1*4))15 8 + 10.4

visual
v11rH13C = 120 + (2 + 2(2 + §2))13.5 2 + 5.0
v11lH13C = 120 + (2 + 2(2 + §2))15.5 1.9 + 4.9
v22rH13C = 120 + (4 + 2(4 + 1))13.5 4.3 + 8.6
v22lH13C = 120 + (4 + 2(4 + 1))15.5 4.1 + 6.9

Remarks:
1. Revision of first peak in v11rH13C (shift to the right in order to give 2 linET).
2. All x11yH13C must use fast mode: (2 + §2) = (2 + 1*2) because of the fast mode of all x22yH13B and all x22yH13A.
3. a22lH13C must use fast mode (4 + 1*4) in spite of the cycEN = *10.4* which proposes slow mode (4 + 2). Perhaps the elementary time is less than 15 in a22lH13C (linEN too small, see nestle procedure).
4. v22rH13C must use (4 + 1) with a cycEN = 8.6 because all precedent v22rH13X use fast mode.

H14A

auditory
a11rH14A = 70 + (8 + 2(2 + §2))15 7.7 + 4.6
a11l H14A = 70 + (8 + 2(2 + §2))14 8.4 + 4.5
a22r H14A = 70 + (9 + 2(4 + 1))15 9.7 + 7.8
a22l H14A = 70 + (9 + 2(4 + 1))14 9.1 + 8.3

visual
v11r H14A = 120 + (3 + 2(2 + §2))12 3 + 5.3
v11l H14A = 120 + (2 + 2(2 + §2))12.5 2 + 6.4
v22r H14A = 120 + (6 + 2(4 + 2))12 6.3 + 8.8
v22l H14A = 120 + (8 + 2(4 + 1))12.5 7.6 + 8.2

H15A

auditory

a11rH15A = 70 + (3 + 2(2 + 1))22	3.2 + 3.4
a11lH15A = 70 + (2 + 2(2 + 1))22	2.1 + 4
a22rH15A = 70 + (12 + 2(4 + 1*4))22	12.9 + 10.4
a22lH15A = 70 + (9 + 2(4 + 1*4))22	8.9 + 11.2

visual

v11rH15A = 120 + (2 + 2(2 + 1))16	2.4 + 4.1
v11lH15A = 120 + (3 + 2(2 + §2))19	3 + 4.6
v22rH15A = 120 + (8 + 2(4 + 1*4))16	7.7 + 12.3
v22lH15A = 120 + (8 + 2(4 + 2))19	7.5 + 10.2

H16A

auditory

a11rH16A = 70 + (4 + 2(2 + §2))14.5	4.2 + 4.5
a11lH16A = 70 + (3 + 2(2 + §2))13	3.4 + 6.6
a22rH16A = 70 + (8 + 2(4 + 1*4))14.5	8.3 + 12.1
a22lH16A = 70 + (9 + 2(4 + 1*4))13	9.2 + 14.3

visual

v11rH16A = 120 + (2 + 2(2 + 1))17	2 + 3.6
v11lH16A = 120 + (2 + 2(2 + §2))14	2 + 5.5
v22rH16A = 120 + (4 + 2(4 + 1*4))17	4.2 + 11
v22lH16A = 120 + (6 + 2(4 + 1))14	6.9 + 8

H16B

auditory

a11rH16B = 70 + (3 + 2(2 + §2))14.5	3.2 + 4.5
a11lH16B = 70 + (4 + 2(2 + §2))13	4.5 + 4.7
a22rH16B = 70 + (7 + 2(4 + 2))14.5	7.3 + 9
a22lH16B = 70 + (9 + 2(4 + 1*4))13	8.4 + 12.1

visual

v11rH16B = 120 + (2 + 2(2 + 1))17	2 + 2.9
v11lH16B = 120 + (2 + 2(2 + §2))14	2 + 4.6
v22rH16B = 120 + (3 + 2(4 + 1))17	3.3 + 7.5
v22lH16B = 120 + (4 + 2(4 + 2))14	4.4 + 9.4

Remarks:

1. Because a22lH16A uses fast mode, a11lH16B must use fast mode too: (2 + §2) = (2 + 1*2)
2. The series a11rH16A(2 + §2), a22rH16A(*4* + *1* *2), a11rH16B(2 + §2), a22rH16B (*4* + 2) is not possible. Either a22rH16A or a22rH16B is wrong. The cycEN (a22rH16B) = 9 may belong to (4 + 1).
3. The series v11lH16A(2 + §2), v22lrH16A(*4* + *1*), v11lH16B(2 + §2), v22lH16B (*4* + 2) is not possible. Either v22lH16A or v22lH16B is wrong. The cycEN (v22lH16A) = 8 may belong to (4 + 2).
4. In this subject there is another possibility: the first series was conducted in November 1998, the second series in October 1999. Is the implicite learning effect reversible within one year?

In any case it can be noticed that the two incompatibilities affect the auditive left hemisphere and the visual right hemisphere. This question can only be answered by other subjects with one year difference between the two series.
An argument against this hypothesis is the implicite learning within the first series but not within the second series. This difference cannot be explained.

H17A

auditory

a11rH17A = 70 + (6 + 2(2 + 2*2))20.5 — 5.6 + 7.8
a11lH17A = 70 + (3 + 2(2 + §2))23.5 — 3.1 + 5
a22rH17A = 70 + (11 + 2(4 + 1))20.5 — 10.4 + 7.9
a22lH17A = 70 + (6 + 2(4 + 1))23.5 — 6.1 + 7.8
a33rH17A = 70 + (11 + 2(6 + 1))20.5 — 10.7 + 9.1
a33lH17A = 70 + (11 + 2(6 + 1))23.5 — 10.6 + 5.5

visual

v11rH17A = 120 + (2 + 2(2 + §2))22 — 2.1 + 4.7
v11lH17A = 120 + (5 + 2(2 + §2))17 — 4.8 + 5.9
v22rH17A = 120 + (8 + 2(4 + 1))22 — 8 + 7.4
v22lH17A = 120 + (6 + 2(4 + 1))17 — 6.5 + 7.8
v33rH17A = 120 + (6 + 2(6 + 1))22 — 6.4 + 7.5
v33lH17A = 120 + (11 + 2(6 + 1))17 — 10.4 + 8.2

Remarks:
1. The implicit learning axiom is fulfilled.

H18A

auditory

a11rH18A = 70 + (3 + 2(2 + §2))13 — 3.4 + 4.6
a11lH18A = 70 + (3 + 2(2 + §2))12 — 3.4 + 4.8
a22rH18A = 70 + (12 + 2(4 + 1*4))13 — 11.5 + 11.2
a22lH18A = 70 + (12 + 2(4 + 2))12 — 11.8 + 9.6
a33rH18A = 70 + (15 + 2(6 + 1))13 — 14.5 + 12.8
a33lH18A = 70 + (18 + 2(6 + 1))12 — 18.8 + 12.2

visual

v11rH18A = 120 + (2 + 2(2 + 1))12 — 2.0 + 2.5
v11lH18A = 120 + (2 + 2(2 + 1))14 — 1.9 + 4.1
v22rH18A = 120 + (5 + 2(4 + 1))12 — 4.7 + 8.4
v22lH18A = 120 + (6 + 2(4 + 1))14 — 6.9 + 7
v33rH18A = 120 + (9 + 2(6 + 1))12 — 9.3 + 9.2
v33lH18A = 120 + (9 + 2(6 + 1))14 — 9.4 + 8

H19A

auditory

a11rH19A = 70 + (7 + 2(2 + §2))15 — 7 + 4.5
a11lH19A = 70 + (4 + 2(2 + §2))15 — 4 + 5.4
a22rH19A = 70 + (11 + 2(4 + 1*4))15 — 10.7 + 11.1
a22lH19A = 70 + (13 + 2(4 + 2))15 — 13 + 9.3

visual

v11rH19A = 120 + (3 + 2(2 + §2))14.5   3.2 + 4.5
v11lH19A = 120 + (2 + 2(2 + §2))14.5   2 + 5.4
v22rH19A = 120 + (5 + 2(4 + 2))14.5   4.9 + 10.3
v22lH19A = 120 + (7 + 2(4 + 1))14.5   7 + 7.7

H19B

auditory

a11rH19B = 70 + (3 + 2(2 + 1))15   3 + 3.9
a11lH19B = 70 + (3 + 2(2 + 1))15   3.3 + 4
a22rH19B = 70 + (9 + 2(4 + 1))15   9.7 + 8.6
a22lH19B = 70 + (9 + 2(4 + 2))15   9 + 10

visual

v11rH19B = 120 + (2 + 2(2 + 1))15   2 + 3.6
v11lH19B = 120 + (3 + 2(2 + 1))15   3 + 4
v22rH19B = 120 + (4 + 2(4 + 1))15   4.3 + 7.1
v22lH19B = 120 + (4 + 2(4 + 1))15   4 + 6.8

Remarks:

1. Either a11lH19B must be slow mode or a22lH19B must be fast mode. The above results are not compatible with the implicit learning axiom.

H20A

auditory

a11rH20A = 70 + (5 + 2(2 + §2))15   5 + 5
a11lH20A = 70 + (6 + 2(2 + §2))13   6.1 + 5.2
a22rH20A = 70 + (14 + 2(4 + 1*4))15   14 + 13.3
a22lH20A = 70 + (14 + 2(4 + 2*4))13   13.8 + 17.2

visual

v11rH20A = 120 + (3 + 2(2 + §2))15   3 + 5.3

Remarks:

1. Subject H20A performed only the above tasks in series A.
2. H20A was performed on 18.11.1998, H20B on 10.08.1999, and H20C on 30.08.1999. The time distance between H20A and H20B may be the cause that a22rH20A uses fast mode and a22rH20B slow mode.

H20B

auditory

a11rH20B = 70 + (5 + 2(2 + §2))15   5 + 5.3
a11lH20B = 70 + (6 + 2(2 + §2))13   5.7 + 5.2
a22rH20B = 70 + (11 + 2(4 + 2))15   10.7 + 10.7
a22lH20B = 70 + (12 + 2(4 + 1*4))13   11.5 + 14.7

visual

v11rH20B = 120 + (5 + 2(2 + §2))15   5.3 + 6
v11lH20B = 120 + (4 + 2(2 + §2))16   4.3 + 6.2
v22rH20B = 120 + (9 + 2(4 + 2))15   8.7 + 10
v22lH20B = 120 + (6 + 2(4 + 1*4))16   6.4 + 12.1

Remarks:

1. The cycEN = 14.7 of task a22lH20B is rather fast mode than slow mode but lies between the two possibilities.

H20C

auditory

a11rH20C = 70 + (5 + 2(2 + §2))15	5 + 4.9
a11lH20C = 70 + (6 + 2(2 + §2))13	6.5 + 5.9
a22rH20C = 70 + (14 + 2(4 + 1))15	13.7 + 8.5
a22lH20C = 70 + (12 + 2(4 + 1*4))13	11.8 + *15.7*

visual

v11rH20C = 120 + (2 + 2(2 + §2))15	2 + 5.2
v11lH20C = 120 + (3 + 2(2 + 1))16	2.7 + 4.3
v22rH20C = 120 + (4 + 2(4 + 1))15	4.3 + 8
v22lH20C = 120 + (5 + 2(4 + 1))16	4.9 + *8.8*

Remarks:

1. v22lH20C must use fast mode because of the irreversibility of implicit learning in the observed period.
2. a22lH20C must use fast mode too because of the same reason.

H21A

auditory

a11rH21A = 70 + (6 + 2(2 + §2))14	5.5 + 4.6
a11lH21A = 70 + (3 + 2(2 + §2))16	3.6 + 5
a22rH21A = 70 + (9 + 2(4 + 2))14	9.1 + 10.1
a22lH21A = 70 + (8 + 2(4 + 2))16	7.4 + 10

visual

v11rH21A = 120 + (2 + 2(2 + 1))14	2 + 3.6
v11lH21A = 120 + (2 + 2(2 + §2))15	2 + 5.3
v22rH21A = 120 + (2 + 2(4 + 1))14	2.3 + 7.6
v22lH21A = 120 + (3 + 2(4 + 2))15	3.7 + 9.2

H21B

auditory

a11rH21B = 70 + (6 + 2(2 + §2))14	6.2 + *4.1*
a11lH21B = 70 + (5 + 2(2 + §2))16	5 + 4.9
a22rH21B = 70 + (9 + 2(4 + 2))14	9.1 + 9.1
a22lH21B = 70 + (9 + 2(4 + 2))16	8.7 + 9.2

visual

v11rH21B = 120 + (2 + 2(2 + 1))14	2 + 3.6
v11lH21B = 120 + (2 + 2(2 + §2))15	2.6 + 6
v22rH21B = 120 + (2 + 2(4 + 1))14	2.6 + 6.6
v22lH21B = 120 + (5 + 2(4 + 1))15	4.8 + 6.2

H22A

auditory

a11rH22A = 70 + (11 + 2(2 + 2*2))14.5 — 10.8 + 7.6
a11lH22A = 70 + (12 + 2(2 + 2*2))15 — 12.3 + 8.2
a22rH22A = 70 + (15 + 2(4 + *1*4*))14.5 — 15.9 + *12.7*
a22lH22A = 70 + (19 + 2(4 + 1*4))15 — 19 + 12.6

visual

v11rH22A = 120 + (3 + 2(2 + 2*2))15 — 3 + 8.9
v11lH22A = 120 + (4 + 2(2 + 2*2))14 — 4.4 + 8.6
v22rH22A = 120 + (5 + 2(4 + 1))15 — 5 + 7.7
v22lH22A = 120 + (6 + 2(4 + 2))14 — 6.6 + 10.2

H22B

auditory

a11rH22B = 70 + (11 + 2(2 + 2*2))14.5 — 10.4 + 9
a11lH22B = 70 + (5 + 2(2 + §2))15 — 5.3 + 6.3
a22rH22B = 70 + (15 + 2(4 + 2*4))14.5 — 14.9 + *16.8*
a22lH22B = 70 + (12 + 2(4 + 1*4))15 — 12 + 13.3

visual

v11rH22B = 120 + (2 + 2(2 + §2))15 — 2 + 5.3
v11lH22B = 120 + (5 + 2(2 + §2))14 — 4.8 + 5.8
v22rH22B = 120 + (5 + 2(4 + 1))15 — 5 + 7.1
v22lH22B = 120 + (6 + 2(4 + 1*4))14 — 6.9 + 11.6

Remarks:

1. The series a11rH22A (2 + 2*2), *a22rH22A (4 + 1*4),* a11rH22B (2 + 2*2), a22rH22B (4 + 2*4) is not possible. The mode of the task a22rH22A is not compatible with the modes of the precedent and subsequent tasks. Perhaps cycEN = (Median-FirstPeak)/linET + 2cycET = 12.7 is too high. This may occur if the median is too high or the the FirstPeak is too low. The median is computed correctly, the first peak consists of three trials standing alone. The distribution starts some 25ms later. It has to be observed whether this mistake occurs frequently. This would be the end of the FirstPeakMethod.
2. In a11lH22B was performed in the morning, the subject estimated an increased attention. Is this the reason why linEN is reduced?
3. a11lH22B and v11yH22B must use (2 + §2) = (2 + 1*2) fast mode because the precedent and the subsequent tasks use fast mode too.

H23A

auditory

a11rH23A = 70 + (3 + 2(2 + §2))15 — 3.3 + 5
a11lH23A = 70 + (4 + 2(2 + §2))16 — 4.3 + 4.7
a22rH23A = 70 + (10 + 2(4 + 1*4))15 — 10 + 12.3
a22lH23A = 70 + (11 + 2(4 + 2))16 — 11.1 + 9.7

visual

v11rH23A = 120 + (3 + 2(2 + §2))14 — 3 + 5.1
v11lH23A = 120 + (5 + 2(2 + §2))13 — 4.9 + 4.5
v22rH23A = 120 + (11 + 2(4 + 1))14 — 10.9 + 7.6
v22lH23A = 120 + (10 + 2(4 + 1))13 — 10.3 + 7.4

Remarks:
1. The fast mode of v22rH23A is an argument for the fast mode of v22rH23B (see below).
2. The slow mode of a22lH23A is an argument for the slow mode of a22lH23B.
3. The elementary times of H23A, H23B, and H23C differ from each other because of the refined methods (FPM26f16, FPM26g(x11y), FPM26g(x22y)).

H23B

auditory

a11rH23B = 70 + (3 + 2(2 + 1))15 3 + 4.2
a11lH23B = 70 + (4 + 2(2 + *1*))14 4.4 + *4.2*
a22rH23B = 70 + (12 + 2(4 + 1))15 12.7 + 7.2
a22lH23B = 70 + (15 + 2(4 + 2))14 15.9 + *9.3*

visual

v11rH23B = 120 + (3 + 2(2 + §2))12 2.6 + 5.5
v11lH23B = 120 + (3 + 2(2 + §2))14 3.4 + 4.8
v22rH23B = 120 + (12 + 2(4 + *1*))12 11.8 + *8.6*
v22lH23B = 120 + (12 + 2(4 + 1))14 11.6 + 6.5

Remarks:
1. I had the following results (FMP26f(x11y):15,14,11,15; FPM26f(x22y): ?,14, 12, 14; FPM26f16(x11y):14,15,12,12). I took 15,14,12,14 regarding the FPM26f (x22y).(see 995808).
2. Because a11yH23B use fast mode, a11yH23C must use fast mode too. That means (2 + 1*2).
3. The difference of the elementary times of H23B and H23C is ± 1ms
4. There is an inconsistency between a11lH23B and a22lH23B: The second cannot use slow mode while the first uses fast mode. Maybe ET = 14 is too small, ET = 15 would solve this puzzle.
5. v22rH23B could be fast mode or slow mode (cycEN = 8.6 lies between these two possibilities).

H23C

auditory

a11rH23C = 70 + (3 + 2(2 + §2))15 3.7 + 5.1
a11lH23C = 70 + (6 + 2(2 + §2))15 6.3 + 5.1
a22rH23C = 70 + (15 + 2(4 + 1))15 15 + 6.4
a22lH23C = 70 + (17 + 2(4 + 1))15 17 + 6.9

visual

v11rH23C = 120 + (5 + 2(2 + §2))13 4.9 + 5.7
v11lH23C = 120 + (6 + 2(2 + §2))13 6.1 + 5.6
v22rH23C = 120 + (12 + 2(4 + 1*4))13 12.6 + 12.2
v22lH23C = 120 + (14 + 2(4 + 1))13 14.2 + 8.1

Remarks:
1. The safety of these results was provided by the program FPM26f16.
2. The error may be + -1ms. This can be seen by help of the three series H23A, H23B, and H23C. The elementary times in these three series deviate about 1 ms.

H24A

auditory

a11rH24A = 70 + (3 + 2(2 + §2))16 — 3.6 + 4.5
a11l H24A = 70 + (3 + 2(2 + §2))16.5 — 3.7 + 4.8
a22r H24A = 70 + (12 + 2(4 + 1*4))16 — 12.1 + 11.3
a22l H24A = 70 + (15 + 2(4 + 2))16.5 — 15.9 + 10.4

visual

v11rH24A = 120 + (2 + 2(2 + §2))12 — 2.2 + 5.9
v11lH24A 120 + (2 + 2(2 + §2))14.5 — 2.5 + 4.8
v22rH24A = 120 + (12 + 2(4 + 1))12 — 11.8 + 6
v22lH24A = 120 + (5 + 2(4 + 1))14.5 — 4.9 + 8.1

H25A

auditory

a11rH25A = 70 + (3 + 2(2 + §2))14 — 3.7 + 5
a11lH25A = 70 + (3 + 2(2 + §2))14 — 2.6 + 5.3
a22rH25A = 70 + (9 + 2(4 + 2))14 — 8.4 + 9.3
a22lH25A = 70 + (9 + 2(4 + 2))14 — 8.7 + 9.7

visual

v11rH25A = 120 + (2 + 2(2 + 1))12 — 2 + 3.2
v11lH25A 120 + (2 + 2(2 + §2))15 — 2 + 4.4
v22rH25A = 120 + (3 + 2(4 + 1))12 — 3 + 7.8
v22lH25A = 120 + (3 + 2(4 + 1))15 — 3.3 + 7.1

H26A

auditory

a11rH26A = 70 + (3 + 2(2 + §2))10.5 — 3.7 + 5.1
a11lH26A = 70 + (2 + 2(2 + §2))12.5 — 2.4 + 5.2
a22rH26A = 70 + (9 + 2(4 + 2*4))10.5 — 9.4 + 17.1
a22lH26A = 70 + (9 + 2(4 + 2*4))12.5 — 8.8 + *13.2*

visual

v11rH26A = 100 + (2 + 2(2 + 1))13.5 — 2 + 3.5
v11lH26A = 100 + (2 + 2(2 + §2))13.5 — 2 + 4.5
v22rH26A = 100 + (3 + 2(4 + *1*4*))13.5 — 3.6 + *10.6*
v22lH26A = 100 + (6 + 2(4 + 2))13.5 — 5.8 + 8.9

Remarks:

1. In this subject the visual constant had to be changed to 100 ms otherwhile the linear pathway and the cyclical pathway would have been too short.
2. a22lH26A has to use slow mode because a11lH26B and a22lH26B use slow mode too.
3. v22rH26A muse use fast mode because v11rH26A uses fast mode.

H26B

auditory

a11rH26B = 70 + (3 + 2(2 + §2))11 — 2.5 + 4.9
a11lH26B = 70 + (3 + 2(2 + §2))11 — 3.9 + 7.9

a22rH26B = 70 + (8 + 2(4 + 2*4))11   8 + 14.7
a22lH26B = 70 + (8 + 2(4 + 2*4))11   8 + 18.2

visual

v11rH26B = 100 + (2 + 2(2 + 1))13   1.9 + 2.7
v11lH26B = 100 + (2 + 2(2 + §2))13   2 + 5.2
v22rH26B = 100 + (2 + 2(4 + 1))13   2.2 + 7.8
v22lH26B = 100 + (4 + 2(4 + 2))13   4.2 + 10.1

Remarks:

1. a22rH26B has cycEN = 14.7. This number is smaller than the cycEN(a22rH26A). This could mean, that there are some (4 + 1*4) too in a22rH26B.

H27A

auditory

a11rH27A = 70 + (15 + 2(2 + 2*2))14.5   15 + 7.6
a11lH27A = 70 + (12 + 2(2 + 2*2))14   12.3 + 8.1
a22rH27A 70 + (15 + 2(4 + 1*4))14.5   15.9 + 12.3
a22lH27A = 70 + (17 + 2(4 + 2))14   16.6 + 10.6

visual

v11rH27A = 120 + (2 + 2(2 + 1))17   2.1 + 2.7
v11lH27A 120 + (2 + 2(2 + §2))16.5   2 + 4.3
v22rH27A = 120 + (3 + 2(4 + 1))17   2.7 + 6.5
v22lH27A = 120 + (3 + 2(4 + 1))16.5   3.9 + 8

H28B

auditory

a11rH28B = 70 + (3 + 2(2 + 1))13   3.4 + 3.9
a11lH28B = 70 + (3 + 2(2 + §2))12.5   3.2 + 4.7
a22rH28B = 70 + (14 + 2(4 + 1))13   13.8 + 8.3
a22lH28B = 70 + (15 + 2(4 + 2*4))12.5   14.4 + *14.1*

visual

v11rH28B = 120 + (2 + 2(2 + 1))16.5   2 + 3.5
v11lH28B = 120 + (2 + 2(2 + §2))16   2 + 4.6
v22rH28B = 120 + (6 + 2(4 + 1))16.5   6.2 + 6.1
v22lH28B = 120 + (6 + 2(4 + 1))16   6.8 + 8

Remarks:

1. The fast modes of H28B are consequences of the fact that H28B is a repetition. H28A is incomplete, therefore its results are not shown here.
2. In a22lH28B slow mode is assumed because of the slow mode of a22lH28C.

H28C

auditory

a11rH28C = 70 + (3 + 2(2 + §2))13   3 + 4.8
a11lH28C = 70 + (2 + 2(2 + §2))13   2.2 + 5.6
a22rH28C = 70 + (14 + 2(4 + *1*))13   13.4 + *9.4*
a22lH28C = 70 + (8 + 2(4 + 2*4))13   8 + 16.3

visual

v11rH28C = 120 + (2 + 2(2 + 1))16.5 2 + 3.9
v11lH28C = 120 + (2 + 2(2 + §2))15 2 + 5.2
v22rH28C = 120 + (6 + 2(4 + 1))16.5 5.6 + 7.3
v22lH28C = 120 + (6 + 2(4 + 1))15 6.7 + 7.3

Remarks:
1. The slow mode of a22rH28C violates the implicit learning axiom, therefore (4 + 1) was taken.

H29A

auditory

a11rH29A = 70 + (4 + 2(2 + §2))12.5 4 + 5.7
a11lH29A = 70 + (6 + 2(2 + §2))12 6.3 + 4.8
a22rH29A = 70 + (9 + 2(4 + 2*4))12.5 9.2 + 17.4
a22lH29A = 70 + (10 + 2(4 + 2*4))12 10.1 + 16.3

visual

v11rH29A = 120 + (2 + 2(2 + §2))14 2 + 4.6
v11lH29A = 120 + (2 + 2(2 + 1))16 2.4 + 4.3
v22rH29A = 120 + (5 + 2(4 + 2))14 4.8 + 8.6
v22lH29A = 120 + (5 + 2(4 + 1))16 4.6 + 7.3

Remarks:
1. In v22rH29A the cycEN = 8.6 permits two pathways: (4 + 1) or (4 + 2)

H29B

auditory

a11rH29B = 70 + (4 + 2(2 + §2))12.5 4.4 + 4.5
a11lH29B = 70 + (4 + 2(2 + §2))12 4.3 + 5.3
a22rH29B = 70 + (9 + 2(4 + 2))12.5 9.2 + 9.8
a22lH29B = 70 + (6 + 2(4 + 1*4))12 6.3 + 11.5

visual

v11rH29B = 120 + (2 + 2(2 + 1))14 2 + 3.9
v11lH29B = 120 + (2 + 2(2 + 1))16 2 + 4
v22rH29B = 120 + (3 + 2(4 + 1))14 3 + 7.1
v22lH29B = 120 + (4 + 2(4 + 1))16 4.3 + 7.6

H30A

auditory

a11rH30A = 70 + (3 + 2(2 + §2))15 3 + 4.7
a11lH30A = 70 + (2 + 2(2 + §2))14 1.9 + 5
a22rH30A = 70 + (8 + 2(4 + 1*4))15 8 + 11.4
a22lH30A = 70 + (8 + 2(4 + 1*4))14 7.6 + 12.3

visual

v11rH30A = 120 + (2 + 2(2 + 1))16 2 + 2.8
v11lH30A = 120 + (2 + 2(2 + 1))14 2 + 4.1
v22rH30A = 120 + (3 + 2(4 + 1))16 2.7 + 6.6
v22lH30A = 120 + (4 + 2(4 + 1))14 4.4 + 6.8

Remarks:

1. In v11rH30A the difference (Fp - con) = 40ms. That means there are only 40/ 14 = 2.9.elemenary times between con and Fp. This is not possible because the minimal pathway needs con + 4ET. That means the first peak is not produced by the fastest possible search process but by local shortcuts. If one takes 4ET for granted, the (Median - Fp) difference diminshes to 1.6 ET. With the 2cycET added one gets 3.6 cycET for the minimal search process and 2 ET for the linear portion.
2. The same is true for v11lH30A, where (Fp-con)/ET = 2.7. That means, Fp is produced by a process which only needs about 3 ET (presumably a linear local shortcut pathway). I therefore take 4ET right of con for the minimal pathway, remaining 2 cycET left of the median. This makes a linear pathway of 2 linET and a cyclical pathway of 2 + 2 = 4 cycET.

H30B

auditory

a11rH30B = 70 + (3 + 2(2 + *1*))15	3 + 4.2
a11lH30B = 70 + (3 + 2(2 + *1*))14	3.4 + *4.8**
a22rH30B = 70 + (6 + 2(4 + 1))15	6 + 8.5
a22lH30B = 70 + (6 + 2(4 + 1*4))14	6.2 + 13.1

visual

v11rH30B = 120 + (2 + 2(2 + 1))16	2 + 2.3
v11lH30B = 120 + (2 + 2(2 + 1))14	2 + *4.6**
v22rH30B = 120 + (2 + 2(4 + 1))16	2.1 + 7.4
v22lH30B = 120 + (4 + 2(4 + 2))14	4.1 + 8.8

Remarks:

1. This is important: After the fast mode in a22yH30A, a11yH30B must use fast mode too because the sensorimotor set SMS1(aNNy) is changed implicitely. In a11lH30B fast mode is proposed though there are 4.8 cycET (the median may lies left of its current position in the distribution of a11lH30B).
2. The equation for v22lH30B must be wrong because the transition back to slow mode is forbidden.

H30A* *(with an universal elementary time of 15 ms for all tasks)*

auditory

a11rH30A = 70 + (3 + 2(2 + §2))15	3 + 4.7
a11lH30A = 70 + (2 + 2(2 + §2))15	1.7 + 4.8
a22rH30A = 70 + (8 + 2(4 + 1*4))15	8 + 11.4
a22lH30A = 70 + (7 + 2(4 + 1*4))15	7 + 11.6

visual

v11rH30A = 120 + (2 + 2(2 + 1))15	2 + 3.2
v11lH30A = 120 + (2 + 2(2 + 1))15	2 + 3.7
v22rH30A = 120 + (3 + 2(4 + 1))15	3 + 6.9
v22lH30A = 120 + (4 + 2(4 + 1))15	4 + 6.5

H30B* *(with an universal elementary time of 15 ms for all tasks)*

auditory

a11rH30B = 70 + (3 + 2(2 + 1))15 — 3 + 4.2
a11lH30B = 70 + (3 + 2(2 + 1))15 — 3 + 4.6*
a22rH30B = 70 + (6 + 2(4 + 1))15 — 6 + 8.5
a22lH30B = 70 + (6 + 2(4 + 1*4))15 — 5.7 + 12.4

visual

v11rH30B = 120 + (2 + 2(2 + 1))15 — 2 + 2.6
v11lH30B = 120 + (2 + 2(2 + 1))15 — 1.7 + 4.4
v22rH30B = 120 + (2 + 2(4 + 1))15 — 2.3 + 7.8
v22lH30B = 120 + (3 + 2(4 + 1))15 — 3.7 + 8.3

Remarks:

1. There are less exeeptions than in then non-symmetrical case. Only a11lH30B needs an exception.
2. The homogenous elementary time contradicts the results of the chronophoresis where slight differences are observable but the nestle method strongly proposes these elementary times of 15 ms.

H31A

auditory

a11rH31A = 70 + (11 + 2(2 + 2*2))17 — 10.6 + 9.7
a11lH31A = 70 + (16 + 2(2 + 2*2))15.5 — 16.1 + 9.9
a22rH31A = 70 + (18 + 2(4 + 1*4))17 — 17.7 + 11.5
a22lH31A = 70 + (26 + 2(4 + 2))15.5 — 26.1 + 10.5

visual

v11rH31A = 120 + (9 + 2(2 + 2*2))16 — 9.9 + 9.6
v11lH31A = 120 + (12 + 2(2 + 2*2))15 — 12 + 10.8
v22rH31A = 120 + (12 + 2(4 + 2))16 — 11.8 + 10.3
v22lH31A = 120 + (15 + 2(4 + 2))15 — 15.7 + 10.2

H31B

auditory

a11rH31B = 70 + (13 + 2(2 + 2*2))16 — 13 + 7.8
a11lH31B = 70 + (13 + 2(2 + 2*2))15 — 13 + 7.9
a22rH31B = 70 + (18 + 2(4 + 1*4))16 — 18.6 + 11.9
a22lH31B = 70 + (20 + 2(4 + 2*4))15 — 20 + 15.4

visual

v11rH31B = 120 + (8 + 2(2 + 2*2))16 — 7.7 + 7.4
v11lH31B = 120 + (11 + 2(2 + 2*2))15 — 11 + 6.9
v22rH31B = 120 + (9 + 2(4 + 2))16 — 9.9 + 9.8
v22lH31B = 120 + (12 + 2(4 + 2))15 — 12.7 + 9.7

Remarks:

1. Either a11rH31B must use fast mode or a22rH31A must use slow mode.
   cycEN(a11rH31B) = 7.8 lies between 6 (2 + 1*2) and 8 (2 + 2*2)
   cycEN(a22rH31A) = 11.5 lies between 10 (4 + 2) and 12 (4 + 1*4).
   I would rather decide a22rH31A to be slow mode.

Symmetries and Statistics

Frequencies of elementary times

The frequencies and the mean values of the four elementary times (ET(aNNr), ET(aNNl), ET(vNNr), and ET(vNNl)) were investigated in the 30 healthy subjects. The auditory elemetary time of the right-handed tasks varies between 10.5 ms and 22 ms, the mean was 14.6 ms. The auditory elementary time of the left-handed tasks lay between 12ms and 22ms, the mean was 14.5ms.

The visual elementary time of the right-handed tasks varies between 12ms and 18ms, the mean lying at 14.7ms and the visual elementary time of the left-handed task lay between 12.5ms and 21ms, the mean was 15.4ms.

All elementary times but the ET(vNNr) showed an approximate normal distribution.

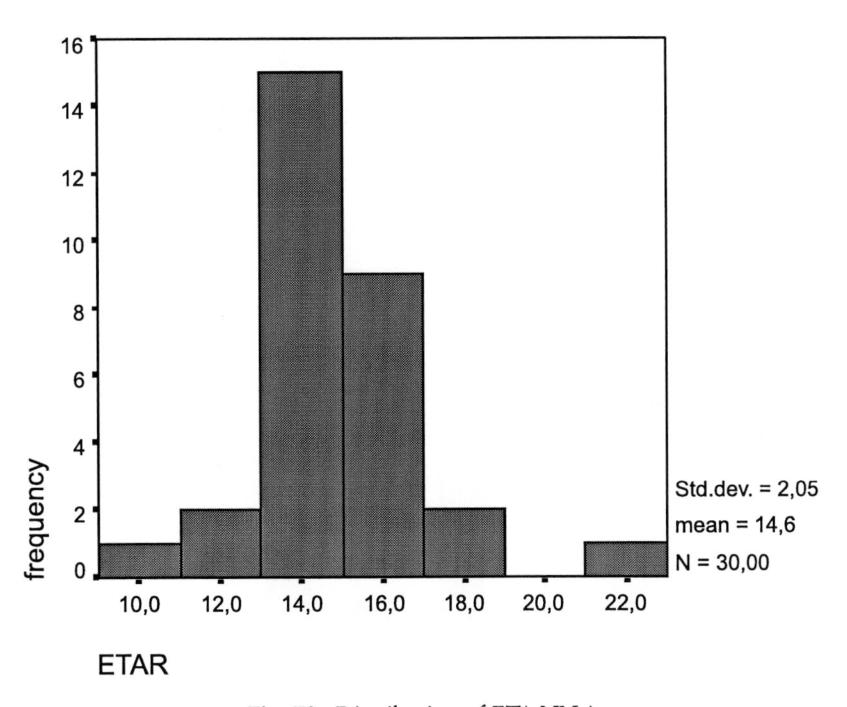

Fig. 70. Distribution of ET(aNNr)

Frequencies of linear pathways

The linear pathways defined by (Fp-con)/ET-2ET tended to have lengths which are multiples of 3: (Fp-con)/ET - 2ET = n*3. These tendencies may be seen in the distributions of the auditory linear pathways in all 30 subjects. The visual linear pathways of the tasks v11y show no such tendencies, that of the tasks v22y do. The

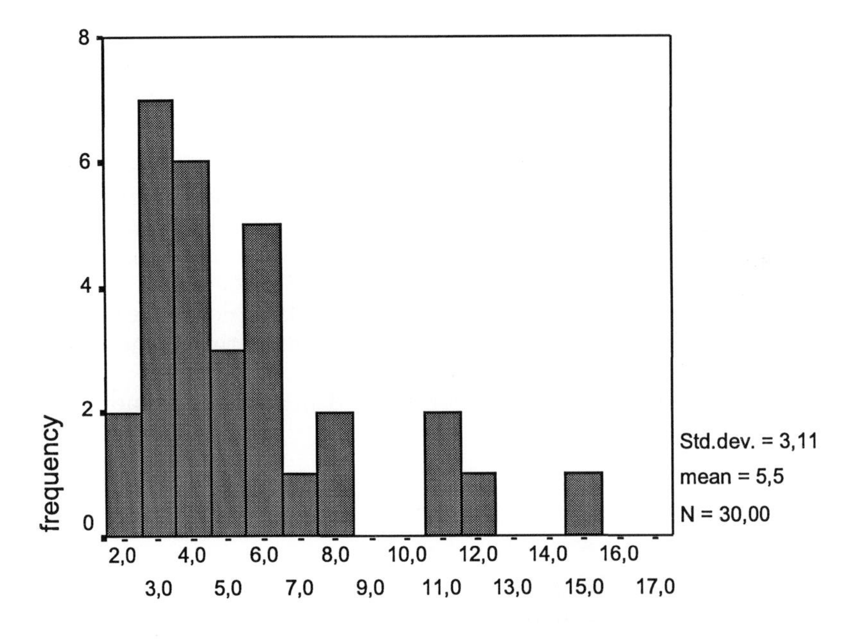

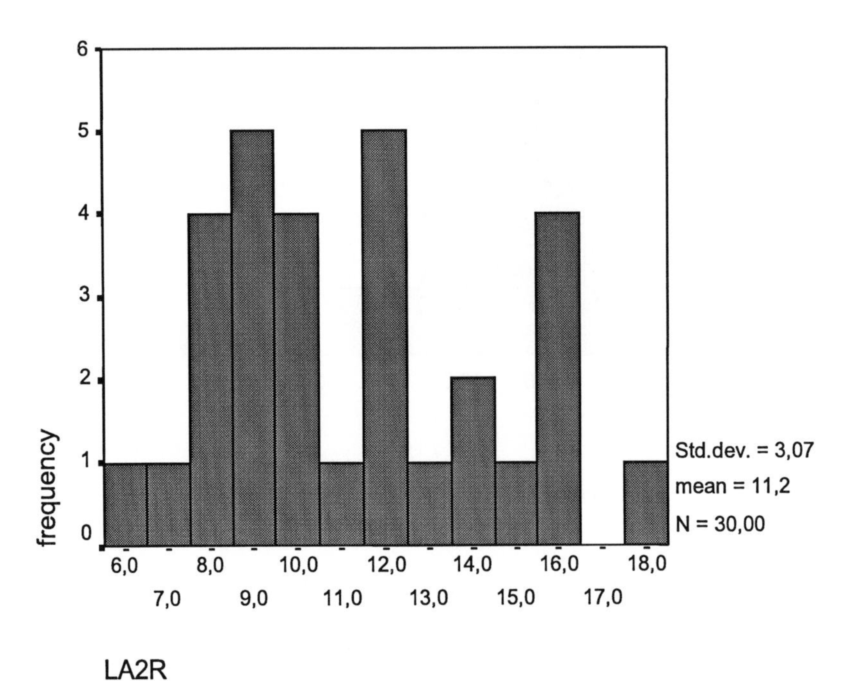

Fig. 71. LA1R= linear pathway of the task a11r, LA2R= linear pathway of the task a22r

other distributions of the linear pathways of the tasks a11r and a22r are shown below. The mean linear pathway of a11r had the length linEN(a11r)=5.5. The other pathways had the *mean length* of linEN(a11l)=5.4, linEN(a22r)=11.2, and linEN (a22l)=12.2.

The visual linear pathways had the *mean values* linEN(v11r)= 3, linEN(v11l)=3, linEN(v22r)=5.8, and linEN(v22l)=6.5.

Frequencies of cyclical pathways

The tasks a11y of the 30 subjects gave some intriguing evidence for the prefered length of the median cyclical pathway. In all subjects, the lengths cycEN=5 and cycEN=8 were prefered in the one stimulus auditory tasks on both sides.

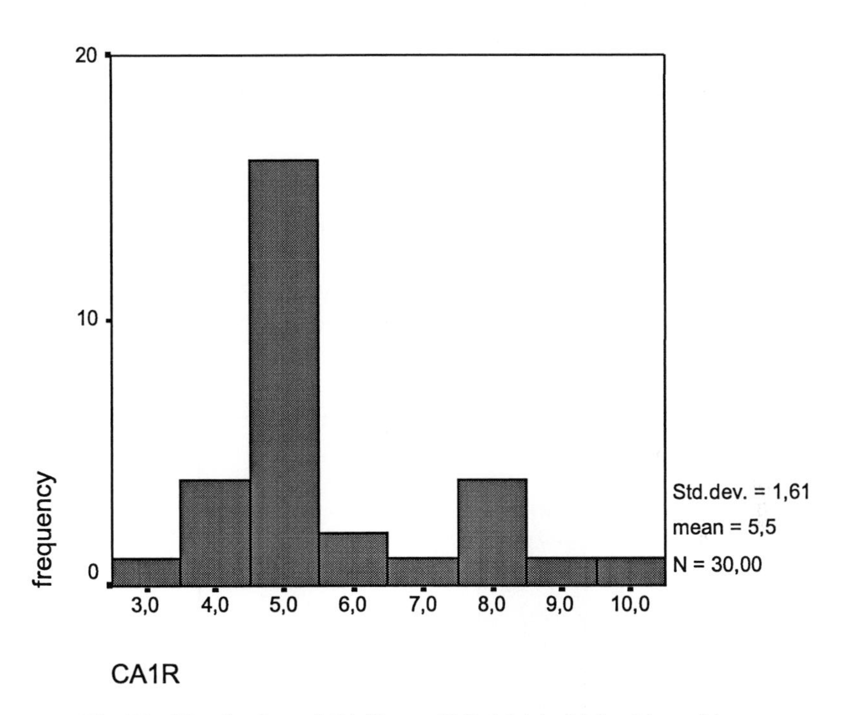

Fig. 72. Distribution of CA1R=cycEN(a11r) in 30 healthy subjects

For the visual task v11l it is cycEN=5 but the task v11r has two peaks at cycEN=3 and cycEN=5:

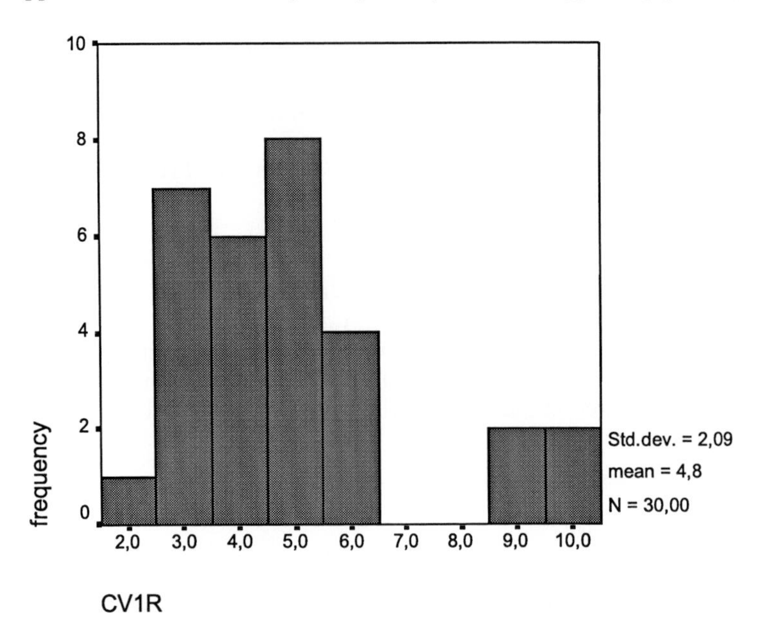

Fig. 73. Distribution of CV1R=cycEN(v11r) in 30 healthy subjects

The tasks v22y clearly have their maximum at cycEN=7 and cycEN=8, with the task v22r having high frequencies in cycEN=9 and cycEN=10 too.

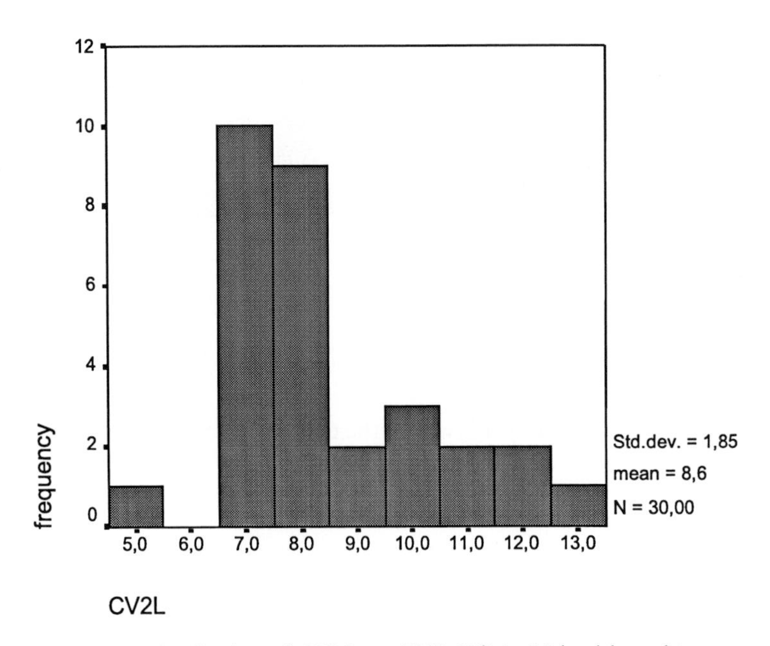

Fig. 74. Distribution of CV2L=cycEN(v22l) in 30 healthy subjects

The most complex pattern can be seen in the frequencies of the a22y cyclical pathways. The task a22l has its relative maxima at cycEN=10 and cycEN=13 while the task a22r has its ones at cycEN=9, cycEN=11, cycEN=12, and cycEN=17.

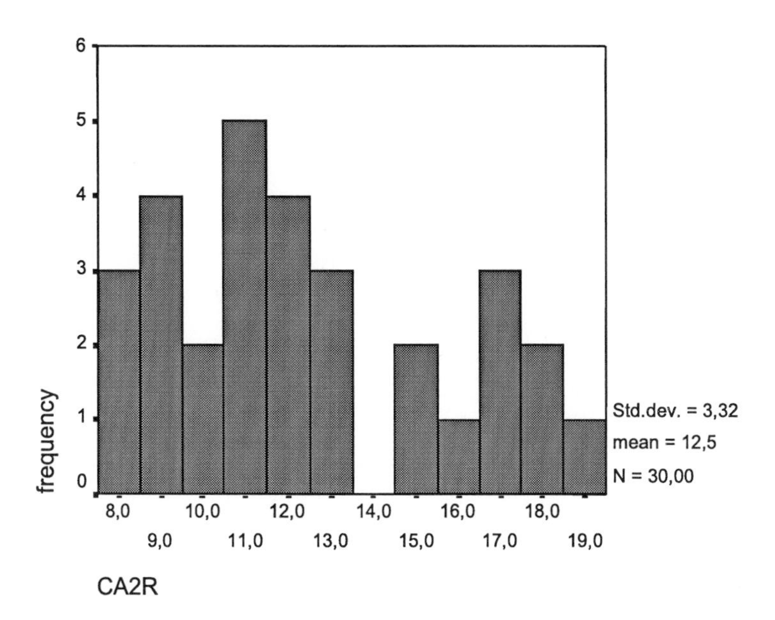

Fig. 75. Distribution of CA2R=cycEN(a22r) in 30 healthy subjects

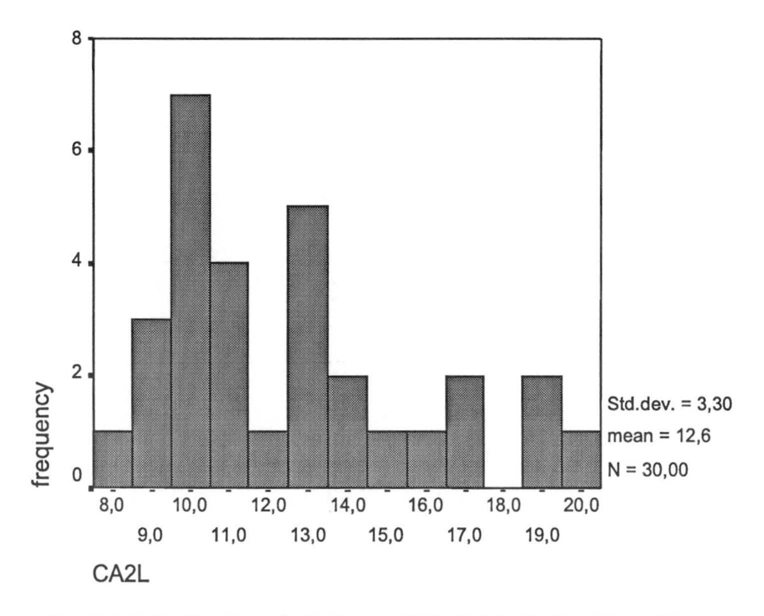

Fig. 76. Distribution of CA2L=cycEN(a22l) in 30 healthy subjects

The Symmetry of linEN(xNNr) and linEN(xNNl)

There is a strong correlation between the lengths of the linear pathways on the two sides. The linEN(a11r) correlates with linEN(a11l), linEN(a22r) with linEN(a22l), linEN(v11r) with linEN(v11l), and linEN(v22r) with linEN(v22l). As an example one of these correlations is shown below.

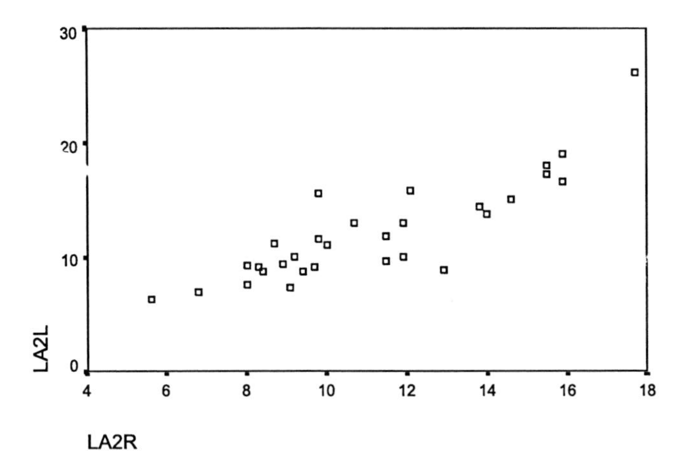

Fig. 77. Scatterplot between LA2R=linear pathway of a22r and LA2L=linear pathway of a22l. The Pearson correlation coefficient is $r= 0.861$, the significance level is $p< 0.01$

The correlation is stronger in auditory tasks than in visual tasks, visible for example in the tasks v22r versus v22l.

The Symmetry of cycEN(xNNr) and cycEN(xNNl)

If one uses scatterplots to investigate the relationship between the different cyclical pathways one sees a strong correlation between cycEN(x11r) and cycEN(x11l), but only a weak correlaton between cycEN(x22r) and cycEN(x22l). As an example the correlaton between cycEN(a11r) and cycEN(a11l) is shown below.

The Symmetry of linEN(xNNy) and cycEN(xNNy)

In the scatterplots there is no correlation between linEN(a22y) and cycEN(a22y). Either for the right side nor for the left. Between linEN(v11y) and cycEN(v11y) there are weak correlations, between linEN(a11y) and cycEN(a11y) there are some weak correlations, too. The same is true between linEN(v22y) and cycEN(v22y). The example below shows the correlation between linEn(a11l) and cycEN(a11l).

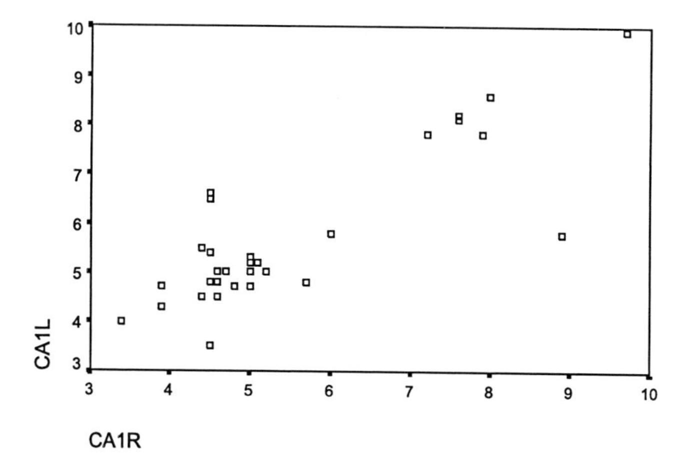

Fig. 78. Scatterplot between CA1R=cyclical pathway of a11r and CA1L=cyclical pathway of a11l. The Pearson correlation coefficient is r= 0.836, the significance level is p< 0.01

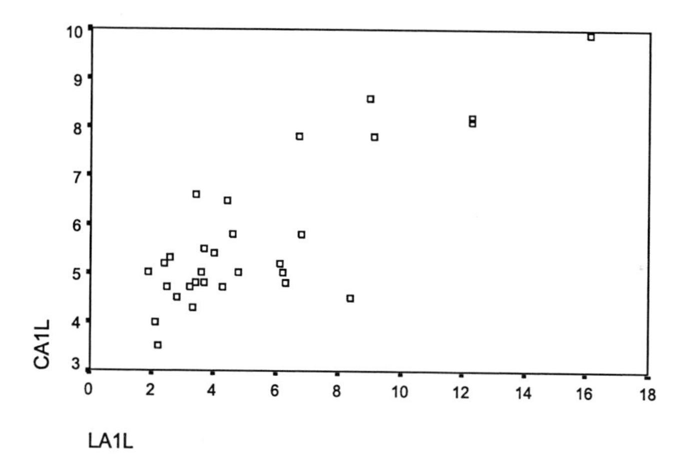

Fig. 79. Scatterplot between LA1l=linear pathway of a11r and CA1L=cyclical pathway of a11l. The Pearson correlation coefficient is r= 0.823, the significance level is p< 0.01

The Symmetry of elementary times

There was no correspondence between the auditory and visual elementary times investigated by the scattering plot with ET(aNNr) and ET(vNNr) as the variable of the x-axis respectively the y-axis. Some subjects showed a striking symmetry: all elementary times had the same value.

The auditory elementary times of the right-handed tasks and the elementary times of the left-handed tasks showed a high correspondence in the scattering plot, their visual counterparts showed a weak correspondence which has to be confirmed.

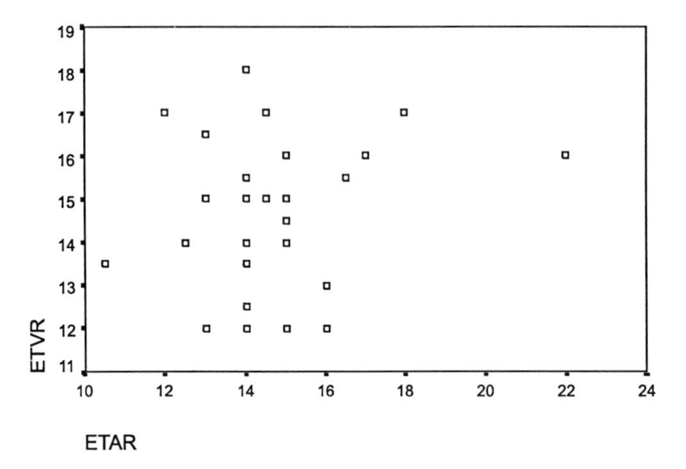

Fig. 80. ETAR=ET(aNNr), ETH30=ET(vNNr). The Pearson correlation coefficient is r= 0.155, there is no significance

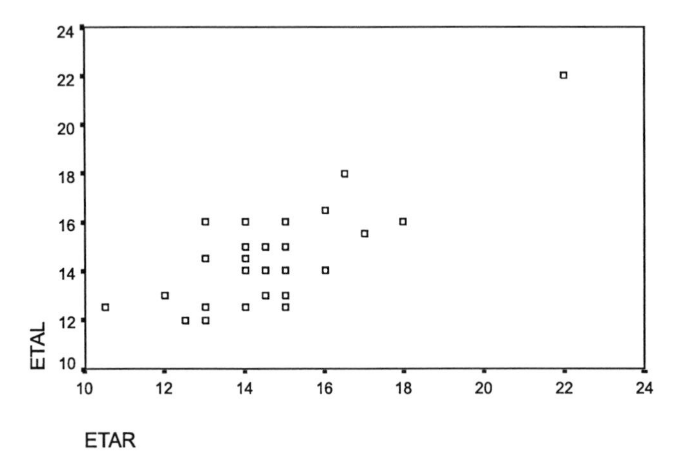

Fig. 81. ETAR=ET(aNNr), ETAL=ET(aNNl). The Pearson correlation coefficient is r= 0.773, the significance level is p< 0.01

Changes of linEN in repetitions

23 of the 30 heahlthy subjects have repeated all eight tasks.If one displays the linEN of each task within a figure, one can compare the linEN of the first series with the linEN of the second series. Principally there are the following possible changes seen in the linEN of two series:

- all linEN remain constant
- all linEN are increased (H01X, H23X, H12X)

- all linEN are decreased (H04X, H08X, H19X, H26X)
- only the auditory linEN are increased, the visual linEN remain constant (H02X)
- only the visual linEN are increased, the auditory linEN remain constant.
- only the auditory linEN are decreased, the visual linEN remain constant (H03X)
- only the visual linEN are decreased, the auditory linEN remain constant.(H13X)
- the auditory linEN are increased, the visual linEN are decreased (H11X, H20X)
- the auditory linEN are decreased, the visual linEN are increased

The increase and the decrease of all linEN shall be visualized in the two subjects H01X and H09X:

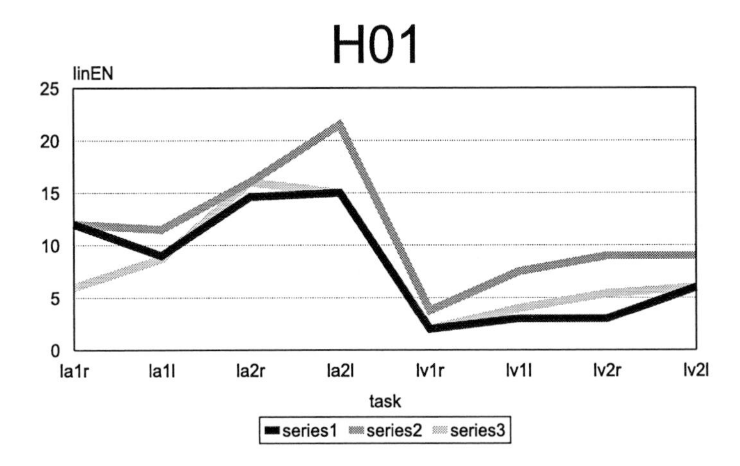

Fig. 82. The length of the linear pathway is coordinated to each task. H01A= dark line, H01B=light line

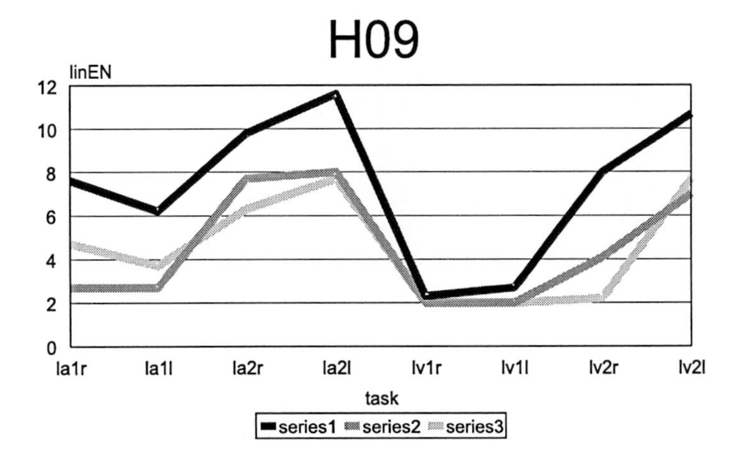

Fig. 83. The length of the linear pathway is coordinated to each task. *H09A* dark line, *H09B* light line

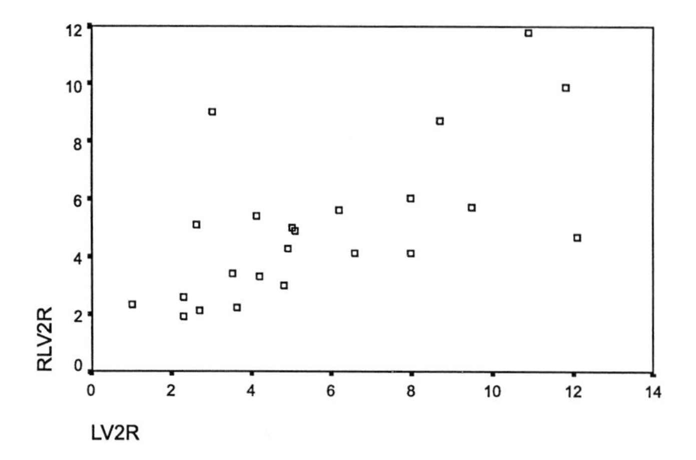

Fig. 84. The length of the linear visual pathway in 23 healthy subjects in series A (LV2R) and series B (RLV2R). The Pearson correlation coefficient is r= 0.641, the significance level is p< 0.01. For one subject the linear pathway increased, in three subjects the linear pathway decreased, the others have approximately the same linear pathway

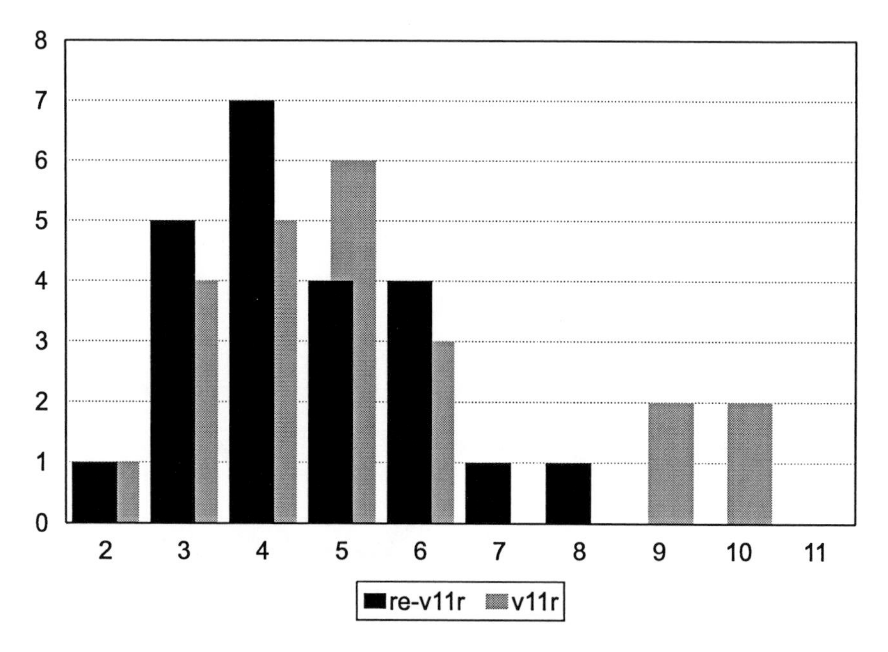

Fig. 85. Distribution of cycEN in 23 healthy subjects. The dark columns represent the repeated task v11r, the light columns represent the task v11r

If one compares the length of the linear visual pathway of the first series A with that of the second series B, one gets the subsequent figure (the length of the linear auditory pathways is treated in Part IV).

This finding is similar in the other auditory and visual tasks (see Part IV for the auditory tasks).

Changes of cycEN in repetitions

The cyclical pathway is reduced in the repeated tasks. As an example, the task v11r and its repetition is shown below.

Conclusions

What can be learned about the stimulus-response pathways from the evidences gathered in the last sections?

Symmetry between elementary times

The strong correspondence between the two auditory elementary times supports the notions that either the two hemispheres work together when performing these tasks or that one hemisphere performs these tasks alone. This observation cannot be made in visual tasks in the same way, here the hemispheres may perform the tasks for themselves. The uncoupling between the auditory and visual elementary times suggests that there is no cooperation between these two pathways at all.

The symmetry of elementary times between different mental pathways may vary:

1. All mental pathways have the same elementary time ET.
2. There are two different elementary times: ET(xNNr) and ET(xNNl). That means all elementary times at the right side are equal and all elementary times at the left side are equal. This is the case in subject H32A.
3. There are four different elementary times: ET(vNNr), ET(vNNl), ET(aNNr), and ET(aNNl). This is the case in subject H01Z.
4. All mental pathways have different elementary times (ET(a11r), ET(a11l), ET(v11r), ET(v11l), ET(a22r), etc.)

   The symmetry decreases from 1 to 4.

The difference between the elementary times of both hemispheres is small normally, but in patients with brain lesions, this difference increases. Subject O01A, for example, has four different elementary times but the side differences between aNNr and aNNl are 3 ms (instead of 1ms).

Distribution of linear pathways

The distribution of the linear pathways shows some preferences of linEN = n*3. This has be seen in the individual equations of the eight pathways before.

Distribution of cyclical pathways

The fact that the cycEN distribution of a11y prefers the values cycEN=5 and cycEN=8 supports the (2+2) and (2+2*2) pathways with observable cycEN=5,6 respectively cycEN=8,9 (see table above). This means that the auditory tasks a11y generally use the *slow mode* in the first series, combined either with a double search or a triple search. This is true also in the task v11l though this task has some lower cycEN=4.

The task v11r has two peaks at cycEN=3 and cycEN=5. That means some subjects use the *fast mode*, double search in these tasks in the first series.

The visual tasks v22y, with their maximum at cycEN=7 and cycEN=8, prefer the *fast mode* with (4+1) and 7 or 8cycET. The task v22r has increased values in cycEN=9 and 10. This means some subjects use the slow mode with (4+2) and cycET=9 or 10 in v22r.

The task a22l uses cycEN=10 and cycEN=13. That means it prefers the *slow mode, double search* (4+2) or the *fast mode, triple search* (4+1*4).

The task a22r uses cycEN=9, cycEN=11,12 and cycEN=17,18. That means it may use either (4+2) or (4+1*4) or (4+2*4).

Symmetry between linear pathways on either side

It has been found that there is a strong correlation between the length of the linear pathways of the tasks xNNr and xNNl. Which conclusions can be drawn from this fact about the route of these two pathways?

Either the two pathways, xNNr and xNNl, run on the same side of the brain or there must be one reason which is responsible for the selection of the length of the linear pathways on both sides. That means it determines both linear pathways without any place for random determination.

Symmetry between the cyclical pathways on either side

The strong correlation between cycEN(x11r) and cycEN(x11l) shows that even the strategy of the searching set system (mode and the number of searching sets) is influenced by some common factor. That means if (2+2*2) is used at one side, the other side uses it too. Here too there is no room for chance but each strategy is part of a larger pattern.

In v11y, the symmetry is broken because some tasks may use fast mode while the other side uses slow mode.

In a22y and v22y there is only a weak correlation. That means that the symmetry relative to mode and number of searching sets must not be as high in these tasks as in the x11y tasks.

Symmetry between linEN and cycEN on the same side

There is a weak correlation between linEN(a11y) and cycEN(a11y), a still weaker one between linEN(v22y) and cycEN(v22y). The other combinations show no certain correlation. Nonetheless, there must be some influence of the determining factor even on the cyclical pathways. If there is a factor which may reduce the reaction times by shortening the linear pathway, the same factor may shorten the cyclical pathway too. That means this factor should favor the fast strategies like double search and fast mode.

Changes in the length of the linear pathway in repetitions

The first observation is the high order of the linEN profile of a subject (see Figs. 82 and 83). This order is approximately preserved if the tasks are repeated. What is the reason for this order, why are the linEN of this size?

There are two reasons for this order: the first is the architecture of the neural-neural relations, the second is the influence of the neuromodulators.

The reduction of cycEN in repeated tasks

The reduction of cycEN is distinctly visible in all tasks. This is caused by the transition from slow mode to fast mode.

The implicit learning axiom

Table 30. Presented is the mean number of internal cycles (internal mean). The external mean = (internal mean - 2). The internal respectively external median is computed by an multiplicative factor. In the brackets, the externally observable median number of searching cycles

	Modes	2 searches	3 searches
x11y	fast	2+1 (3-4 cycET)	2+1*2 (6 cycET)
	slow	2+2 (5-6 cycET)	2+2*2 (8-9 cycET)
x22y	fast	4+1 (7-8 cycET)	4+1*4 (12-14cycET)
	slow	4+2 (9-10 cycET)	4+2*4 (20 cycET)
x33y	fast	6+1 (10-12 cycET)	6+1*6 (20 cycET)
	slow	6+2 (13-14 cycET)	6+2*6 (30 cycET)

A (2+1*2) search can only be assumed in a repeated task if the preceding task had used a (2+2*2) search.

This axiom states that if a pathway uses fast mode, all subsequent pathways of this kind of task have to use fast mode too. *Implicit learning is not reversible* in the time between the first series and the second series. For example if a subject uses fast mode in the task a11rINA then all subsequent tasks a22rINA, a11rINB, a22rINB, a11rINC, a22rINC etc. have to use fast mode, too.

In some cases, the implicit learning axiom is not fulfilled. In these cases, mistakes either of first peak recognition or of median computation can be found. In still other cases, it is difficult to decide whether certain cycEN belong to one category or the other in the above table. For example a cycEN(x11y) = 7 may belong to the category "slow triple search" with (2+2*2) searching cycles or to the categories "slow double search" with (2+2) respectively "fast triple search" with (2+1*2) searching cycles.

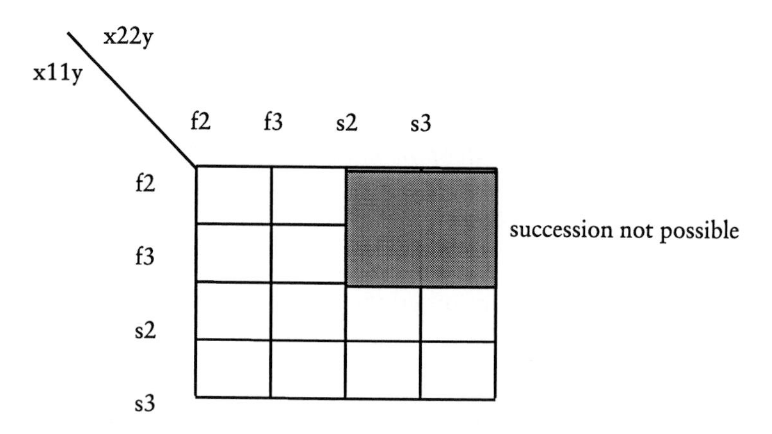

Fig. 86. Possible and not possible successions of modes and number of searches for the response element. f2 means for example: fast mode, double search

Double search/triple search

The number of double searches or triple searches has not been established yet.

The pathways of patients with monohemispheric brain lesions

Convergence tables

Question

Investigating patients with a defined (by CT or MRI) monohemispheric brain lesion has advantages for the present study. The reaction time data of these patients are well suited to test the proposed equations and evaluate their asymmetries caused by the brain lesion.

The patients with brain lesions had all different lesions. It could not be expected that all these patients show disturbed stimulus-response pathways. Because of the different locations of the brain lesions, different influences on the pathways had to be expected. The question, therefore, was to correlate the localisation of brain lesion with one or more disordered xNNy pathways.

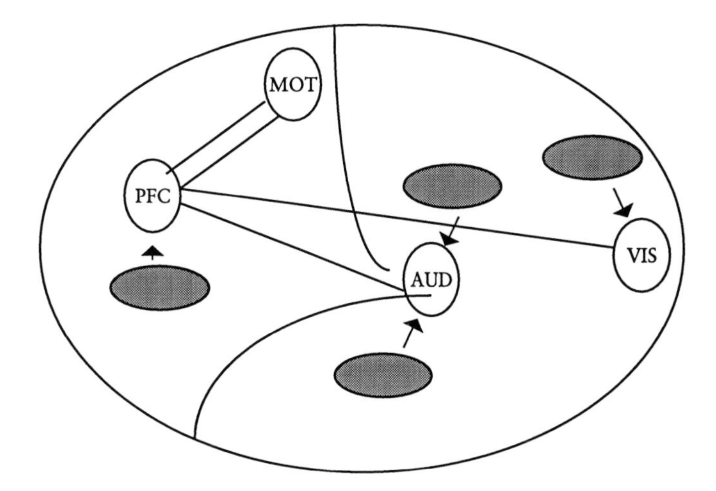

Fig. 87. Intracortical connections used by the auditory and visual stimulus-response pathway

There are frontal pattern where the lesions disturbs the PFC (FOP), the temporo-partietal pattern (TEP) where the lesion disturbs the auditory sensory areas and the parieto-occipital pattern (POP), where the lesion disturbs the visual sensory areas. A combination of the two last patterns is caused by a lesion lying between them (combinded pattern or TEP/POP). At last the lesion may not affect the stimulus-response pathway at all (neutral pattern or NEP). In principle there may be also a central pattern (COP) which affects the motor areas.

Summing up, the patterns of disturbation of mental pathways in patients with monohemispheric brain lesions can be caused by three possible reasons:

- An increased elementary time can be found in both sensory modalities in the lesioned hemisphere. This pattern can be caused by a frontal monohemispheric lesion.
- The elementary time of only one sensory modality is increased. This pattern can be caused by a monohemispheric lesion in the non-frontal brain.
- The lesion does not affect the elementary times of the mental pathways at all.

Methods

Methods of Measuring

The patients came from several hospitals, two neurosurgery departments and one neurorehabilitative unit. All subjects were naive (unpracticed) when doing the tasks for the first time. The methods of measuring the reaction times were the same as for healthy subjects (see above) but the number of trials was reduced in order to treat them with care. The patients performed 50 trials of x11y, 200 trials of x22y, and 100 trials of m22y (m=mixed, ie. auditory and visual stimuli). The sequence of tasks was always the same within one sensory modality: x11y, x22y, x33y. Whether the auditory tasks or the visual tasks were performed first was variable.

Methods of Evaluation

Previous method

The following sequence of procedures was applied to the reaction time data of these patients:

1. Chronophoresis by SINGorg01 and SINGorg02
2. Evaluation of the x11y distribution by FPMorg18
3. Convergence tables, fixation of the elementary time ET(x11y),establishing the equations of the x11y pathways
4. Evaluation of the x22y distribution by FPMorg18
5. Comparing the ET(x11y) with Nestle(x22y)
6. Confirming the consistency of mode(x11y) with mode(x22y)
7. Confirming the consistency of asymmetrical elementary times with the side of the brain lesion.

Present method

The data of the last patients (O06A, O11A, O16A, O14A) were evaluated by the sequential processes used in the healthy and the schizophrenic group. The program versions FPM31o and FPM26o58, and SINGOR07 were used to compensate the fact, that only 50 trials of x11y were performed. That alternative way was the sequence CHANGE3, FPM31f, FPM26f59, SING106n, and SING104r.

The results of these different programs were united into convergence tables (see below).

Convergence tables

One has to use a common representation: the elementary time which occurs most frequently is typed in bold letters, the second frequent elementary time is underlined or written in normal letters. If both elementary times have the same frequency, then both are written in normal letters. In the column "Results" the first number is the elementary time, the second number the "second elementary time" or "extra elementary time". Both are separated by their respective symmetry between xNNr and xNNl.

This double representation of the elementary time and the "extra elementary time" prevents from confounding the two values. This is necessary because sometimes the "extra elementary time" is more frequent than the regular one. This would signal an incorrect increase of the regular elementary time.

O01A rft	FPM31e (x11y)	FPM31e (x22y)	FPM26f58 (x11y)	FPM26f58 (x22y)	Results	Chrono-phoresis	Median-Fp Method
aNNr	15,17	?	17	15	16, ?	17.5	
aNNl	20,25	11,17,20	23,31?	17.5	17,20	17.5	
vNNr	20	16	15,27,34	13.5	13.5,16	13.5, 22	
vNNl	23	24	15,27	15,25	13, ?	10	

O02A rpol	FPM31e (x11y)	FPM31e (x22y)	FPM26f58 (x11y)	FPM26f58 (x22y)	Results	Chrono-phoresis	Median-Fp Method
aNNr	15	14,21,25	15,20	16.5	15,20.5		
aNNl	19,25	16,21	19,24	15	15.5,20		
vNNr	17	21	16,23	14,16,18	17,22		
vNNl	15,28	12,24	?	20,23	14.5,23		

O03A rf	FPM31e (x11y)	FPM31e (x22y)	FPM26f58 (x11y)	FPM26f58 (x22y)	Results	Chrono-phoresis	Median-Fp Method
aNNr	18	10,20	?	17,30	17.5	14	
aNNl	18, 21	19	?	20,25	19.5	11.5, 23	
vNNr	15	12,17,19	13,16,19	13,28,32	15.5	15	
vNNl	21	13, 19	19,21,24	16,22,24	21.5	17.5	

O04A lp	FPM31e (x11y)	FPM31e (x22y)	FPM26f58 (x11y)	FPM26f58 (x22y)	Results	Chrono-phoresis	Median-Fp Method
aNNr	15	14, 17	23,26	18	16,24.5	16.5,21,27	
aNNl	19, 23	22	23, 27,33	26	11.5,23	14,25	
vNNr	14, 22	17,22	16,21	14.5, 25.5, 30	15, 22	21,5	
vNNl	26, 28?	15, 26	?	27	14,?	21,5	

O05A rpo	FPM31e (x11y)	FPM31e (x22y)	FPM26f58 (x11y)	FPM26f58 (x22y)	Results	Chrono-phoresis	Median-Fp Method
aNNr	13, 24	18, 20	17,20,29	17,30	13.5, 18.5	14	
aNNl	24	18, 25	12	14,22	12,16	17.5	
vNNr	10, 15	18, 24	23.5, 30	20, 27,31	15,22.5	14.5, 21	
vNNl	15, 19	16,20,23, 28	17, 30.5	14.5, 31	15,19.5	15,5	

O06A lt	FPM31e (x11y)	FPM31e (x22y)	FPM26f58 (x11y)	FPM26f58 (x22y)	Results	Chrono-phoresis	Median-Fp Method
aNNr	18,22	25	23,29	13	14,23	12.5, 23.5	
aNNl	16	23	24,33	?	16,23.5	19.5	
vNNr	?	17,21	15	29	15.5,21	15	
vNNl	17	24,11	17	17	17,23	17.5	

Remarks:
If a duplication of elementary times showed up both in the FPM programs and in the chronophoresis (SINGLE program), a real duplication has been assumed and has been related to the model.

O07A lf	FPM31e (x11y)	FPM31e (x22y)	FPM26f58 (x11y)	FPM26f58 (x22y)	Results	Chrono-phoresis	Median-Fp Method
aNNr	16	15	15, 26.5	22,26,29	15,26	12, 16.5	
aNNl	20, 24	15, 18	21, 26.5, 35	12,14	13, 17 20.5	14.5	
vNNr	15,19	15, 16	17,19,34	16,28,33	16, 19	17	
vNNl	15, 21	16	26	14,19	14.5,20	15, 23.5	

O08A lt	FPM31e (x11y)	FPM31e (x22y)	FPM26f58 (x11y)	FPM26f58 (x22y)	Results	Chrono-phoresis	Median-Fp Method
aNNr	19?	18,27	15	16, 23	17, 25	16.5	
aNNl	15, 23	15,23	14,23	12, 14.5	14.5,23	14.5	
vNNr	13,20,24,29	15	24.5, 33	?	15.5,23	16.5	
vNNl	19	17	17,25,35	20, 23, 26	18, 25	13.5	

O09A l(k)	FPM31e (x11y)	FPM31e (x22y)	FPM26f58 (x11y)	FPM26f58 (x22y)	Results	Chrono-phoresis	Median-Fp Method
aNNr	(15)	18	?	?	15,18	10.5	
aNNl	12,15,24	12,15, 19.5	?	?	12,15	17.5	
vNNr	16	16.5, 25	16.5, 20	?	16.5	19	
vNNl	15,21	17,21,25	18, 21	16.5, 21	16,5,21	16	

O10A rt	FPM31e (x11y)	FPM31e (x22y)	FPM26f58 (x11y)	FPM26f58 (x22y)	Results	Chrono-phoresis	Median-Fp Method
aNNr	17	16,25	26,29,33	19	13.5, 16.5	13	
aNNl	16, 20.5	15,19	32	14,20,29	15, 20	20	
vNNr	12,21	13,15, 17.5	19.5, 29	?	14, 19	12,17.5	
vNNl	20	14,27	22,24	14	14,22	12.5,22	

Remarks:

1. It is unclear whether 13.5 or 27 is the correct value in aNNrO10A. There is a similar case in vNNlO08A where either 25 or 12.5 is the second value.

O11A lt	FPM31e (x11y)	FPM31e (x22y)	FPM26f58 (x11y)	FPM26f58 (x22y)	Results	Chrono-phoresis	Median-Fp Method
aNNr	15	?	14	16,21,26, 28	15,27	–	
aNNl	16,25	21,28	19	24,30	15,24.5	32	
vNNr	13,17,22	14,21,24	26,29,33	?	13.5,22	–	
vNNl	20,25,29	15,21,27	31	13	13,20.5	12.5	

O12A lf	FPM31e (x11y)	FPM31e (x22y)	FPM26f58 (x11y)	FPM26f58 (x22y)	Results	Chrono-phoresis	Median-Fp Method
aNNr	14,17,19	14,20,27	16,24,32	?	14,17	17.5	
aNNl	11,13,17	24	12,16	?	12,16.5	13, 21	
vNNr	15,17,21	15	16.5, 20.5	?	16, 21	17	
vNNl	15,23,25	17,24	?	?	16, 24	12, 20.5	

O13A lft	FPM31e (x11y)	FPM31e (x22y)	FPM26f58 (x11y)	FPM26f58 (x22y)	Results	Chrono-phoresis	Median-Fp Method
aNNr	25	25,27	?	15, 19.5	13.5,19.5	16	
aNNl	11.18.24	15	21,24	?	11.5,16.5	12,24	
vNNr	15	14,22	17,25,30	14	14.5, 23.5	10,23.5	
vNNl	10,19	14,25	25, 30.5	12,18.5	12, 19	11,18	

O14A rs	FPM31e (x11y)	FPM31e (x22y)	FPM26f58 (x11y)	FPM26f58 (x22y)	Results	Chrono-phoresis	Median-Fp Method
aNNr	16,25	15,16	12	28	16,26.5	19	
aNNl	15,29	14,16	15	?	15,?	18,27	
vNNr	26	13	19	?	13,19	15,23.5	
vNNl	20	14,18,24	19,26,28	?	14, 19	13,25	

O15A lp	FPM31e (x11y)	FPM31e (x22y)	FPM26f58 (x11y)	FPM26f58 (x22y)	Results	Chrono-phoresis	Median-Fp Method
aNNr	19	18,26	?	18	1826		
aNNl	14,21	18,21,26	17,25	18,20.5,25	18,21,25		
vNNr	18,21	18,23,26?	18	14.5	18, 22		
vNNl	17,20	18,23,26?	22,24	17	17, 22		

O16A lf	FPM31e (x11y)	FPM31e (x22y)	FPM26f58 (x11y)	FPM26f58 (x22y)	Results	Chrono-phoresis	Median-Fp Method
aNNr	?	17, 21	16	?	16.5,21	10.5, 14.5	
aNNl	13	23	13	?	13,23	13,17,22.5	
vNNr	15,21	15	13	?	15,21	14.5,18.5	
vNNl	?	22	10	?	10.5,22	11,16,24.5	

O17A rfp	FPM31e (x11y)	FPM31e (x22y)	FPM26f58 (x11y)	FPM26f58 (x22y)	Results	Chrono-phoresis	Median-Fp Method
aNNr	12,14,17, 21	13,26	17,20,26, 33	21.5,26	13, 17, 21	11.5, 17	
aNNl	13	13	22	22.5, 35	13, 22	23	
vNNr	17	17,23	17,20,34	18,20	17,21	12.5, 19	
vNNl	15	24,26	28,35	22	15, 23	14.5	

O18A rf	FPM31e (x11y)	FPM31e (x22y)	FPM26f58 (x11y)	FPM26f58 (x22y)	Results	Chrono-phoresis	Median-Fp Method
aNNr	20	17,22	19	16,22	16.5,21	16	
aNNl	14, 23,26	15,20	24,28	21,28	14,22	14,21.5	
vNNr	15,17,25	14,21	17,20	15	15.5,22	15	
vNNl	11,15,17, 24	16,27	19,22,28	16,17	16,23	17.5, 33	

O19A Lfp	FPM31e (x11y)	FPM31e (x22y)	FPM26f58 (x11y)	FPM26f58 (x22y)	Results	Chrono-phoresis	Median-Fp Method
aNNr	18.5, 24	20,25	?	28,31	13.5,19	17	
aNNl	15,18,20	15,21,28	?	15	14,20.5	21.5	
vNNr	16,22,25	16,21,26, 29	33	20,28	15.5,20	17	
vNNl	16	15,17,26, 28	?	?	16,?	19.5, 23.5	

O20A rt(k)	FPM31e (x11y)	FPM31e (x22y)	FPM26f58 (x11y)	FPM26f58 (x22y)	Results	Chrono-phoresis	Median-Fp Method
aNNr	12,18,28	16,26,28	19	19	14,19	15.5, 18.5	
aNNl	15,25	19,25	13	14	13.5,19	12,18	
vNNr	28	20,24	?	13,19	13, 19.5	16,22.5	
vNNl	15,21,26, 28	19,24	?	19,26,32	14,19	16.5, 23	

Remarks:
1. The lower values are in bold letters, the higher values are underlined.

The meaning of the first elementary time and the "extra elementary time" in the above convergence tables

In rrinciple there could be three different elementary times within one task: an elementary time of the linear pathway, an elementary time of the cyclical pathway searching on the sensory area and an elementary time of the cyclical pathway searching on the motor area. As is proven in the chapter about simulation, the nestle procedure is able to detect different elementary times hidden within reaction times. The existence of the above "extra elementary time" could be due to reality. The other possibility is the artificial charakter of this additional elementary time. The evaluation programs tend to produce additional values standing in a mathematical relation with the real elementary time (see "extra plots" within the chronophrosis). The final decision has not been made whether these additional values represent some reality or are artificial products of the evaluation programs.

It is an important observation that these values usually are less frequent in the convergence tables than the primary values. That means if there would exist some real periods they should vary less than the periods known so far.

If one tries to get some mathematical relation between the normal elementary times and the "extra elementary times" one does not see any relation but in the comparison between ETvl and ETvl2. This lack of any relation could be due to the difficulty in measuring the second values. Because they are less frequent in the convergence tables it is more difficult to get a useful mean value.

Many "extra elementary times" (without any statistical significancy) may be calculated by the term: extra elementary time = primary elementary time * 4/3 eg 20 = 15 * 4/3

This would mean that 4 primary elementary times are equivalent to 3 extra elementary times.

The Fig. below shows that besides this relation another mathematical relation exists:

extra elementary time = primary elementary time * 3/2
eg 24 = 16 * 3/2

This would mean that 3 primary elementary times are seen as 2 extra elementary times.

These 4/3 and 3/2 multiples are typical for the extra plots artificially produced by the program.

That evidence supports the notion that the extra elementary times are not real but by-products of the evaluating programs.

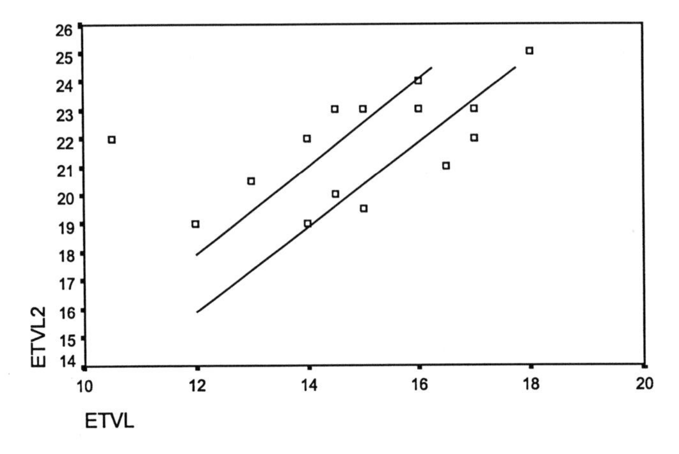

Fig. 88. The relation between ETVL and ETVL2 shows that the latter may be expressed by two hypothetical mathematical relations: ETVL2=ETVL*4/3 and ETVL2=ETVL*3/2. The Pearson correlation coefficient is $r=0.816$, the significance level is $p< 0.01$ ($n=15$)

Collection of equations

Findings from patients with monohemispheric organic brain lesions in alphabetical order.

O01A (right, fronto-temporal meningeoma)

auditory

a11rO01A = 70 + (3 + 2(2 + §2))16 3.3 + 6.1
a11lO01A = 70 + (3 + 2(2 + 2*2))17 3.9 + 7.9
a22rO01A = 70 + (12 + 2(4 + 1))16 11.8 + 5.4
a22lO01A = 70 + (12 + 2(4 + 1))17 11.5 + 6.8

visual

v11rO01A = 120 + (6 + 2(2 + §2))13.5 5.4 + 5
v11lO01A = 120 + (9 + 2(2 + §2))13 8.4 + 5.5
v22rO01A = 120 + (15 + 2(4 + 1))13.5 14.3 + 8.5
v22lO01A = 120 + (17 + 2(4 + 2))13 16.5 + 10.5

O02A (right polar astrocytoma)

auditory

a11rO02A = 70 + (5 + 2(2 + 1))15 4.7 + 4.1
a11lO02A = 70 + (6 + 2(2 + §2))15.5 5.4 + 4.7
a22rO02A = 70 + (12 + 2(4 + 1*4))15 12 + 15.3
a22lO02A = 70 + (13 + 2(4 + 1*4))15.5 13.2 + 11.3

visual

v11rO02A = 120 + (3 + 2(2 + 2*2))17 3.3 + 6.8
v11lO02A = 120 + (6 + 2(2 + §2))14.5 6.3 + 6.1
v22rO02A = 120 + (9 + 2(4 + 1))17 9.2 + 6.6
v22lO02A = 120 + (12 + 2(4 + 2))14.5 11.8 + 9.4

O03A (right frontal tumor at the wing of the sphenoid, operated)

auditory

a11rO03A = 70 + (3 + 2(2 + 1))17.5 2.6 + 3.9
a11lO03A = 70 + (3 + 2(2 + 2*2))19.5 3.9 + 10.3
a22rO03A = 70 + (15 + 2(4 + 1*4))17 14.6 + 11.4
a22lO03A = 70 + (15 + 2(4 + 1))19.5 14.4 + 7.6

visual

v11rO03A = 120 + (4 + 2(2 + §2))15.5 4.1 + 4.8
v11lO03A = 120 + (4 + 2(2 + 1))21.5 4.3 + 4
v22rO03A = 120 + (6 + 2(4 + 1))15.5 6.7 + 8.3
v22lO03A = 120 + (6 + 2(4 + 1))21.5 5.7 + 7.4

O04A (left parietal tumor, right pyramidal meningeoma, operated)

auditory

a11rO04A = 70 + (6 + 2(2 + 2*2))16 5.5 + 7.9
a11lO04A = 70 + (9 + 2(2 + 2*2))11.5 9.3 + 7.4
a22rO04A = 70 + (21 + 2(4 + 1*4))16 21.1 + 12.8
a22lO04A = 70 + (24 + 2(4 + 2*4))11.5 23.7 + 20.3

visual

v11rO04A = 120 + (9 + 2(2 + 2*2))15 8.7 + 11.5
v11lO04A = 120 + (18 + 2(2 + 2*2))14 18 + 8.4
v22rO04A = 120 + (20 + 2(4 + 2*4))15 20 + 20.1
v22lO04A = 120 + (21 + 2(4 + 2*4))14 21.6 + 16.5

O05A (right parieto-occipital meningeoma, operated)

auditory

a11rO05A = 70 + (5 + 2(2 + 2*2))13.5 4.7 + 7.3
a11lO05A = 70 + (12 + 2(2 + §2))12 11.3 + 5.5
a22rO05A = 70 + (12 + 2(4 + 1*4))13.5 12.8 + 11.6
a22lO05A = 70 + (18 + 2(4 + 2))12 17.2 + 9

visual

Equation	Value
v11rO05A = 120 + (2 + 2(2 + §2))15	2 + 6.3
v11lO05A = 120 + (5 + 2(2 + §2))15	5.3 + 4.9
v22rO05A = 120 + (7 + 2(4 + 1))15	7.3 + 6.7
v22lO05A = 120 + (5 + 2(4 + 2))15	4.7 + 9.8

O06A (left temporal lesion after ADEM)

auditory

Equation	Value
a11rO06A = 70 + (4 + 2(2 + 1))14	4.1 + 3.9
a11lO06A = 70 + (2 + 2(2 + 2*2))16	1.8 + 6.9
a22rO06A = 70 + (12 + 2(4 + 1*4))14	12.6 + 14.3
a22lO06A = 70 + (14 + 2(4 + 2))16	13.6 + 9.1

visual

Equation	Value
v11rO06A = 120 + (2 + 2(2 + 1))15.5	2 + 2.4
v11lO06A = 120 + (3 + 2(2 + §2))17	3.6 + 5.6
v22rO06A = 120 + (3 + 2(4 + 1))15.5	3.5 + 6
v22lO06A = 120 + (4 + 2(4 + 2))17	4.2 + 8.9

O07A (recidiv of a left-frontal astrocytoma)

auditory

Equation	Value
a11rO07A = 70 + (3 + 2(2 + §2))15	3 + 5.1
a11lO07A = 70 + (5 + 2(2 + §2))13	4.9 + 4.7
a22rO07A = 70 + (15 + 2(4 + 2*4))15	14.7 + 22.1
a22lO07A = 70 + (9 + 2(4 + 2*4))13	9.5 + 23.8

visual

Equation	Value
v11rO07A = 120 + (3 + 2(2 + 2*2))16	2.7 + 8
v11lO07A = 120 + (4 + 2(2 + §2))14.5	4.2 + 6.3
v22rO07A = 120 + (5 + 2(4 + 2*4))16	5.2 + 18.8
v22lO07A = 120 + (5 + 2(4 + 1*4))14.5	4.9 + 14.3

O08A (left temporal intra-cranial hematoma)

auditory

Equation	Value
a11rO08A = 70 + (4 + 2(2 + §2))17	4.2 + 5.5
a11lO08A = 70 + (6 + 2(2 + 1))14.5	5.6 + 3.8
a22rO08A = 70 + (12 + 2(4 + 1*4))17	11.8 + 14.2
a22lO08A = 70 + (12 + 2(4 + 2*4))14.5	11.8 + 16.4

visual

Equation	Value
v11rO08A = 120 + (2 + 2(2 + 1))15.5	2 + 3.6
v11lO08A = 120 + (2 + 2(2 + §2))18	2.2 + 4.7
v22rO08A = 120 + (2 + 2(4 + 1))15.5	2.2 + 8
v22lO08A = 120 + (5 + 2(4 + 1))18	5.2 + 5.4

O09A (left wing of the sphenoid meningeoma)

auditory

Equation	Value
a11rO09A = 70 + (2 + 2(2 + 1))18	2.2 + 2.8
a11lO09A = 70 + (6 + 2(2 + §2))15	6 + 4.3
a22rO09A = 70 + (15 + 2(4 + 1))18	14.7 + 7.6
a22lO09A = 70 + (15 + 2(4 + 2))15	15.7 + 10.3

visual

v11rO09A = 120 + (3 + 2(2 + §2))16.5	2.5 + 4.5
v11lO09A = 120 + (3 + 2(2 + §2))16.5	3.2 + 6
v22rO09A = 120 + (8 + 2(4 + 2))16.5	8 + 10.5
v22lO09A = 120 + (8 + 2(4 + 1*4))16.5	8 + 12.2

O10A (right temporal tumor, operated, recidiv?)

auditory

a11rO10A = 70 + (6 + 2(2 + 2*2))13.5	6.5 + 10.5
a11lO10A = 70 + (17 + 2(2 + §2))15	17 + 6
a22rO10A = 70 + (21 + 2(4 + 2))13.5	20.6 + 9.9
a22lO10A = 70 + (21 + 2(4 + 2*4))15	21.7 + 17

visual

v11rO10A = 120 + (10 + 2(2 + 2*2))14	10.1 + 9.5
v11lO10A = 120 + (9 + 2(2 + 2*2))14	9.8 + 9.8
v22rO10A = 120 + (15 + 2(4 + 2))14	14.8 + 9.4
v22lO10A = 120 + (21 + 2(4 + 2))14	21.2 + 10.8

O11A (left temporal resection, epilepsy surgery)

auditory

a11rO11A = 70 + (5 + 2(2 + §2))15	5 + 4.3
a11O11A = 70 + (4 + 2(2 + §2))15	4.3 + 6
a22rO11A = 70 + (25 + 2(4 + 2*4))15	25.3 + 22.7
a22lO11A = 70 + (30 + 2(4 + 2*4))15	30 + 20.9

visual

v11rO11A = 120 + (9 + 2(2 + 2*2))13.5	9.5 + 6.6
v11lO11A = 120 + (6 + 2(2 + 2*2))13	6.5 + 11.3
v22rO11A = 120 + (12 + 2(4 + 1*4))13.5	12.8 + 15.1
v22lO11A = 120 + (18 + 2(4 + 2*4))13	18 + 24.1

O12A (left frontal tumor, glioma?)

auditory

a11rO12A = 70 + (17 + 2(2 + 2*2))17	17.1 + 7.4
a11lO12A = 70 + (15 + 2(2 + 2*2))16.5	15.9 + 10
a22rO12A = 70 + (42 + 2(4 + 2*4))17	42.7 + 21.7
a22lO12A = 70 + (28 + 2(4 + 2*4))16.5	28.9 + 16.3

visual

v11rO12A = 120 + (14 + 2(2 + 2*2))16	13.6 + 7.6
v11lO12A = 120 + (15 + 2(2 + 2*2))16	14.6 + 10.6
v22rO12A = 120 + (15 + 2(4 + 1))16	15.8 + 8.4
v22lO12A = 120 + (13 + 2(4 + 1*4))16	13.3 + 11.8

O13A (left fronto-temporal and left frontal-superior tumors)

auditory

a11rO13A = 70 + (3 + 2(2 + 1))13.5	3.9 + 4.2
a11lO13A = 70 + (5 + 2(2 + 2*2))11.5	5.4 + 6.8
a22rO13A = 70 + (15 + 2(4 + 1*4))13.5	15 + 14.3
a22lO13A = 70 + (19 + 2(4 + 2*4))11.5	19.3 + 19.8

visual

v11rO13A = 120 + (3 + 2(2 + §2))14.5	3.5 + 7.5
v11lO13A = 120 + (6 + 2(2 + §2))12	6.8 + 6
v22rO13A = 120 + (12 + 2(4 + 2))14.5	12.1 + 8.9
v22lO13A = 120 + (14 + 2(4 + 1*4))12	14.3 + 13.5

O14A (right lateral ventricle tumor)

auditory

a11rO14A = 70 + (6 + 2(2 + 2*2))16	6 + 10.3
a11lO14A = 70 + (13 + 2(2 + 2*2))15	13.3 + 10.7
a22rO14A = 70 + (22 + 2(4 + 1*4))16	22.4 + 13.7
a22lO14A = 70 + (21 + 2(4 + 2*4))15	21.7 + 20.3

visual

v11rO14A = 120 + (9 + 2(2 + 2*2))13	9.5 + 11.5
v11lO14A = 120 + (9 + 2(2 + 2*2))14	8.7 + 8.3
v22rO14A = 120 + (17 + 2(4 + 2*4))13	17.2 + 19.5
v22lO14A = 120 + (15 + 2(4 + 2*4))14	15.9 + 15.7

O15A (left parietal tumor, operated)

auditory

a11rO15A = 70 + (7 + 2(2 + 2*2))18	7.4 + 7
a11O15A = 70 + (9 + 2(2 + §2))18	8.6 + 5.4
a22rO15A = 70 + (16 + 2(4 + 1*4))18	16.3 + 12
a22lO15A = 70 + (23 + 2(4 + 1))18	22.7 + 7.1

visual

v11rO15A = 120 + (5 + 2(2 + §2))18	4.9 + 5.8
v11lO15A = 120 + (10 + 2(2 + 2*2))17	10.1 + 8.5
v22rO15A = 120 + (15 + 2(4 + 1*4))18	14.9 + 13.8
v22lO15A = 120 + (16 + 2(4 + 1*4))17	16.2 + 13.1

O16A (left frontal tumor)

auditory

a11rO16A = 70 + (7 + 2(2 + 2*2))16.5	7 + 6.8
a11lO16A = 70 + (14 + 2(2 + 2*2))13	13.8 + 8.5
a22rO16A = 70 + (24 + 2(4 + 2*4))16.5	23.8 + 17.5
a22lO16A = 70 + (36 + 2(4 + 2*4))13	36.5 + 19.7

visual

v11rO16A = 120 + (7 + 2(2 + 2*2))15	7.3 + 9
v11lO16A = 120 + (8 + 2(2 + 2*2))10.5	8 + 9.9
v22rO16A = 120 + (12 + 2(4 + 1*4))15	11.3 + 15.3
v22lO16A = 120 + (18 + 2(4 + 2*4))10.5	18.5 + 20.3

O17A (right fronto-parietal chronic subdural hematoma, medial meningeoma)

auditory

a11rO17A = 70 + (3 + 2(2 + §2))13	3 + 4.7
a11lO17A = 70 + (5 + 2(2 + §2))13	4.5 + 5.6
a22rO17A = 70 + (17 + 2(4 + 2*4))13	17.2 + 16.7
a22lO17A = 70 + (24 + 2(4 + 1*4))13	23.4 + 11

visual

v11rO17A = 120 + (2 + 2(2 + §2))17 2.1 + 6.1
v11lO17A = 120 + (3 + 2(2 + 2*2))15 2.7 + 6.7
v22rO17A = 120 + (3 + 2(4 + 1))17 3.9 + 8.2
v22lO17A = 120 + (9 + 2(4 + 1))15 9.7 + 7.4

O18A (right frontal, white matter glioma)

auditory

a11rO18A = 70 + (5 + 2(2 + §2))16.5 4.4 + 4.4
a11lO18A = 70 + (6 + 2(2 + §2))14 6.6 + 5
a22rO18A = 70 + (12 + 2(4 + 2))16.5 11.6 + 9.8
a22lO18A = 70 + (20 + 2(4 + 1*4))14 19.8 + 12.9

visual

v11rO18A = 120 + (3 + 2(2 + §2))15.5 3.5 + 4.9
v11lO18A = 120 + (3 + 2(2 + §2))16 2.7 + 5.4
v22rO18A = 120 + (11 + 2(4 + 2))15.5 10.9 + 9.9
v22lO18A = 120 + (10 + 2(4 + 2))16 9.6 + 9.1

O19A (left fronto-parietal, chronic subdural hematoma)

auditory

a11rO19A = 70 + (14 + 2(2 + 2*2))13.5 13.9 + 9.9
a11lO19A = 70 + (5 + 2(2 + 2*2))14 4.8 + 8.3
a22rO19A = 70 + (26 + 2(4 + 1*4))13.5 26.1 + 12.9
a22lO19A = 70 + (24 + 2(4 + 1*4))14 24.1 + 14.6

visual

v11rO19A = 120 + (9 + 2(2 + 2*2))15.5 9.3 + 8.9
v11lO19A = 120 + (13 + 2(2 + §2))16 13 + 6.3
v22rO19A = 120 + (15 + 2(4 + 2))15.5 15.7 + 10.2
v22lO19A = 120 + (17 + 2(4 + 1))16 17.1 + 6.9

O20A (right temporal meningeoma at the wing of the sphenoids, operated)

auditory

a11rO20A = 70 + (18 + 2(2 + 2*2))14 17.6 + 9.9
a11lO20A = 70 + (18 + 2(2 + 2*2))13.5 18 + 12.1
a22rO20A = 70 + (24 + 2(4 + 2*4))14 24.4 + 22.6
a22lO20A = 70 + (27 + 2(4 + 2*4))13.5 27.3 + 20.2

visual

v11rO20A = 120 + (27 + 2(2 + 2*2))13 27.6 + 10.9
v11lO20A = 120 + (18 + 2(2 + 2*2))14 18.7 + 8.9
v22rO20A = 120 + (21 + 2(4 + 2*4))13 21.8 + 16.5
v22lO20A = 120 + (21 + 2(4 + 2))14 21.9 + 10.4

Deviations from normal pathways in patients with monohemispheric brain lesion

The statistical method used in this chapter is to compare the pathway of the side contralateral to the lesion (diseased pathway) with the pathway of the side homolateral to the lesion (healthy pathway).

Comparison of the median reaction times homolateral (xNNh) and contralateral (xNNd) to the lesion

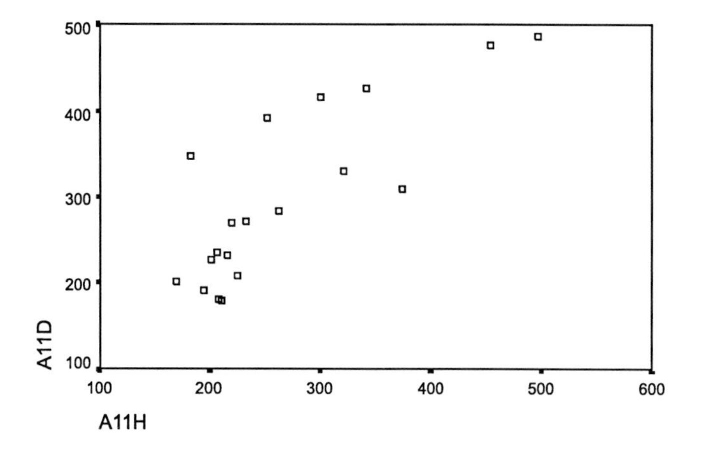

Fig. 89. Comparison of the median reaction times of the healthy side (a11h) with the median reaction times of the diseased side (a11d). The Pearson correlation coefficient is $r=0.816$, the significance level is $p< 0.01$

In the task a11y most of the patients show a median reaction time which is near the symmetry line or above. This means *an increased median reaction time in the tasks contralateral to the lesion.* There are four patients (O03A, O19A, O10A, and O14A) which deviate most from the symmetry axis by an increased median reaction time at the contralateral side and one patient (O16A) which shows an increased median reaction time in the task homolateral to the lesion.

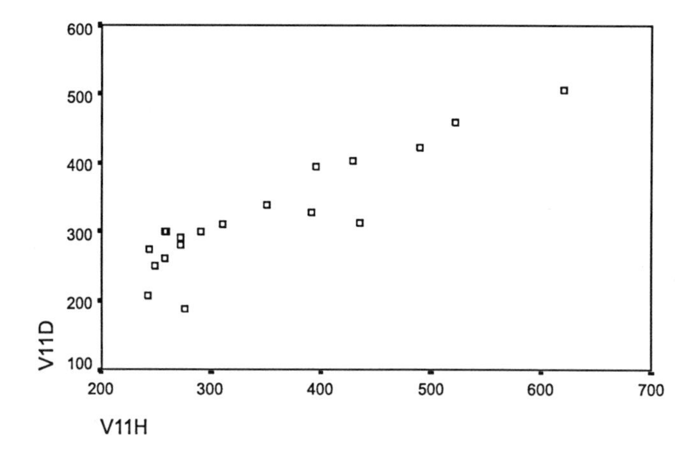

Fig. 90. Comparison of the median reaction times of the healthy side (v11h) with the median reaction times of the diseased side (v11d). The Pearson correlation coefficient is $r=0.861$, the significance level is $p< 0.01$

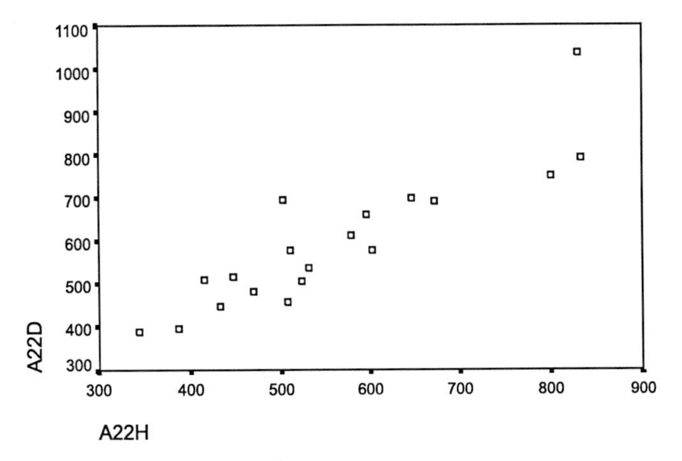

Fig. 91. Comparison of the median reaction times of the healthy side (a22h) with the median reaction times of the diseased side (a22d). The Pearson correlation coefficient is r=0.895, the significance level is p< 0.01

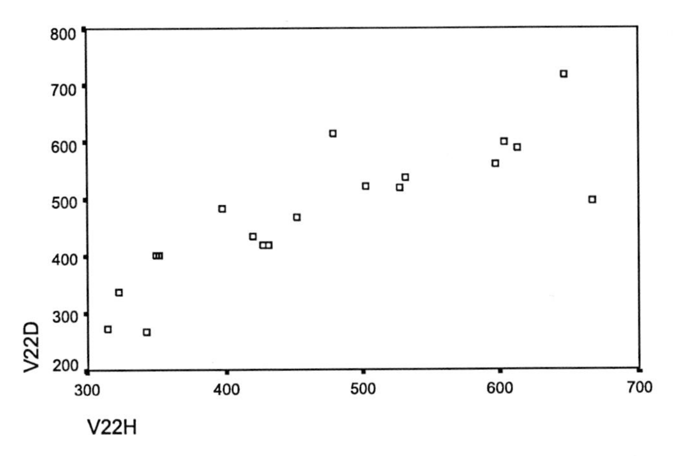

Fig. 92. Comparison of the median reaction times of the healthy side (v22h) with the median reaction times of the diseased side (v22d). The Pearson correlation coefficient is r=0.844, the significance level is p< 0.01

In the visual tasks, v11y most of the patients show symmetrical median reaction times or a slight increase in the tasks homolateral to the lesion. Three of them (O06A, O15A, and O20A) have clearly increased median reaction times at the tasks *homolateral* to the lesion.

In the tasks x22y, the results are similar: most of the patients show symmetrical median reaction times in the tasks which are homolateral and contralateral to the lesion. In the auditory tasks a22y, the subjects O10A and O12A have increased median reaction times in the task contralateral to the lesion, in the visual tasks v22y the subject O10A has an increased median reaction time contralateral to the lesion and the patient O11A homolateral to the lesion.

These findings should be compared to the findings in healthy subjects.

Comparison of the elementary times homolateral (ET(xNNh)) and contralateral (ET(xNNc)) to the lesion

This section uses the first or regular elementary times. If one compares the auditory elementary time of the homolateral and the contralateral side of the lesion one sees that most of the *contralateral elementary times are slightly increased* relative to their homolateral counterparts. 11 patients lie above the symmetry axis (especially O04A, O16A, O08A, and O03A), 5 lie below (eg O05A, O06A, and O18A).

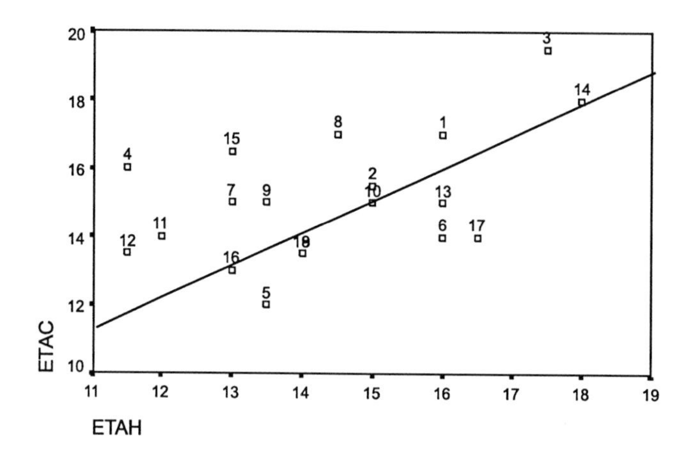

Fig. 93. Relation between the auditory elementary times homolateral to the lesion (ETAH) and contralateral to the lesion (ETAC).12=O13AF, 11=O12AF, 4=O04AP, 15=O16AF, 8=O08AT, 3=O03AF, 7=O07AF, 9=O10AT, 5=O05AP, 6=O06AT, 17=O18AF The Pearson correlation coefficient is r=0.515, the significance level is $p< 0.05$

In the next figure, which compares the visual elementary time of the task homolateral to the lesion (ETVH) with the visual elementary time of the task contralateral to the lesion (ETVC), the difference is not as clear as in the auditory case. Two patients (O16A and O03A) show a clear increase of visual elementary time for the tasks contralateral to the lesion. These two patients with frontal lesions have also a clear increase of auditory elementary time in the contralateral tasks. The other two patients who have an increased auditory elementary time in contralateral tasks without having an equivalent finding in visual elementary times have parietal (O04A) and temporal (O08A) lesions.

The model predicts different kinds of disturbing the pathways. A frontal lesion may affect the auditory and visual tasks on one side, a temporal, parietal, or occipital lesion can affect only one sensory modality (unless the lesion lies between the auditory and visual sensory areas and affects both of them). Are there cases which fulfil this prediction?

In the figures which compare ETAH to ETAC and ETVH to ETVC four patients with frontal lesions (O16AF, O13AF, O07AF, and O03AF) show an increase of auditory *and* visual elementary time in the contralateral hemisphere.

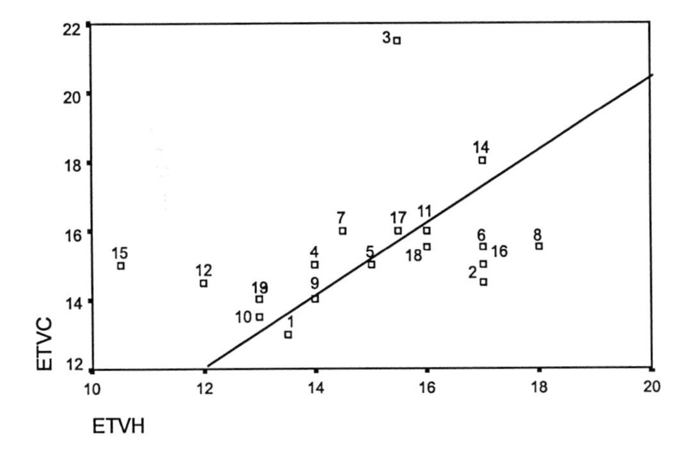

Fig. 94. Relation between the visual elementary times homolateral to the lesion (ETVH) and contralateral to the lesion (ETVC). 15= O16AF, 12=O13AF, 7=O07AF, 3=O03AF. The Pearson correlation coefficient is r=0.385, there is no significance

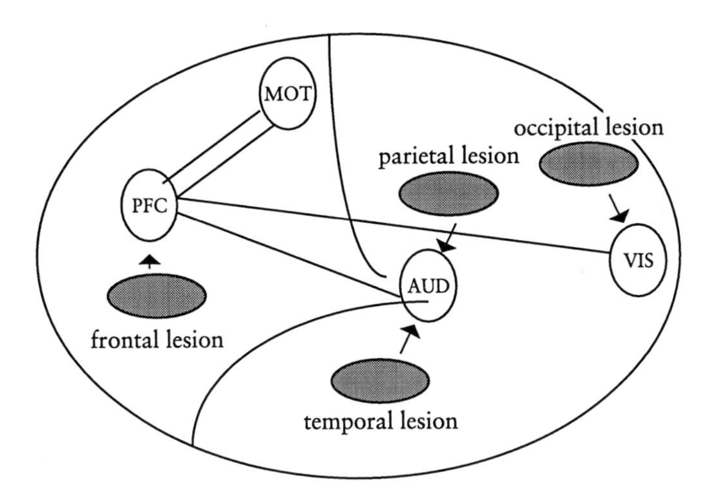

Fig. 95. Intracortical connections used by the auditory and visual stimulus-response pathway. A frontal lesion may disturb both the auditory and the visual areas, the temporal and parietal lesions may disturb the auditory area without disturbing the visual area

Three patients with temporal and parietal lesion (O04AP, O08AT, O10AT) show *only* an increase of auditory elementary time in the contralateral hemisphere.

In one patient with a frontal lesion (O12AF), only the auditory elementary time of the contralateral task is increased. Perhaps the lesion attaches only the top element of the auditory task set but not the top element of the visual task set.

If one compares the elementary times of healthy subjects with that of patients with monohemispheric brain lesions, one gets the following table.

Table 31. Comparison between the mean elementary time in 20 patients with monohemispheric brain lesions and 30 healthy subjects. *aET* auditory elementary time, *vET* visual elementary time

	aET	vET
homolat	14.4	14.8
contralat	**15.1**	**15.3**
healthy r	14.6	14.7
healthy l	14.5	15.4

The elementary times of the contralateral side are slightly increased, the visual elementary time in healthy subjects is not symmetrical.

Comparing the length of the linear and cyclical pathway homolateral (linENh, cycENh) and contralateral (linENc, cycENc) to the lesion

There are no patients which show an increase of the linear pathway both in a11c and a22c relative to their homolateral counterparts. The linear pathway of O05A, O10A, O14A, and O19A are prolonged in a11c relative to a11h but this is not the case in a22c. One patient (O16A) shows an increase of linear pathway in the homolateral tasks of a11h and a22h.

In the visual tasks two patients (O04A, O20A) show an increase in the homolateral tasks v11h. This is not the case in v22h.

Altogether, the linear pathways are astonishingly symmetric between the homolateral and the contralateral tasks. As an example of this symmetry, the relation between the length of the linear pathway of v22h and v22c is shown below.

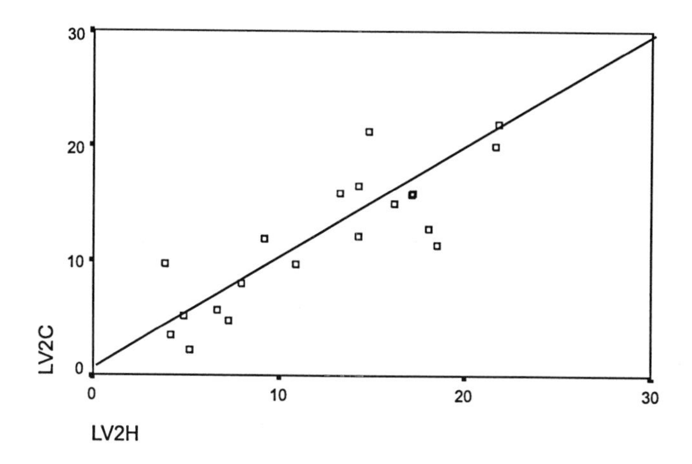

Fig. 96. Linear relationship between LV2H = the length of the linear pathway of the homolateral side and LV2C = linear pathway at the contralateral side of the lesion in a visual two stimuli task. The Pearson correlation coefficient is $r=0.848$, the significance level is $p< 0.01$

The same is true for the length of the cyclical pathway. There is no tendency of cyclical pathways being increased at the contralateral side of the lesion relative to the homolateral side.

As an example the relation between ca2h/ ca2c and cv2h/cv2c are shown.

The Pearson correlation coefficient is r=0.692, the significance level is p< 0.01.

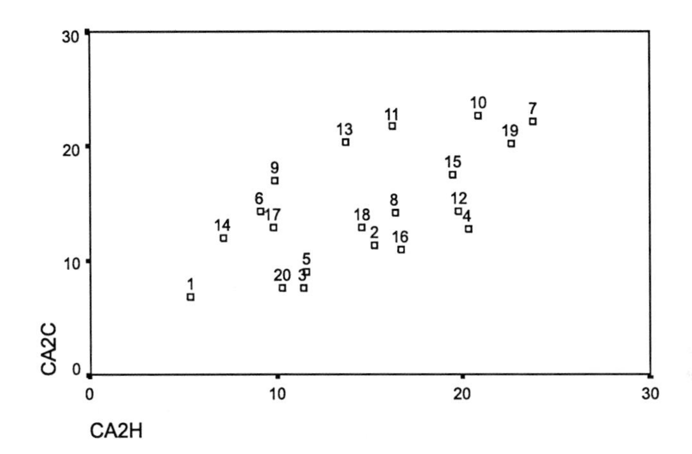

Fig. 97. Linear relationship between CA2H = the length of the cyclical pathway of the homolateral side and CA2C = linear pathway at the contralateral side of the lesion in a visual two stimuli task. The Pearson correlation coefficient is r=0.636, the significance level is p< 0.01

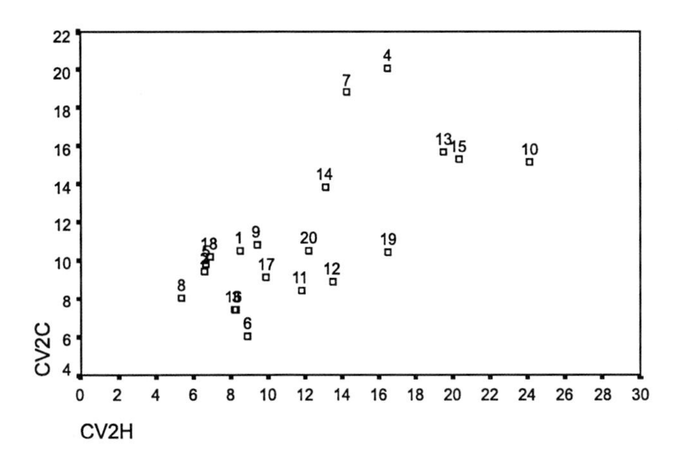

Fig. 98. Relation between CV2H = cycEN(v22h) and CV2C=cycEN(v22c) at the contrelateral side

This symmetry is astonishing because there are much longer linear and cyclical pathways as the comparison between the pathways of healthy and that of patients with brain lesions show.

Table 32. Comparison between the mean length of the linear pathway in 20 patients with monohemispheric brain lesions and 30 healthy subjects. The values of the patients are without considering the side of the lesion

	LA1R	LA1L	LA2R	LA2L	LV1R	LV1L	LV2R	LV2L
Lesion	6.3	8.2	19	20	7	7.9	11.4	12.9
Healthy	5.5	5.4	11.2	12.2	3.0	3.0	5.8	6.5

Table 33. Comparison between the mean length of the cyclical pathway in 20 patients with monohemispheric brain lesion and 30 healthy subjects

	CA1R	CA1L	CA2R	CA2L	CV1R	CV1L	CV2R	CV2L
Lesion	6.3	7.0	14.4	14.7	7.1	7.2	11.4	11.9
Healthy	5.5	5.7	12.5	12.6	4.8	5.4	8.8	8.6

Because the prolongation of the linear and the cyclical pathway is irrespective of the side of the lesion and rather general, one may assume that there is another influencing factor. The mean age of the patients with brain lesions is 48 years (14 patients) whereas the mean age of the healthy sample is 28.5 years, this could be the reason for these general differences.

Conclusions

What can be learned from the findings in these patients about the normal and abnormal structure of stimulus-response pathways?

Comparison of the median reaction times homolateral (xNNh) and contralateral (xNNd) to the lesion

If one looks at the elementary times of the patients deviating from the symmetry axis in the median reaction time plot, one sees that in most cases this deviation is *not due to a changed elementary time* but to a changed length of the linear or the cyclical pathway. And the change of a pathway x11y does not imply a change of the respective x22y pathway (only the patient O10A shows that fact).

In all tasks, 7 patients show an increased median reaction time in the task contralateral to the lesion and 5 patients in the task homolateral to the lesion. Therefore until now no general conclusion between the side of the lesion and the side of the altered pathway may be drawn.

Either the alterations are independent from the lesion or a task must not use the contralateral side constantly or the patients have used the same hemisphere in both sides of the task (by not fixating the arrow).

The length of the linear and the cyclical pathway are due to the formation process which takes place before the performance of the task. In this pre-phase the memory set is shaped. Deviations must therefore be caused by influences to this shaping process.

Comparison of the elementary times homolateral (ET(xNNh)) and contralateral (ET(xNNd)) to the lesion

If one compares the findings from healthy subjects with the findings from patients having monohemispheric brain lesions one does not see any clear difference beside *the slight increase of auditory elementary time in contralateral tasks.*

Patterns of elementary time asymmetries

If one considers the error of ±2ms for elementary time only a few patients with monohemispheric brain lesions have a sure asymmetry of elementary time if one compares the homolateral elementary time with the contralateral elementary time (relative to the lesion). The remaining patients show symmetrical elementary times.

Right frontal lesion
O03A: ET(xNNl) ↑ 2 and 6
O18A: ET(aNNr) ↑ 2.5 symmetrical

Left frontal lesion
O13A: ET(xNNr) ↑ 2 and 2.5
O12A: symmetrical
O16A: ET(xNNr) ↑ *3.5 and 4.5*
O07A: ET(xNNr) ↑ 2 and 1.5

Right temporo-parietal lesion
O01A: ET(aNNl) ↑ 1 symmetrical
O10A: ET(aNNl) ↑ 1.5 symmetrical
O17A: symmetrical
O20A: symmetrical

Left temporo-parietal lesion
O06A ET(aNNl) ↑ 2 symmetrical
O15A: ET(vNNr) ↑ 1 symmetrical
O08A: ET(aNNr) ↑ 2.5 ET(vNNl) ↑ 2.5 symmetrical
O11A: symmetrical
O19A: (fronto-parietal) - symmetrical

Right parieto-occipital lesion
O05A: ET(aNNr) ↑ 1.5 symmetrical

Left parieto-occipital lesion
O04A: ET(aNNr) ↑ *4.5* ET(vNNr) ↑ 1 symmetrical

Right lateral ventricle
O14A: symmetrical

Right fronto-polar lesion
O02A: ET(vNNr) ↑ 2.5 symmetrical

Left temporo-basal (KBFM)
O09A: ET(aNNr) ↑ *3*

Comparing the length of the linear and cyclical pathway homolateral (linENh, cycENh) and contralateral (linENc, cycENc) to the lesion

If one looks at the elementary times and the length of the linear or the cyclical pathway of these patients with an increased median reaction time one sees that the length of the pathway contributes mostly to the increased median reaction time. This length is responsible for the increase in the auditory median reaction times and the decrease in vsual median reaction times in tasks conralateral to the lesion.

But most patients show more or less symmetry between the linear and cyclical pathways of the homolateral side compared with the contralateral side. Whether the few cases with asymmetrical pathways are due to the lesion has to be confirmed by a greater sample.

Testing the model by mixed tasks in patients with monohemispheric brain lesion

Can the mixed tasks be used to test the validity of the proposed model? In a patient with a temporal lesion only *one* basic element in a mixed task is damaged In mixed tasks, the elementary time of one stimulus elements should therefore be different from the others.

How does a set system function with *one* different elementary time.

How does the chronophoresis measure the different elementary times? The simulation shows that independently varying elementary times (in a wide range) can be separated by the chronophoresis and co-varying elementary times (in a narrow range) produce one mean value in the chronophoresis. The studies with artificial data show that the chronophoresis gives the *average* elementary time of a set system if there are two different elementary times which covariate tightly.

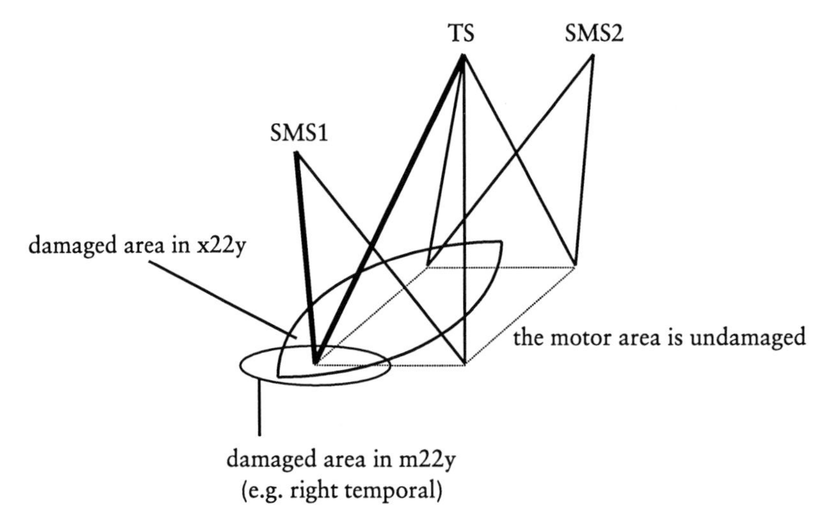

Fig. 99. Memory set system of the task x22y with one task set TS and two sensorimotor sets SMS1/2. Each sensorimotor set connects a stimulus element with an response element. In a mixed task only one stimulus element may be affected, in a task x22y both stimulus elements are affected

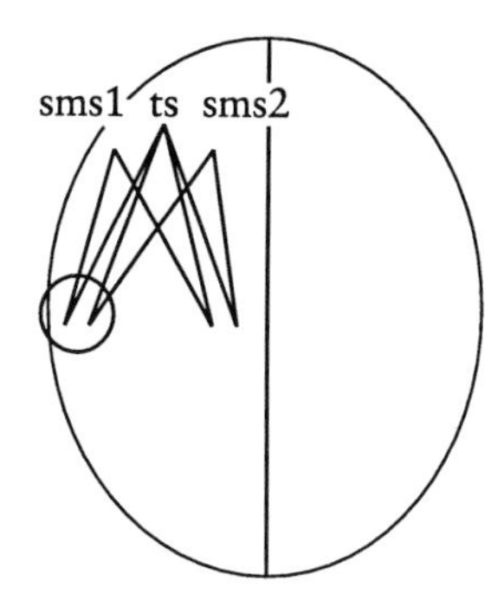

Fig. 100. The task a22r with a left temporal tumor The task set and the sensorimotor set1 have one half of their basic elements affected. If the affected elements are scanned this takes more time than scanning the healthy elements. In *a22l, two* of the four task set elements are impaired, but in the mixed task *m22r only one* of the task set elements is impaired

If the different elementary times produce one mean value in the chronophoresis, one should get a smaller increase of elementary time in the task m22y than in the task a22y because a22y has *two* affected stimulus elements out of four contrary to *one* out of four affected stimulus elements in m22y.

A temporal lesions cannot affect the response elements in the motor area nor the visual stimulus elements. Only the auditory stimulus elements are impaired.

The damaged basic element implies that all relations which are attached to it are slowed (bold lines). The remaining relations work normally.

One mean value in chronophoresis

The auditory elementary time of O10A is ET(aNNl)=15ms. Because of the above considerations, the task elements of a22l are (D,D,N,N) with D=damaged and N=normal and 15ms being the average of the damaged and the non damaged portions. How large is the normal elementary time of the healthy portions of O10A's brain? Because of the symmetry principle, one may assume an healthy elementary time similar to ET(aNNr)=13.5 ms.

If one takes this value for the elementary time of the undamaged motor elements in the task a22l, one gets:

$$(D+D+N+N)/4 = (D+D+13.5+13.5)/4 = 15$$
$$2D = 80-27 = 53$$
$$D = 26.5$$

This would be the true damaged elementary time in the task a22lO10A with the two motor elementary times being normal, ie. 13.5.

As we have seen, the mixed task m22rO10A has the composition (D,N,N,N). This would predict an average elementar time ET(m22rO10A) = (28+13.5+13.5+13.5)/4 = 17. Indeed the chronophoresis of m22rO10A shows a plot P16 (and P22 and P32). These considerations have to be tested with a greater sample.

The splitting of elementary times in the chronophoresis of patients with temporo-parietal and parieto-occipitial pattern

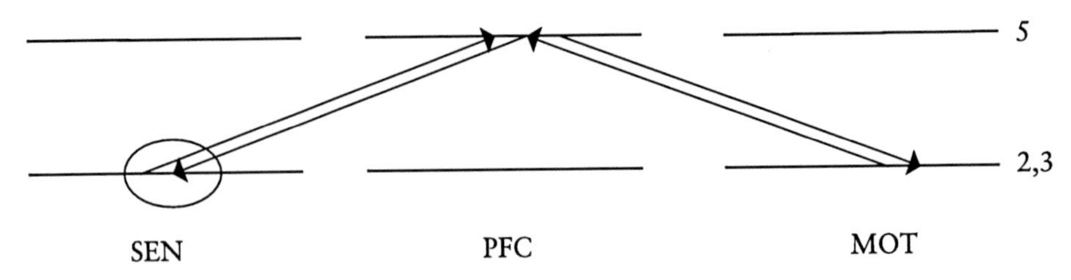

Fig. 101. If the sensory area is lesioned, the elementary time between layer 5 of PFC and layer 2,3 of the sensory area is increased but the elementary time between layer 5 of PFC and layer 2,3 of the motor area remains the same (splitting of elementary time)

If the sensory area, used by the memory set, is affected only these searches on this area are disturbed. The searches on the motor area remain normal. One has to assume that the input and the output behavior of the sensory area is affected. Therefore both directions of the connection between the top element and its sensory basic element have a prolonged elementary time. Because the task set contains an equal number of sensory and motor elements, on the average half of the searches are prolonged and half are normal. That means the average elementary time is (ET1 + ET2)/2.

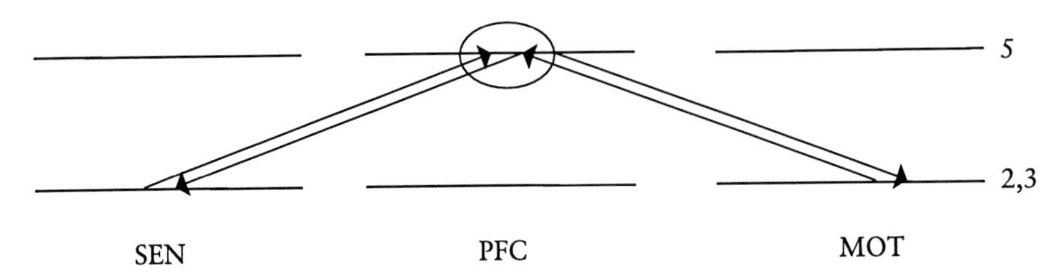

Fig. 102. If the layer 5 of the PFC is lesioned, both elementary times are increased

If the prefrontal area is affected by the lesion, both the searches on the sensory and the motor areas are disturbed (showing an prolonged elementary time). This means there is *no splitting* of elementary times and *both modalities* are affected. This last phenomenon may be absent if the lesion specifically affects one modality which are differently located in the prefrontal cortex (eg O12A).

It remains the question whether the prolonged elementary time is caused by an extra plot. These extra plots are methodical artifacts of the nestle procedure in the FPM and SINGLE programs. In order to answer that question, the prolonged elementary time should not have a mathematical relationship to the normal elementary time like the extraplots to the normal plot. And control tasks (either the contralateral task of the patient or the comparable task of a healthy subject) should not have this splitting.

Finally one has to examine the chronophoreses of patients with non-frontal lesions for this splitting. *Patient O04A has such a splitting in a11r* (see chronophoresis) This has to be tested in a greater sample of patients.

The problem of different elementary times in mixed tasks

How does the set system deal with asymmetrically increased elementary times in simultaneous search? It is no serious fault having some slowed elementary times if the memory sets search successively (one after another). But simultaneous searches may be affected. The slowed elementary times may cause asynchronous activation of the response element and therefore delay the finding of this element.

Subjects who differ markedly in their auditory and visual elementary times should have difficulties in performing mixed tasks if they use simultaneous searches.

This problem emerges also in subjects who have a asymmetry of elementary times caused by disease. In these cases, the disturbed areas cause increased elementary times, whereas the healthy areas use normal elementary times.

How can the increased elementary time be explained at the molecular level?

This question cannot be answered to date. Generally, the conduction of activity from the generation of the action potential at the axon hillock down the axon to the synapses, the dendrites and the body of the subsequent cell has to be affected in an unknown way.

Discussion of the method

The side of the lesion should not be known to the evaluating scientist.

At best there are two independent evaluations of one subject which give the same results.

The critical issue of the evaluation of patients with monohemispheric brain lesions is the definition of the first peak (Fp). Because these subjects did only 50 trials of the tasks x11y, the early trials are rather rare and it is difficult therefore to determine the first peak (the minimal pathways).

Implicit learning axiom

Is it possible that a fast mode search in a11y implies such a strong tSMS relation that a third search in a22y is not needed? Then a22y must work with fast mode too:

a22y = (con + n*linET + 4cycCT + *1*cycCT)

The pathways of patients with schizophrenia

Convergence tables

Preliminary remarks

The patients with schizophrenia performed 100 x11y tasks, 200 x22y tasks and eventually 300 x33y tasks. The reason of this design was that the index finger performed approximately 100 trials within each task. This was necessary to make the evaluations of the various tasks comparable to each other. The patients with schizophrenia had first rank symptoms in their history but were without productive symptoms now. All had negative symptoms like deficiency of emotion and concentration. All were treated with neuroleptics. The age of the patients lay between 20 and 40 years.

The evaluation of the reaction time results was done in the same way as in the healthy subjects. The programs FPM31e and FPM26f58 were used to get the four elementary times of the right-handed and the left-handed auditory pathways as well as the visual ones. The elementary times were selected by rules of agreement stated below.

Rules of agreement:

1. I take the integer number *occuring in both* FPM27d(x11y) and FPM27d(x22y).
2. If they have no common number, I take the *mean* of both methods (a:if two numbers are proposed) or the mean of the two adjacent numbers (b:if more than two numbers are proposed).
3. If there is no number in one method, I take the common number of xNNr and xNNl.
4. If there is no common number of xNNr and xNNl, I take the mean of both.

Previous agreements:

1. I take the integer number *occuring in both* FPM26g(x11y) and FPM26g(x22y).
2. If they have no common number, I take the *mean* of both methods.
3. If there is no number in one method, I take the common number of xNNr and xNNl.
4. If there is no common number of xNNr and xNNl, I take the mean of both.

Convergence tables

From 13 of the 22 patients with schizophrenia the convergence tables are shown below as examples for the methods used to get the elementary times.

S02A	FPM31e (x11y)	FPM31 (x22y)	FPM26f58 (x11y)	FPM26f58 (x22y)	Results	Chrono-phorese	Median-Fp Method
aNNr	23?	13,20	13	?	13		
aNNl	11?	15,20	18,21	?	10.5,20		
vNNr	25	20,25,27	22	12.5, 25	12.5		
vNNl	?	14,23	?	?	13		

S02B	FPM31e (x11y)	FPM31 (x22y)	FPM26f58 (x11y)	FPM26f58 (x22y)	Results	Chrono-phorese	Median-Fp Method
aNNr	15	13,15	?	12,13	14		
aNNl	11,23	19	20,25	?	10		
vNNr	11	15,27	14,16	24	12.5		
vNNl	10,20	12	22,24	19	11.5		

S03A	FPM31e (x11y)	FPM31 (x22y)	FPM26f58 (x11y)	FPM26f58 (x22y)	Results	Chrono-phorese	Median-Fp Method
aNNr	13,22	29	?	15	14	13.5	
aNNl	?	24	24	19,21,27	12	15	
vNNr	?	25	26	?	13	15	
vNNl	22,29	20,27	?	?	10.5,14	12	

S05A	FPM31e (x11y)	FPM31 (x22y)	FPM26f58 (x11y)	FPM26f58 (x22y)	Results	Chrono-phorese	Median-Fp Method
aNNr	23	13,15	24	12,26	12.5		
aNNl	26	13,15,26	18	13	13		
vNNr	?	20	11	15	15		
vNNl	17	14,16,21	?	17	17		

S07A	FPM31e (x11y)	FPM31 (x22y)	FPM26f58 (x11y)	FPM26f58 (x22y)	Results	Chrono-phorese	Median-Fp Method
aNNr	15	11,15,25	12,15,18		12,15	12	15
aNNl	15	27	13,17		13.5,15	15	15
vNNr	15,21	24	11-14		11	11.5	11.4
vNNl	12,25	?	?		12.5	15.5	12,17

The (Median-Fp) Method is not used to compute the results. In healthy subjects only the FPM results were used to gain the ETs.

Remarks:
1. The occurrence of ET and 2ET is more important than the occurrence of single ET.
2. In this subject the chronophoresis results have influenced the choice of the elementary times.
   The ET(aNNr)=12 and ET(aNNl)=13.5 was used because of the chronophoresis results.
   The ET(aNNl) could be 15 too.

S08A	FPM31e (x11y)	FPM31 (x22y)	FPM26f58 (x11y)	FPM26f58 (x22y)	Results	Chrono- phorese	Median- Fp Method
aNNr	18	12,23,16	18,25	?	12,17.5	16	18.3
aNNl	13,26,14	15,25	?	24,18?	14,12	15	13.2
vNNr	11,22,16	23?	14,22,28	23,27?	15,22.5	17.5	14.9
vNNl	22	24	?	13,22,24	23	21	22

S08B	FPM31e (x11y)	FPM31 (x22y)	FPM26f58 (x11y)	FPM26f58 (x22y)	Results	Chrono- phorese	Median- Fp Method
aNNr	?	12,25	11	?	12	16.5	
aNNl	13?	16	?	17	16.5	17.5	
vNNr	27	13,28	?	?	13.5	16	
vNNl	15	25	23	?	24	23.5	

Remarks.
1. Because of symmetry reasons with S08A, the ET(aNNr)=16.5.

S10A	FPM31e (x11y)	FPM31 (x22y)	FPM26f58 (x11y)	FPM26f58 (x22y)	Results	Chrono- phorese	Median- Fp Method
aNNr	11,22	11	?	?	11	13,20.5	
aNNl	21	17,25,28	?	17,21,34	17, 21	21.5	
vNNr	21	14	?	26	13.5	12,19	
vNNl	11,26	13,22,27	26	?	13	18.5	

S10B	FPM31e (x11y)	FPM31 (x22y)	FPM26f58 (x11y)	FPM26f58 (x22y)	Results	Chrono- phorese	Median- Fp Method
aNNr	18,27	23?	27	?	13.5	20.5	
aNNl	?	20	?	?	20	14.5, 22	
vNNr	15,21	14,21	?	?	14.5,21	11,21	
vNNl	10	14	12,24	?	13	10,14,18	

S10	Results of S10A	Results of S10B	Results of S10	Intersect. chronoph.
aNNr	11	13.5	12	20.5
aNNl	17, 21	20	20.5	22
vNNr	13.5	14.5,21	14	11,20.5
vNNl	13	13	13	15,18.5

Remarks:

1. The intersection of chronophoreses was taken between SING106n(x11y,S10B) and SING104r(x11y, S10A).

S11A	FPM31e (x11y)	FPM31 (x22y)	FPM26f58 (x11y)	FPM26f58 (x22y)	Results	Chrono-phorese	Median-Fp Method
aNNr	16	14	16,19	15	15	14.5	
aNNl	11,13	14,22,24	11	?	11	19.5	
vNNr	?	15	16	?	16	18	
vNNl	?	13-15	15	?	13.5	13.5	

S11B	FPM31e (x11y)	FPM31 (x22y)	FPM26f58 (x11y)	FPM26f58 (x22y)	Results	Chrono-phorese	Median-Fp Method
aNNr	27	19,29	?	?	14	18.5	
aNNl	?	11,14,23?	12	?	11.5	13	
vNNr	23	25	15,17	?	12,16	13	
vNNl	15,29	?	?	12,21	13.5	17	

S12A	FPM31e (x11y)	FPM31 (x22y)	FPM26f58 (x11y)	FPM26f58 (x22y)	Results	Chrono-phorese	Median-Fp Method
aNNr	22,29	16,22	?	20	15,21		11.1
aNNl	14,28	11,21	?	13	13.5		10.7, 18.4
vNNr	?	19	28,33	11,16,22	15.5		10.7
vNNl	17	?	11	11	11,17		12.9

S12B	FPM31e (x11y)	FPM31 (x22y)	FPM26f58 (x11y)	FPM26f58 (x22y)	Results	Chrono-phorese	Median-Fp Method
aNNr	18,24	15,26	14	14,27	13.5	14	
aNNl	27	13,26	23,28	18-20?	13	16	
vNNr	16,19	?	?	12,14	15	12.5	
vNNl	20?	15	18	12,14?	16.5	19	

S13A	FPM31e (x11y)	FPM31 (x22y)	FPM26f58 (x11y)	FPM26f58 (x22y)	Results	Chrono-phorese	Median-Fp Method
aNNr	15,27	15	14,22	16,22	15	15.5	
aNNl	17	?	30?	20,28?	14.5,17	15	
vNNr	15,(25)	12,19	17,34	12,35	17	15	
vNNl	14	18	?	15,16	15	14	

S14A	FPM31e (x11y)	FPM31 (x22y)	FPM26f58 (x11y)	FPM26f58 (x22y)	Results	Chrono-phorese	Median-Fp Method
aNNr	19,23 ?	15	17?	16	15.5		13.5
aNNl	14,26	?	26	30,32 ?	13.5		16.1
vNNr	12,19 ?	19,22,28	14	16,32	15		13.8
vNNl	11,20,28 ?	23	18,21	15,25	21		20.9

S15A	FPM31e (x11y)	FPM31 (x22y)	FPM26f58 (x11y)	FPM26f58 (x22y)	Results	Chrono-phorese	Median-Fp Method
aNNr	22	16?	34	23	11.5	12	12.6, 17.6
aNNl	20	16?	31	15	15.5	17.5	10.6, 18.2
vNNr	15	19,26	23	19,21	19	17.5	11.6, 17.6
vNNl	17	27	13,19	?	13.5,18	13,5	11,9, 17.8

Remarks:
1. The chronophoresis is produced by SING104r(x11y).
2. The proposals of the diverse FPM procedures may be equal to 2ET.

S15B	FPM31e (x11y)	FPM31 (x22y)	FPM26f58 (x11y)	FPM26f58 (x22y)	Results	Chrono-phorese	Median-Fp Method
aNNr	23?	13,28	13,18,26	13,24	13	14.5	12.2, 20,9
aNNl	25	13,17	16,25	17,24	17,25	16	13.7, 19.2
vNNr	15	15,24	20,24	11	15,24	14.5	10.3, 14.5
vNNl	14,17,18	17	?	?	14,17	14.5	11.5, 17.3

Remarks:
1. The agreements propose two elementary times in three of the four pathways. The results of the chronophoresis (SING104r(x11y)) are used to select one of these as the real value.

2. The chronophoreses of S15A and S15B (not the middle of the plots as shown above) have a common intersection. These are the elementary times: 13, 17.5, 16, 14.5.

S15	Results of S15A	Results of S15B	Results of S15	Intersect. chronoph.
aNNr	11.5	13	13	13
aNNl	15.5	17,25	17	17.5
vNNr	19	15,24	15.5	16
vNNl	13.5,18	14,17	14	14.5

Remarks:

1. Until now in subjects with repetitions I have used common elementary times to generate the equations (with the exception of subject S08A/S08B). I have to give rules how to proceed in future subjects.
2. Until now the intersection of chronophoreses is taken as additional evidence but is not used to generate the common elementary times.
3. The best possibility *to get the common elementary times is to apply the agreements* onto the results of the different series, including the intersection of chronophoreses. If one takes the neighboring values and computes their mean value one gets: 13, 17.25, 15.5, 14.25. If one rounds these values to integers or half-integers one gets the above results (17.25 ->17 and 14.25 -> 14).
4. In the case of having no intersection of chronophoreses, the agreements are only applied to the results of the FPM procedures.

S17A	FPM31e (x11y)	FPM31 (x22y)	FPM26f58 (x11y)	FPM26f58 (x22y)	Results	Chrono-phorese	Median-Fp Method
aNNr	19?	15,29	?	?	15	16	
aNNl	18	18,19	18,25	22-25	18	14,21	
vNNr	22	14,18,24	17	17	17	16	
vNNl	15,24	15,25	31,32	14	15	12.5	

S17B	FPM31e (x11y)	FPM31 (x22y)	FPM26f58 (x11y)	FPM26f58 (x22y)	Results	Chrono-phorese	Median-Fp Method
aNNr	14	17	13	26	13.5	12.5	
aNNl	17	19	17	19,35	18	17.5	
vNNr	17	?	16	?	16.5	13	
vNNl	15	16	?	17	16	18	

S20A	FPM31e (x11y)	FPM31 (x22y)	FPM26f58 (x11y)	FPM26f58 (x22y)	Results	Chrono-phorese	Median-Fp Method
aNNr	15,26	14,29	?	?	14.5		
aNNl	16	16,25,28	?	?	16		
vNNr	15	18,25	29	?	15		
vNNl	?	14,29	?	11,13	13.5		

Collection of Equations

S01A

auditory

a11rS01A = 70 + (6 + 2(2 + §2))17   6.2 + 5.1
a11lS01A = 70 + (10 + 2(2 + 2*2))15   10.3 + 10
a22rS01A = 70 + (12 + 2(4 + 1*4))17   11.5 + 13.8
a22lS01A = 70 + (18 + 2(4 + 1*4))15   18.7 + 13.6

visual

v11rS01A = 120 + (6 + 2(2 + 2*2))17   6.2 + 8.4
v11lS01A = 120 + (5 + 2(2 + 2*2))17   4.8 + 8.1
v22rS01A = 120 + (12 + 2(4 + 1*4))17   11.8 + 12.2
v22lS01A = 120 + (13 + 2(4 + 2))17   13 + 10.4

S02A

auditory

a11rS02A = 70 + (8 + 2(2 + 2*2))13   8 + 9.8
a11lS02A = 70 + (15 + 2(2 + 2*2))10.5   15.1 + 8.4
a22rS02A = 70 + (*24* + 2(4 + 2*4))13   24.2 + 26.6
a22lS02A = 70 + (*39* + 2(4 + 2*4))10.5   39.4 + 26.5

visual

v11rS02A = 120 + (3 + 2(2 + 2*2))12.5   3.6 + 7.1
v11lS02A = 120 + (9 + 2(2 + 2*2))13   8.4 + 8.2
v22rS02A = 120 + (10 + 2(4 + 2*4))12.5   10 + *14.5*
v22lS02A = 120 + (15 + 2(4 + 2*4))13   14.9 + 15.8

Remarks:

1. The cycEN(v22r) = 14.5 may be slow mode or fast mode. The slow mode of v22rS02B supports the slow mode of v22rS02A (implicit learning axiom).

S02B

auditory

a11rS02B = 70 + (9 + 2(2 + 2*2))*14*   9.4 + 12.7
a11lS02B = 70 + (14 + 2(2 + 2*2))10   14 + 15.2
a22rS02B = 70 + (*20* + 2(4 + 2*4))*14*   20.1 + 24.5
a22lS02B = 70 + (*36* + 2(4 + 2*4))10   36.5 + 26.1

visual

v11rS02B = 120 + (4 + 2(2 + 2*2))12.5   4.4 + 12
v11lS02B = 120 + (12 + 2(2 + 2*2))11.5   11.5 + 14.9
v22rS02B = 120 + (15 + 2(4 + 1*4))12.5   15.2 + 13.6
v22lS02B = 120 + (15 + 2(4 + 2*4))11.5   15.4 + 21.5

S03A

auditory

a11rS03A = 70 + (5 + 2(2 + §2))14   4.4 + 5.3
a11lS03A = 70 + (5 + 2(2 + 2*2))12   4.7 + 7.1
a22rS03A = 70 + (*23* + 2(4 + 2*4))14   22.5 + 17.4
a22lS03A = 70 + (*24* + 2(4 + 2*4))12   23.8 + 23.9

visual

v11rS03A = 120 + (3 + 2(2 + 2*2))13	3.4 + 10
v11lS03A = 120 + (4 + 2(2 + 2*2))14	4.1 + 11.7
v22rS03A = 120 + (16 + 2(4 + 2*4))13	16 + 1 + 18.9
v22lS03A = 120 + (15 + 2(4 + 1*4))14	15.1 + 11.6

Remarks:

1. In a22yS03A the first peak (Fp) gives wrong cycEN. These are too high in so far, that cyclenumbers of 28.4 or 34.3 cannot be explained by the current theory. Therefore the first peak has to be shifted to the right side. The next first peak Fp´ is taken to compute the structure of the pathway.

S04A

auditory

a11rS04A = 70 + (6 + 2(2 + §2))11	5.7 + 5
a11lS04A = 70 + (6 + 2(2 + §2))13	6.1 + 5
a22rS04A = 70 + (24 + 2(4 + 2*4))11	23.9 + 26.5
a22lS04A = 70 + (27 + 2(4 + 2*4))13	26.5 + 19.9

visual

v11rS04A = 120 + (2 + 2(2 + §2))17	2 + 6
v11lS04A = 120 + (6 + 2(2 + §2))11	6.1 + 6
v22rS04A = 120 + (8 + 2(4 + 1))17	7.7 + 6
v22lS04A = 120 + (12 + 2(4 + 2))11	12 + 10.7

Remarks:

1. The linEN of a22y are prolonged.
2. The cycEN of a22r is unusual high.
3. The subject shows visual side differences.

S05A

auditory

a11rS05A = 70 + (6 + 2(2 + 2*2))12.5	6.4 + 8.2
a11lS05A = 70 + (9 + 2(2 + 2*2))13	9.2 + 7.2
a22rS05A = 70 + (15 + 2(4 + 2*4))12.5	15.6 + 22.6
a22lS05A = 70 + (18 + 2(4 + 1*4))13	18.4 + 11.2

visual

v11rS05A = 120 + (3 + 2(2 + §2))15	3.3 + 6.3
v11lS05A = 120 + (3 + 2(2 + 2*2))17	3.9 + 8.1
v22rS05A = 120 + (9 + 2(4 + 1*4))15	9 + 15.1
v22lS05A = 120 + (12 + 2(4 + 1*4))17	12.4 + 13.2

S06A (3.98)

auditory

a11rS06A = 70 + (4 + 2(2 + §2))16	4.3 + 4.9
a11lS06A = 70 + (4 + 2(2 + §2))16.5	4.1 + 5.5
a22rS06A = 70 + (12 + 2(4 + 1))16	11.4 + 8.4
a22lS06A = 70 + (15 + 2(4 + 2))16.5	15 + 10.8

visual

v11rS06A = 120 + (3 + 2(2 + §2))14	3 + 5.9
v11lS06A = 120 + (5 + 2(2 + §2))14.5	5 + 5.8
v22rS06A = 120 + (7 + 2(4 + 2))14	7.3 + 9.9
v22lS06A = 120 + (8 + 2(4 + 1*4))14.5	8.3 + 11

S06B(6.98)

auditory

a11rS06B = 70 + (4 + 2(2 + §2))17.5	4 + 6.6
a11lS06B = 70 + (3 + 2(2 + §2))16.5	3.5 + 5.2
a22rS06B = 70 + (9 + 2(4 + 1*4))17.5	8.9 + 10.9
a22lS06B = 70 + (9 + 2(4 + 2))16.5	9.5 + 9.9

visual

v11rS06B = 120 + (3 + 2(2 + §2))14	3.6 + 3.8
v11lS06B = 120 + (3 + 2(2 + §2))14.5	2.5 + 4.6
v22rS06B = 120 + (2 + 2(4 + 2))14	1.6 + 10
v22lS06B = 120 + (6 + 2(4 + 2))14.5	5.6 + 9.9

Remarks

1. Because a22rS06A uses fast mode, the equation of a11rS06B must contain (2 + 1*2). or the implicit learning has been reversed. This hypothesis is supported by the slow mode of v22lS06B.
2. Either v11rS06B has to use slow mode or v22rS06B has to use fast mode in order to fulfil the implicit learning axiom.

S07A

auditory

a11rS07A = 70 + (5 + 2(2 + §2))12	5.1 + 4.7
a11lS07A = 70 + (5 + 2(2 + §2))13.5	5 + 4.8
a22rS07A = 70 + (15 + 2(4 + 2*4))12	15.1 + 15.6
a22lS07A = 70 + (12 + 2(4 + 1*4))13.5	12.4 + 13.8

visual

v11rS07A = 120 + (5 + 2(2 + 2*2))11	5.3 + .9.8
v11lS07A = 120 + (9 + 2(2 + 2*2))12.5	8.8 + 8.3
v22rS07A = 120 + (15 + 2(4 + 1*4))11	15.3 + 14.7
v22lS07A = 120 + (12 + 2(4 + 1*4))12.5	12.4 + 14.5

Remarks:

1. S07A has a beautiful chronophoresis and evoked responses.
2. Is it possible that S07A can compensate the increased linear pathway in one hemisphere by an normal pathway in the other hemisphere (while subject S04A has increased pathways in both hemispheres)

S08A

auditory

a11rS08A = 70 + (3 + 2(2 + §2))17.5	3.1 + 6.8
a11lS08A = 70 + (5 + 2(2 + 2*2))14	4.4 + 7.6
a22rS08A = 70 + (9 + 2(4 + 2))17.5	9.7 + 9.8
a22lS08A = 70 + (18 + 2(4 + 1*4))14	11.9 + 17.6

visual

v11rS08A = 120 + (2 + 2(2 + §2))15 — 2 + 5.9
v11lS08A = 120 + (2 + 2(2 + §2))*23* — 2 + 4.8
v22rS08A = 120 + (11 + 2(4 + 2))15 — 10.3 + 9.5
v22lS08A = 120 + (5 + 2(4 + 1))*23* — 4.7 + 6.8

S08B

auditory

a11rS08B = 70 + (5 + 2(2 + §2))16.5 — 5.3 + 5.7
a11lS08B = 70 + (3 + 2(2 + §2))16.5 — 3.7 + 6.2
a22rS08B = 70 + (11 + 2(4 + 1*4))16.5 — 10.4 + 11
a22lS08B = 70 + (12 + 2(4 + 2))16.5 — 11.6 + 9.6

visual

v11rS08B = 120 + (2 + 2(2 + §2))13.5 — 2 + 6.7
v11lS08B = 120 + (2 + 2(2 + 1))*24* — 2 + 3.5
v22rS08B = 120 + (9 + 2(4 + 1*4))13.5 — 9.5 + 11.1
v22lS08B = 120 + (5 + 2(4 + 1))*24* — 5.1 + 6.5

Remarks:
1. It is possible, that ET(vNNl) = 24/2.
2. The fast mode of v11lS08B is due to the repetition of the task.
3. The prolongation of linEN(a22lS08A) is reversed in a22lS08B (therapy?).
4. This is the first subject, where the ETs of the repetition are measured apart from the ETs of the precedent tasks.

S09A

auditory

a11rS09A = 70 + (5 + 2(2 + §2))15 — 4.7 + 4.9
a11lS09A = 70 + (3 + 2(2 + §2))15.5 — 3.5 + 6.3
a22rS09A = 70 + (12 + 2(4 + 1*4))15 — 12.7 + 13
a22lS09A = 70 + (14 + 2(4 + 1*4))15 — 13.8 + 13.9

visual

v11rS09A = 120 + (5 + 2(2 + 2*2))23 — 5 + 13.9
v11lS09A = 120 + (9 + 2(2 + 2*2))28 — 8.9 + 9.1
v22rS09A = 120 + (8 + 2(4 + 1*4))23 — 8.2 + 10.7
v22lS09A = 120 + (8 + 2(4 + 1*4))28 — 7.3 + 11.3

S10A

auditory

a11rS10A = 70 + (6 + 2(2 + §2))12 — 5.5 + 5
a11lS10A = 70 + (3 + 2(2 + §2))*20.5* — 2.9 + 5.9
a22rS10A = 70 + (13 + 2(4 + 2*4))12 — 13 + 18.9
a22lS10A = 70 + (11 + 2(4 + 2))*20.5* — 10.4 + 10.7
a33rS10A = 70 + (*30* + 2(6 + 2*6))12 — 30.1 + 24.8*
a33lS10A = 70 + (17 + 2(6 + 2))*20.5* — 16.8 + 14.9*

visual

v11rS10A = 120 + (9 + 2(2 + 2*2))14 — 9,1 + 7.8
v11lS10A = 120 + (6 + 2(2 + §2))13 — 5.3 + 6.8

v22rS10A = 120 + (15 + 2(4 + 2))14 15.5 + 9.7
v22lS10A = 120 + (12 + 2(4 + 1*4))13 12.6 + 13.8
v33rS10A = 120 + (15 + 2(6 + 1))14 14.4 + 11.2
v33lS10A = 120 + (21 + 2(6 + 1))13 21.5 + 12.1

Remarks:
1. The cycEN(a33rS10A) = 24.8 lies between fast mode (6 + 1*6) and slow mode (6 + 2*6).
2. The cycEN(a33lS10A) = 14.9 lies between slow mode (6 + 2) and fast mode (6 + 1*6).

S10B

auditory

a11rS10B = 70 + (8 + 2(2 + 2*2))12 8 + 7.8
a11lS10B = 70 + (3 + 2(2 + §2))*20.5* 3.1 + 6
a22rS10B = 70 + (*24* + 2(4 + 1*4))12 23.4 + 14.3
a22lS10B = 70 + (9 + 2(4 + 2*4))*20.5* 8.7 + 16.3*

visual

v11rS10B = 120 + (6 + 2(2 + 2*2))14 5.5 + 6.9
v11lS10B = 120 + (6 + 2(2 + 2*2))13 6.5 + 7.5
v22rS10B = 120 + (9 + 2(4 + 1*4))14 9.8 + 11.6
v22lS10B = 120 + (15 + 2(4 + 2*4))13 14.9 + 16.3*

Remarks:
1. The cycEN(v11rS10A) = 6.9 may be (2 + §2) or (2 + 2*2).
2. The cycEN(v22lS10B) = 16.3 may be (4 + 1*4) or (4 + 2*4). Because v22lS10A and v33lS10A use fast mode, v22lS10B may use fast mode too.

S11A

auditory

a11rS11A = 70 + (2 + 2(2 + §2))*15* 2.3 + 5.5
a11lS11A = 70 + (*12* + 2(2 + §2))11 11.6 + 6.4
a22rS11A = 70 + (11 + 2(4 + 1*4))*15* 10.7 + 12.1
a22lS11A = 70 + (*18* + 2(4 + 1*4))11 17.5 + 14.6
a33rS11A = 70 + (16 + 2(6 + 6))*15* 16 + 14.2
a33lS11A = 70 + (27 + 2(6 + 1*6))11 26.6 + 17.4

visual

v11rS11A = 120 + (2 + 2(2 + §2))16 2.1 + 4.9
v11lS11A = 120 + (6 + 2(2 + §2))13.5 5.4 + 4.6
v22rS11A = 120 + (5 + 2(4 + 2))16 4.9 + 8.7
v22lS11A = 120 + (8 + 2(4 + 2))13.5 8 + 9.7
v33rS11A = 120 + (6 + 2(6 + 1))16 6.4 + 9.8
v33lS11A = 120 + (12 + 2(6 + 1))13.5 11.7 + 10.7

Remarks:
1. The x33y tasks have been omitted because they show an *inreased* elementary time.
2. Remarkable are the side differences and the relative long linEN(a22l).
3. The subject hears voices especially if real noises are present.

S11B

auditory

a11rS11B = 70 + (6 + 2(2 + §2))14 6.6 + 4.9
a11lS11B = 70 + (9 + 2(2 + §2))11.5 9.7 + 6.1
a22rS11B = 70 + (11 + 2(4 + 1*4))14 10.7 + 12.1
a22lS11B = 70 + (17 + 2(4 + 1*4))11.5 17.1 + 13.7

visual

v11rS11B = 120 + (2 + 2(2 + §2))16 2.1 + 5.4
v11lS11B = 120 + (5 + 2(2 + §2))13.5 4.7 + 5.6
v22rS11B = 120 + (8 + 2(4 + 2))16 7.7 + 9.6
v22lS11B = 120 + (12 + 2(4 + 2))13.5 11.7 + 9

S12A

auditory

a11rS12A = 70 + (3 + 2(2 + §2))15 3.7 + 5
a11lS12A = 70 + (6 + 2(2 + §2))13.5 5.4 + 5
a22rS12A = 70 + (9 + 2(4 + 2))15 9.7 + 9.8
a22lS12A = 70 + (11 + 2(4 + 2*4))13.5 11.3 + 16.3

visual

v11rS12A = 120 + (3 + 2(2 + 2*2))15.5 3.5 + 7.2
v11lS12A = 120 + (3 + 2(2 + 2*2))17 3.3 + 8.1
v22rS12A = 120 + (12 + 2(4 + 2))15.5 11.9 + 9.2
v22lS12A = 120 + (12 + 2(4 + 2))17 11.5 + 9.9

Remarks:
1. The cycEN(a22lS12A) = 15.8 lies between fast mode and slow mode.

S12B

auditory

a11rS12B = 70 + (5 + 2(2 + §2))13.5 4.7 + 4.7
a11lS12B = 70 + (5 + 2(2 + §2))13 5.3 + 4.8
a22rS12B = 70 + (11 + 2(4 + 1*4))13.5 11.3 + 11.5
a22lS12B = 70 + (15 + 2(4 + 1*4))13 14.5 + 12.8

visual

v11rS12B = 120 + (4 + 2(2 + §2))15 4.3 + 6.3
v11lS12B = 120 + (4 + 2(2 + §2))16.5 4.1 + 5.4
v22rS12B = 120 + (12 + 2(4 + 2))15 12.7 + 9.3
v22lS12B = 120 + (11 + 2(4 + 2))16.5 10.4 + 10.4

S13A

auditory

a11rS13A = 70 + (3 + 2(2 + 2*2))15 2.7 + 5.5
a11lS13A = 70 + (3 + 2(2 + §2))14.5 3.9 + 5.9
a22rS13A = 70 + (5 + 2(4 + 2*4))15 5 + 17.7
a22lS13A = 70 + (21 + 2(4 + 1*4))14.5 21.8 + 13.7

visual

v11rS13A = 120 + (2 + 2(2 + 2*2))17   2.4 + 7.3
v11lS13A = 120 + (12 + 2(2 + 2*2))15   12 + 12.8
v22rS13A = 120 + (9 + 2(4 + 1*4))17   8.9 + 13.9
v22lS13A = 120 + (15 + 2(4 + 2*4))15   14.7 + 16.5

Remarks:

1. a22lS13A has an earlier first peak (Fp) at 195 ms. This means a linEN = 6.6 and a cycEN = 28.9. Such a cycEN is not possible, therefore a later first peak (Fp´) has been assumed at 415ms.

S14A

auditory

a11rS14A = 70 + (4 + 2(2 + 2*2))15.5   4.1 + 8.2
a11lS14A = 70 + (6 + 2(2 + §2))13.5   5.8 + 6.1
a22rS14A = 70 + (15 + 2(4 + 1*4))15.5   14.8 + 11.2
a22lS14A = 70 + (21 + 2(4 + 1*4))13.5   20.6 + 14.5

visual

v11rS14A = 120 + (2 + 2(2 + 1))15   2 + 4.1
v11lS14A = 120 + (2 + 2(2 + 1))21   2 + 3.5
v22rS14A = 120 + (6 + 2(4 + 1*4))15   6.3 + 11.7
v22lS14A = 120 + (6 + 2(4 + 1*4))21   5.4 + 12.4

Remarks:

1. The fast mode of v11y is due to the fact that S14A is a replication of S14. The data of S14 are incomplete therefore S14 is not presented here.

S15A

auditory

a11rS15A = 70 + (5 + 2(2 + §2))13   5.3 + 6.7
a11lS15A = 70 + (6 + 2(2 + §2))17   5.6 + 6.5
a22rS15A = 70 + (15 + 2(4 + 1*4))13   14.9 + 12.2
a22lS15A = 70 + (15 + 2(4 + 1*4))17   15.1 + 13.2
a33rS15A = 70 + (23 + 2(6 + 1*6))13   22.6 + 15
a33lS15A = 70 + (14 + 2(6 + 1*6))17   13.3 + 14.4

visual

v11rS15A = 120 + (3 + 2(2 + §2))15.5   3.2 + 4.8
v11lS15A = 120 + (3 + 2(2 + 2*2))14   3.4 + 7.9
v22rS15A = 120 + (6 + 2(4 + 1))15.5   5.7 + 7.6
v22lS15A = 120 + (6 + 2(4 + 1))14   6.9 + 8.4
v33rS15A = 120 + (8 + 2(6 + 1))15.5   8.3 + 8.5
v33lS15A = 120 + (8 + 2(6 + 1))14   7.6 + 10.9

Remarks:

1. In a33lS15A the cycEN = 14.4 is taken as fast mode in order to fulfil the implicit learning axiom.

S15B

auditory

a11rS15B = 70 + (9 + 2(2 + 2*2))13   8.4 + 8.5
a11lS15B = 70 + (3 + 2(2 + 2*2))_17_   3.9 + 7.8
a22rS15B = 70 + (15 + 2(4 + 2*4))13   15.7 + 18.8
a22lS15B = 70 + (14 + 2(4 + 2))_17_   13.6 + 10.5

visual

v11rS15B = 120 + (2 + 2(2 + 2))15.5   2 + 5.6
v11lS15B = 120 + (3 + 2(2 + 2*2))14   3.4 + 9.1
v22rS15B = 120 + (6 + 2(4 + 2))15.5   5.7 + 9
v22lS15B = 120 + (9 + 2(4 + 1))14   8.4 + 8.2

Remarks:

1. There is an unexplained violation of the implicit learning axiom from S15A to S15B. I have to look for some reason for this violation of this axiom (which constantly holds in healthy subjects).

S16A

auditory

a11rS16A = 70 + (6 + 2(2 + 2*2))15   5.6 + 8.4
a11lS16A = 70 + (3 + 2(2 + 2*2))15   2.6 + 9.7
a22rS16A = 70 + (6 + 2(4 + 2*4))15   5.3 + 22.7
a22lS16A = 70 + (17 + 2(4 + 2*4))15   17.6 + 15.4

visual

v11rS16A   missing
v11lS16A   missing
v22rS16A = 120 + (8 + 2(4 + 1*4))15   8 + 13.6
v22lS16A = 120 + (12 + 2(4 + 2))15   12.3 + 10.1

S17A

auditory

a11rS17A = 70 + (6 + 2(2 + §2))15   5.7 + 5.9
a11lS17A = 70 + (5 + 2(2 + §2))_18_   4.4 + 6.3
a22rS17A = 70 + (15 + 2(4 + 2*4))15   15 + 19.3
a22lS17A = 70 + (17 + 2(4 + 1*4))_18_   16.9 + 14
a33rS17A = 70 + (27 + 2(6 + 1))15   27 + 12.5
a33lS17A = 70 + (16 + 2(6 + 1*6))_18_   16.1 + 19.6

visual

v11rS17A = 120 + (3 + 2(2 + 2*2))17   3.6 + 6.9
v11lS17A = 120 + (4 + 2(2 + 2*2))15   4 + 7.6
v22rS17A = 120 + (12 + 2(4 + 2))17   12.7 + 9.4
v22lS17A = 120 + (8 + 2(4 + 2))15   8 + 9.9
v33rS17A = 120 + (9 + 2(6 + 1))17   8.6 + 11.9
v33lS17A = 120 + (9 + 2(6 + 1))15   9.3 + 10.1

S17B

auditory

a11rS17B = 70 + (6 + 2(2 + §2))13.5   6.9 + 5.8
a11lS17B = 70 + (5 + 2(2 + §2))_18_   5.2 + 5.3

a22rS17B = 70 + (18 + 2(4 + 2*4))13.5	18.3 + 16.4
a22lS17B = 70 + (9 + 2(4 + 2*4))18	8.3 + 18.3

visual

v11rS17B = 120 + (4 + 2(2 + 2*2))16.5	4.1 + 7.5
v11lS17B = 120 + (3 + 2(2 + 2*2))16	3.6 + 6.6
v22rS17B = 120 + (11 + 2(4 + 2))16.5	11.3 + 9.9
v22lS17B = 120 + (10 + 2(4 + 2))16	9.3 + 9.8

Remarks:

1. a22y use slow mode in spite of the fast mode of a33y. The same is valid for v22y and v33y. This means that the subject does not save the implicit learning results of xNNyS17A. Perhaps the period between S17A and S17B is too long.

S18A

auditory

a11rS18A = 70 + (2 + 2(2 + 1))17.5	2 + 3.6
a11lS18A = 70 + (6 + 2(2 + 2*2))13.5	6.9 + 6.8
a22rS18A = 70 + (12 + 2(4 + 1*4))17.5	11.4 + 12.7
a22lS18A = 70 + (12 + 2(4 + 2*4))13.5	12.4 + 22.9

visual

v11rS18A = 120 + (2 + 2(2 + 1))20	2 + 2.4
v11lS18A = 120 + (3 + 2(2 + §2))18.5	3.4 + 6.2
v22rS18A = 120 + (3 + 2(4 + 1))20	2.8 + 8.4
v22lS18A = 120 + (3 + 2(4 + 1))18.5	3.9 + 8.2

S19A

auditory

a11rS19A = 70 + (3 + 2(2 + §2))17	3.3 + 5.8
a11lS19A = 70 + (5 + 2(2 + §2))14	5.1 + 5.4
a22rS19A = 70 + (9 + 2(4 + 2*4))17	8.9 + 15.6
a22lS19A = 70 + (11 + 2(4 + 2*4))14	11.2 + 16

visual

v11rS19A = 120 + (2 + 2(2 + §2))18	2 + 5.1
v11lS19A = 120 + (2 + 2(2 + 1))18	1.6 + 3.9
v22rS19A = 120 + (4 + 2(4 + 2))18	4.1 + 9.3
v22lS19A = 120 + (4 + 2(4 + 1))18	4.1 + 6.9

S20A

auditory

a11rS20A = 70 + (10 + 2(2 + 2*2))14.5	10.1 + 6.7
a11lS20A = 70 + (6 + 2(2 + §2))16	6.8 + 6
a22rS20A = 70 + (27 + 2(4 + 1*4))14.5	27.3 + 11.4
a22lS20A = 70 + (21 + 2(4 + 2*4))16	20.5 + 19.3

visual

v11rS20A = 120 + (12 + 2(2 + 2*2))15	12.3 + 9.8
v11lS20A = 120 + (6 + 2(2 + 2*2))13.5	5.4 + 7.1
v22rS20A = 120 + (18 + 2(4 + 1*4))15	17.7 + 13.8
v22lS20A = 120 + (12 + 2(4 + 2*4))13.5	11.3 + 20.2

S21A

auditory

a11rS21A = 70 + (6 + 2(2 + 2*2))20.5	5.6 + 7.8
a11lS21A = 70 + (3 + 2(2 + §2))23.5	3.1 + 5
a22rS21A = 70 + (11 + 2(4 + 1))20.5	10.4 + 7.9
a22lS21A = 70 + (6 + 2(4 + 1))23.5	6.1 + 7.8
a33rS21A = 70 + (11 + 2(6 + 1))20.5	10.7 + 9.1
a33lS21A = 70 + (11 + 2(6 + 1))23.5	10.6 + 5.5

visual

v11rS21A = 120 + (2 + 2(2 + §2))22	2.1 + 4.7
v11lS21A = 120 + (5 + 2(2 + §2))17	4.8 + 5.9
v22rS21A = 120 + (8 + 2(4 + 1))22	8 + 7.4
v22lS21A = 120 + (6 + 2(4 + 1))17	6.5 + 7.8
v33rS21A = 120 + (6 + 2(6 + 1))22	6.4 + 7.5
v33lS21A = 120 + (11 + 2(6 + 1))17	10.4 + 8.2

S22A

auditory

a11rS22A = 70 + (3 + 2(2 + 2*2))13.5	2.8 + 9.4
a11lS22A = 70 + (9 + 2(2 + 1*2))14	9.1 + 6.5
a22rS22A = 70 + (12 + 2(4 + 2*4))13.5	12.8 + 15.2
a22lS22A = 70 + (15 + 2(4 + 1*4))14	14.8 + 14.3

visual

v11rS22A = 120 + (2 + 2(2 + 1))16	2 + 3.8
v11lS22A = 120 + (2 + 2(2 + §2))15	2.3 + 5.7
v22rS22A = 120 + (5 + 2(4 + *1*))16	4.6 + 9.4
v22lS22A = 120 + (5 + 2(4 + 1*4))15	5.3 + 11.8

Remarks:

1. v22rS22A must be fast mode because of the fast mode of v11rS22A and the implicit learning axiom. This implicates, that the first peak should lie right of its current position and the cycEN would be smaller.

Discussion

Many patients with schizophrenia show a typical a22y pattern: left of the first peak lies a small number of reaction times. If one takes the first of these low reaction times as the first peak, the cycle number would be prolonged, if one takes the usual first peak with height>1 then the linear portion would be prolonged. The subject O16A shows a remarkable rhythm of this low reaction times with a period of 3ET. This would suggest the use of fewer areas in tasks with this low reaction times.

Another striking feature of the distributions of patients with schizophrenia is the high variability of reaction times. Therefore it could be possible that there is *no constant linEN* in these patients.

The increased length of the linear pathway in patients with schizophrenia means that the PFC cannot use the simple perceptions of the early sensory areas but has to use later areas with more complex meaning. Maybe this is the reason why simple noises cannot be perceived as what they are but give rise to experiences in higher areas like language areas. Maybe this is the reason why they are experienced as voices. The subject S11A was the first which reportend these voices being dependent from a noisy environment (vanishing in silent places).

The gravity of the disease may be dependent on the occurrence of the increased linear pathway in one hemishpere or in both hemispheres. In the case that only one hemisphere is being altered, the healthy pathway can compensate this phenomenon.

Is the prolongation of the linear portion of visual pathways compatible with delusional perceptions?

Deviations from normal pathways of healthy subjects

Frequency of elementary times in patients with schizophrenia

The frequencies of the elementary times approximately showed a normal distribution with nearly the same mean values in the two auditory elementary times ET(aNNr) and ET(aNNl) with mean value of ET(aNNr)=14.8ms and mean value of ET(aNNl)=14.9ms.

The two visual elementary times ET(vNNr) and ET(vNNl) showed a slightly increased mean value compared with the healthy subjects with mean value of ET(vNNr) = 16.1ms and mean value of ET(vNNl)=16.2ms.

The small increase of mean values may be due to the occurrence of higher elementary times in some subjects (values from 11 to 28 ms in ET(vNNl) and 11 to 23 ms in ET(vNNr)).

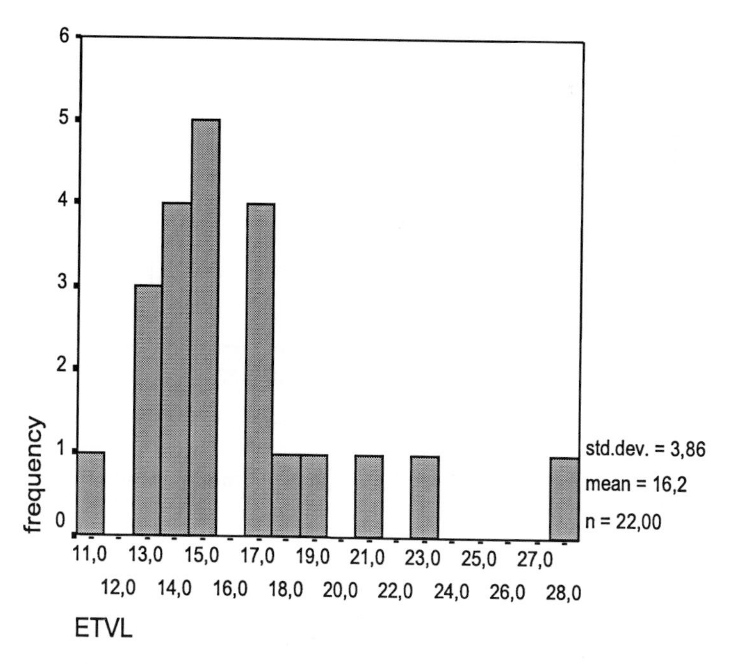

Fig. 103. Frequencies of elementary times in the tasks vNNl in 22 patients with schizophrenia

The lengths of linear pathways in patients with schizophrenia

The linear pathways of these patients tend to be longer than in healthy subjects. The table shows the comparison of the means values of the two groups.

Table 34. Comparison between the mean length of the linear patway in 22 patients with schizophrenia and 30 healthy subjects

	LA1R	LA1L	LA2R	LA2L	LV1R	LV1L	LV2R	LV2L
Schizo.	4.8	6.2	13.9	17	3.8	5.0	9.4	9.6
Healthy	5.5	5.4	11.2	12.2	3.0	3.0	5.8	6.5

The distribution of the linear pathways shows the same preference of multiples of 3 as in healthy subjects. As an example the distribution of the linear pathway of a22l is shown below.

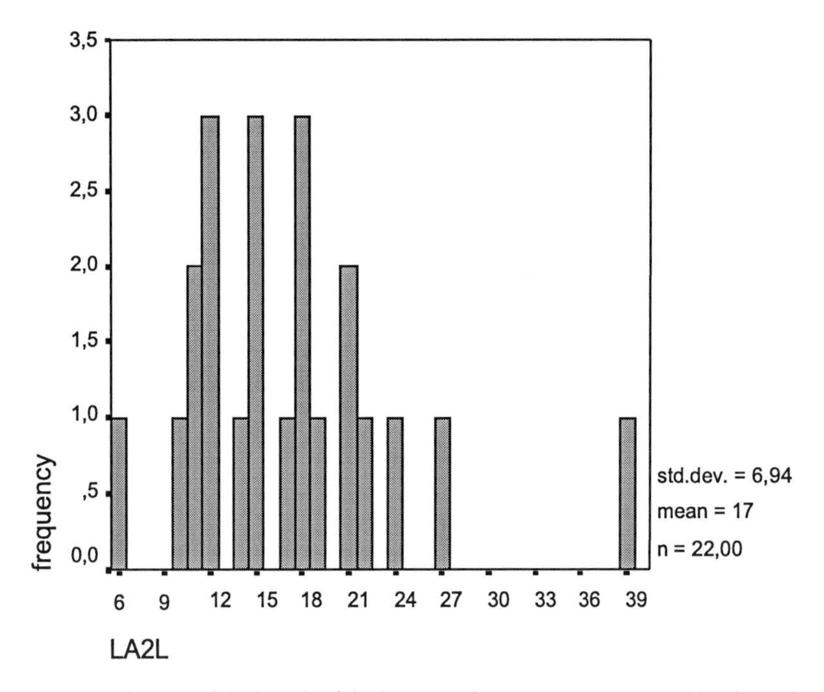

Fig. 104. Distribution of the length of the linear pathway in 22 patients with schizophrenia

The lengths of cyclical pathways in patients with schizophrenia

The cyclical pathways of these patients are also longer than in healthy subjects. The table shows the comparison of the means values of the two groups.

Table 35. Comparison between the mean length of the cyclical patway in 22 patients with schizophrenia and 30 healthy subjects

	CA1R	CA1L	CA2R	CA2L	CV1R	CV1L	CV2R	CV2L
Schizo.	6.3	6.5	**15**	**15.6**	6.8	7.2	**11.1**	**11.4**
Healthy	5.5	5.7	12.5	12.6	4.8	5.4	8.8	8.6

The distribution of cycEN(a11r) shows a preference of the (2+2) or (2+1*2) and slighthly the (2+2*2) pathway. The task a11l prefers the (2+2) or (2+1*2) pathway, the task a22r shows no clear preferences The task a22l tends to use the (4+1*4) strategy and the task v11r is similar to its auditory counterpart. The task v11l prefers (2+2) or (2+1*2) and (2+2*2), the tasks v22r and v22l use mostly the (4+2) pathway.

The symmetry of linear pathways in patients with schizophrenia

In contrast to the symmetry of linear pathways in healthy subjects, the patients with schizophrenia show no clear correlation between the two sides.

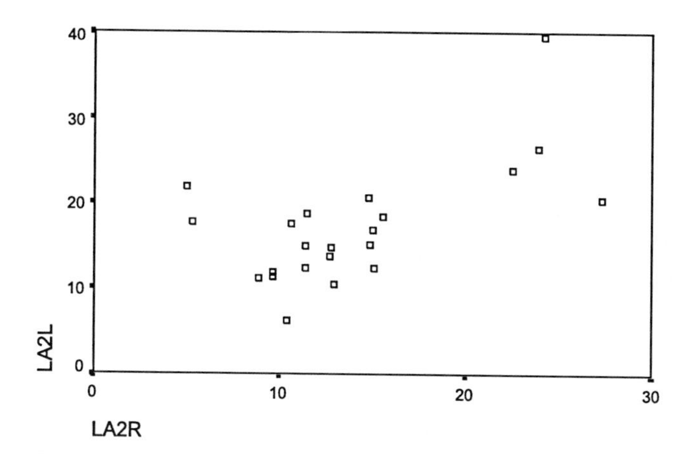

Fig. 105. Scatterplot between LA2R=linear pathway of a22r and LA2L=linear pathway of a22l. The Pearson correlation coefficient is r=0.604, the significance level is $p < 0.01$

If one looks at this Fig., one sees that the values of linEN(a22l) lie above the values of linEN(a22r). The next figure visualizes this relationship between the two linear pathways. This is different from the situation in healthy subjects (see figure below the next figure).

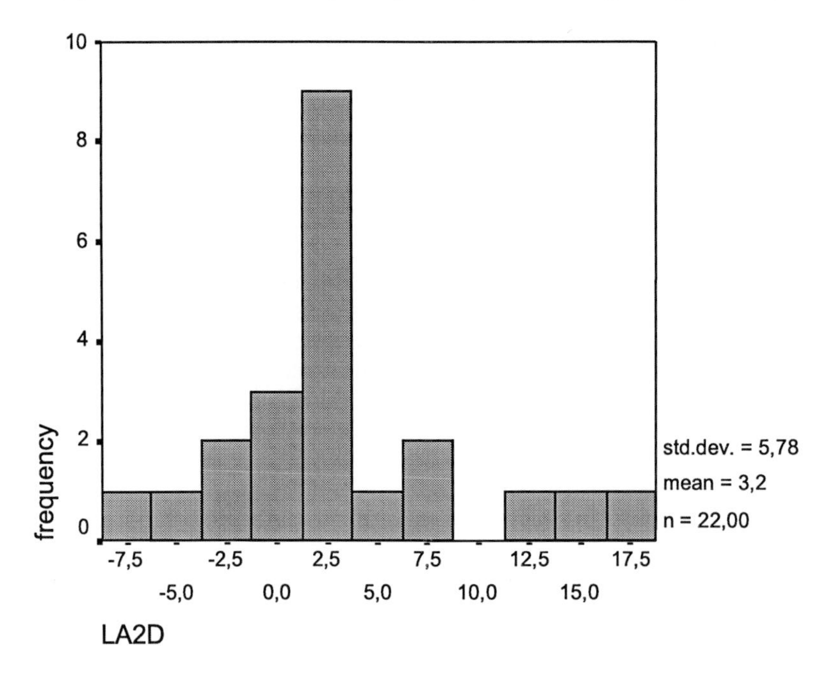

Fig. 106. Frequency of LA2D=linEN(a22l)-linEN(a22r) in 22 patients with schizophrenia

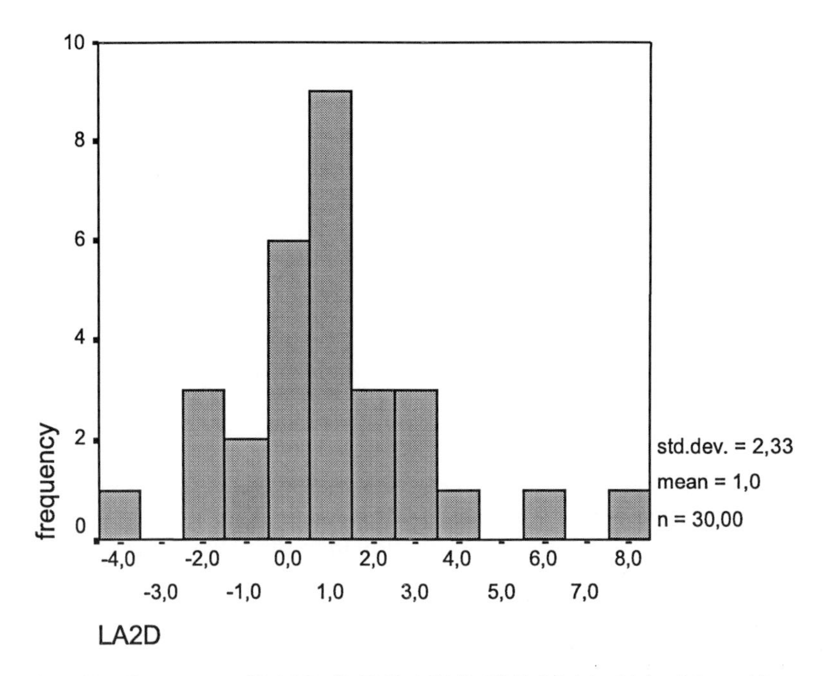

Fig. 107. Frequency of LA2D=linEN(a22l)-linEN(a22r) in 30 healthy subjects

The symmetry of cyclical pathways in patients with schizophrenia

The cyclical pathways are not as symmetrical in patients with schizophrenia as in healthy subjects. As an example, the relationship between cycEN(a11r) and cycEN(a11l) is used to compare the cyclical pathways.

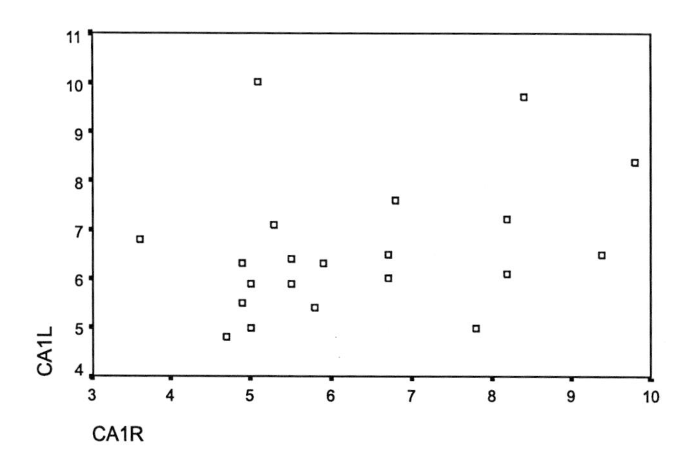

Fig. 108. Scatterplot between CA1R=cyclical pathway of a11r and CA1L=cyclical pathway of a11l. The Pearson correlation coefficient is r=0.338, there is no significant correlation

The symmetry of elementary times in patients with schizophrenia

In contrast to the healthy control group, patients with schizophrenia show an asymmetry of auditory elementary times. The visual elementary times remain symmetrical (see figures below).

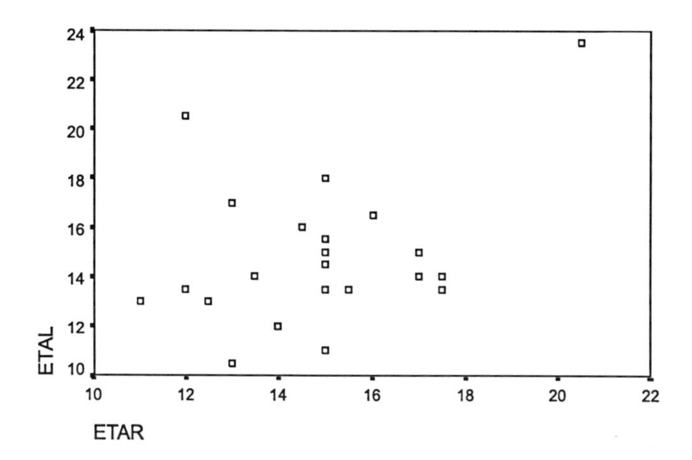

Fig. 109. Relatonship between ETAR=ET(aNNr) and ETAL=ET(aNNl) in 22 patients with schizophrenia. The Pearson correlation coefficient is r=0.354, there is no significanct correlation

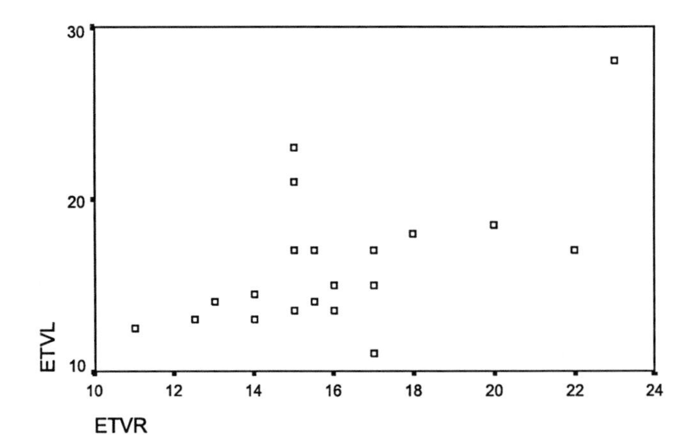

Fig. 110. Relatonship between ETH30=ET(vNNr) and ETVL=ET(vNNl) in 22 patients with schizophrenia. The Pearson correlation coefficient is $r=0.571$, the significance level is $p<0.01$

The increased elementary times do not correlate with the prolonged pathways as the following figure shows.

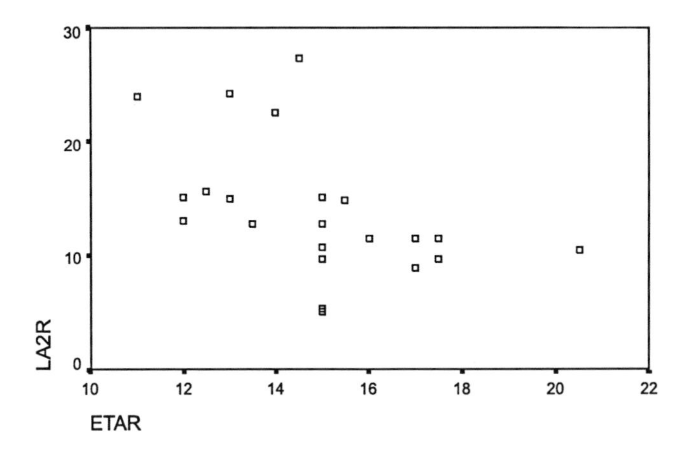

Fig. 111. Relationship between ETAR = ET(aNNr) and LA2R = linEN(a22r) in 22 patients with schizophrenia. The Pearson correlation coefficient is $r=-0.491$, the significance level is $p<0.05$

The increased number of patients with asymmetrical auditory elementary times is shown by the next two figures. Note that the possible error in healthy subjects lies within the range of ± 2ms.

The visual and auditory elementary times of the patients with schizophrenia lie within the error range of ±2ms (see next figure).

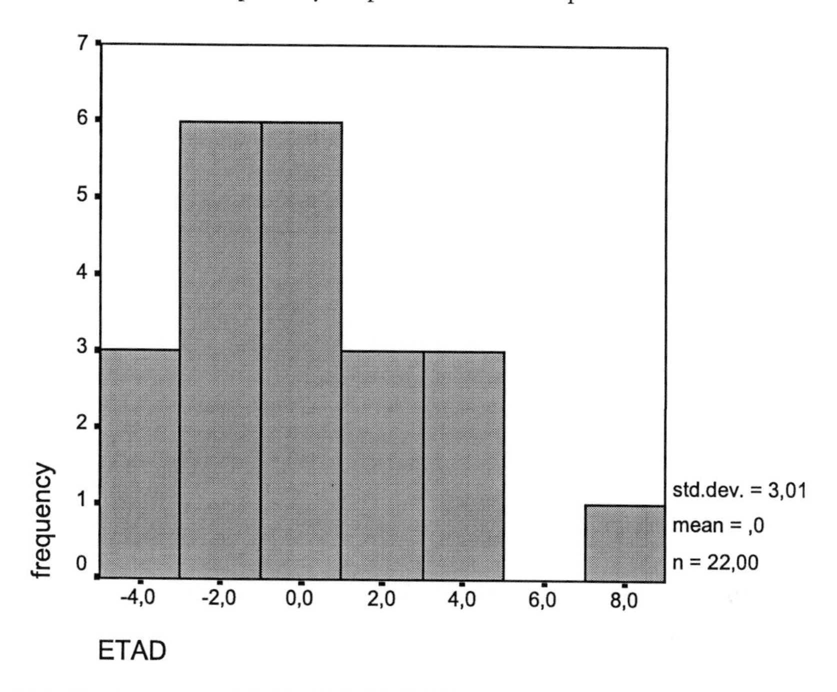

Fig. 112. The frequency of ETAD=ET(aNNl)-ET(aNNr) in 22 patients with schizophrenia

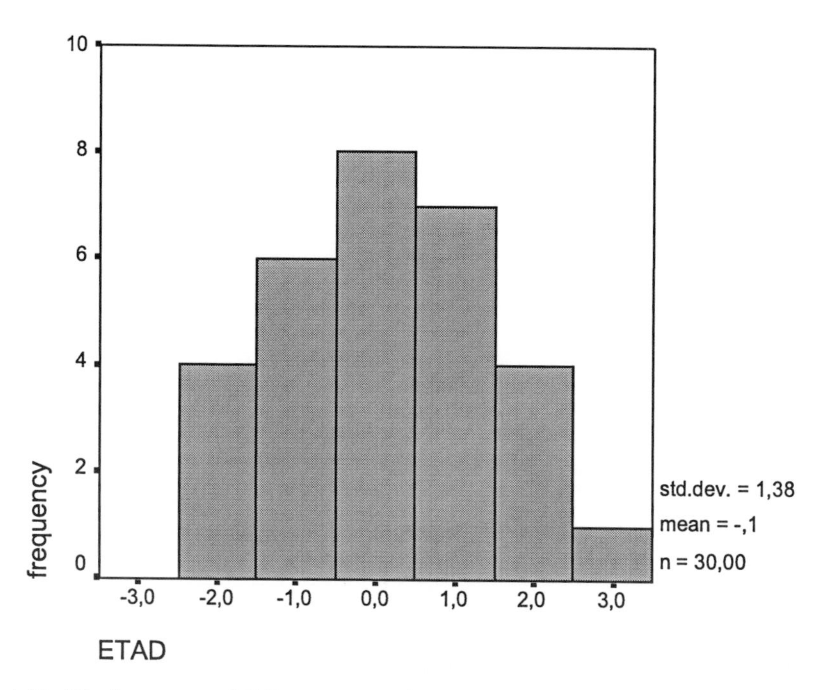

Fig. 113. The frequency of ETAD=ET(aNNl)-ET(aNNr) in 30 healthy subjects. Note the different scale of the x-axis in the above two figures

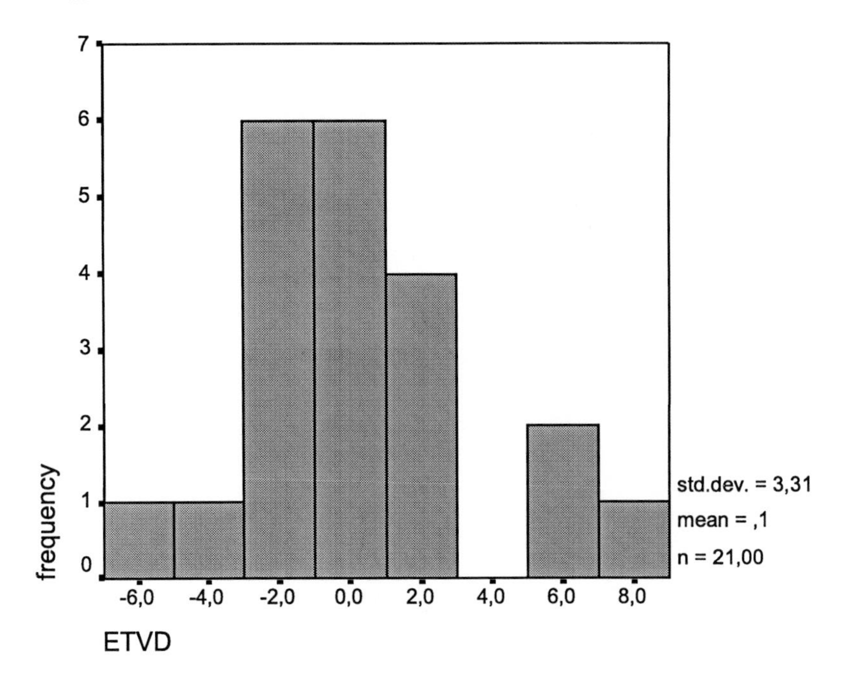

Fig. 114. The frequency of ETVD=ET(vNNl)-ET(vNNr) in 22 patients with schizophrenia

The changes of linear pathways in repeated tasks

Some of the patients with schizophrenia were ready to repeat the reaction tasks. Because of the small number of repeats no general statements can be made. Some

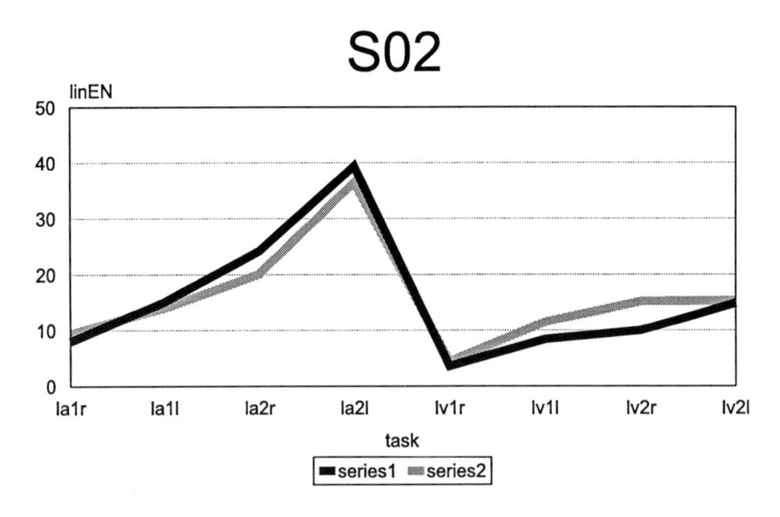

Fig. 115. Two series of the tasks a11r to v22l repeated by the patient S02

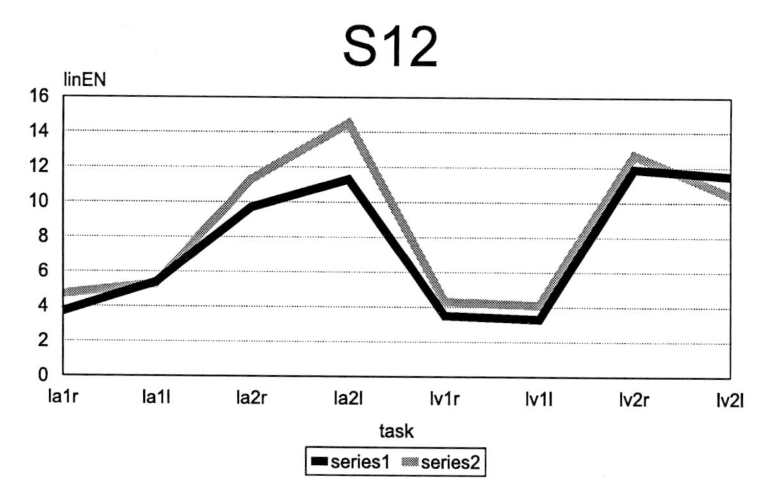

Fig. 116. Two series of the tasks a11r to v22l repeated by the patient S12

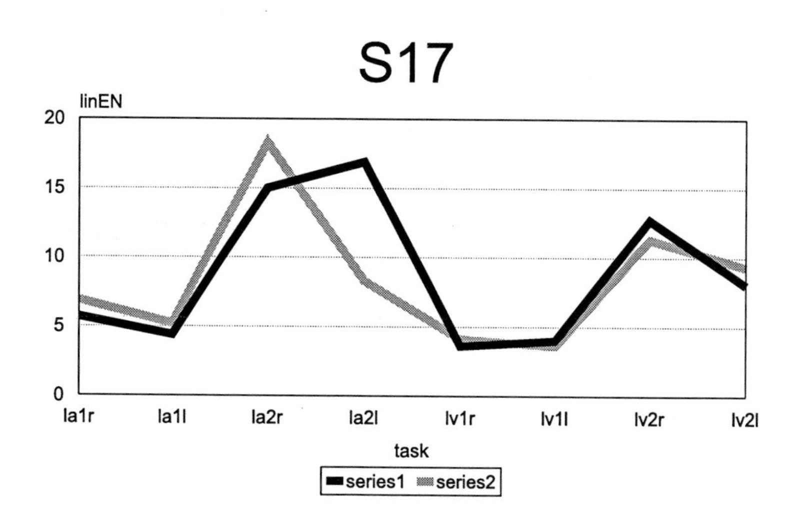

Fig. 117. Two series of the tasks a11r to v22l repeated by the patient S17

patients showed a striking stability in the lengths of the linear pathways (eg S02), some show a homogenous translation (eg S12), and some show an unusual reversal of the asymmetry between linEN(a22r) and linEN(a22l) (eg S17).

Conclusions

What can we learn about schizophrenia from these findings?

Frequency of elementary times

The reason for the occurrence of longer elementary times in some subjects is not known. As has been said earlier, the elementary time is the time an activated neuron (or set of neurons) needs in order to activate its successor neuron (or set of neurons). This time may be caused by the spreading of the action potential along the axon, the crossing of the synaptical cleft at the target neuron, the spreading of the postsynpatic potentials along the dendrites and the cell body of the target cell until the new action potential is generated at the hillock of the succeeding neuron.

Where in this chain of events the slowing down occurs is not known (is not subject of this work).

Length of the linear pathway

The length of the linear pathway is determined during the instruction phase. In this phase, the undifferentiated precursor memory set gets the structure of the mature task set. The mechanisms of this structuring processes are highly hypothetical. The basic elements, out of which the task elements are selected, may have the ability to activate some top elements. In this way, a number of competing memory sets is activated. Their top elements compete with each other in order to select one memory set which becomes the task set.

The available (existing) architecture of memory sets, the neuromodulatory influence, and the externally activated basic elements determine the selection of the memory set which becomes the task set.

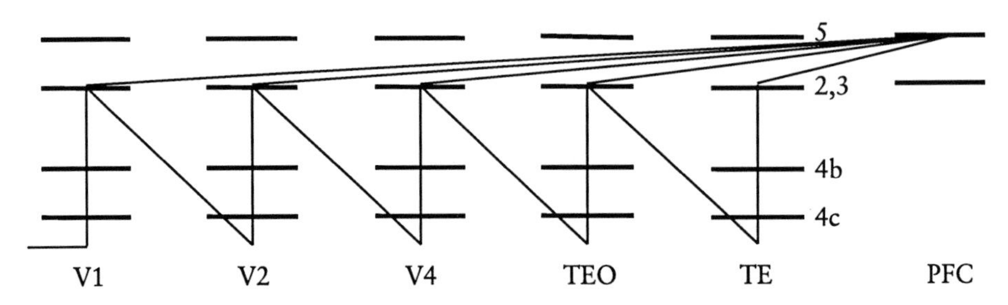

Fig. 118. The cortical structure of the sensory part of the visual pathway. The sequence of sensory areas has been taken from Reid (1999)

Each of the sensory areas may provide the stimulus element for the mature memory set (=task set). The available pre-existing memory sets compete with each other in order to select one successful memory set which becomes the task set. This competition may be influenced by neuromodulators (norepinephrine within the sensory areas and dopamine within the prefrontal cortex). If the concentration of the neuromodulators is high, short pathways are prefered, guaranteeing a fast reaction in emergency situations. If the concentration of the neuro-modulators is low, longer pathways are permitted, providing the chance of unusual creative behavior. In healthy subjects the transitions between these states may be seen by comparing the lengths of linear pathways of different series.

Some patients with schizophrenia do not use these standard pathways but use non-standard pathways (relative to their length). Either the neuromodulators are very low in these patients or they have lost their standard memory sets.

The mature memory set (=task set) freezes the length of its linear pathway and dominates the future trials of a task. If the patients would forget the selected task set, they should have the chance of selecting a task set with a shorter linear pathway. This is not the case. In these patients the task set with the long linear pathway dominates the subsequent trials of the task.

The notion of a low neuromodulatory level is contradicted by investigations which found an even higher level of neuromodulators. This finding would rather support the idea of lacking memory sets. Then the high neuromodulator level would be an attempt to compensate this lack. But because of the non-existence of standard memory sets, the high neuromodulatory level would have consequences on the non-standard pathways.

If a patient must use non-standard pathways he or she cannot experience a realistic view of the world, but has to experience it by areas distant to the sensory areas, associated with unusual emotions and thoughts (this may be experienced as paranoia etc.).

Length of the cyclical pathway

The cyclical pathways are increased in the tasks x22y analogous to the linear pathways.

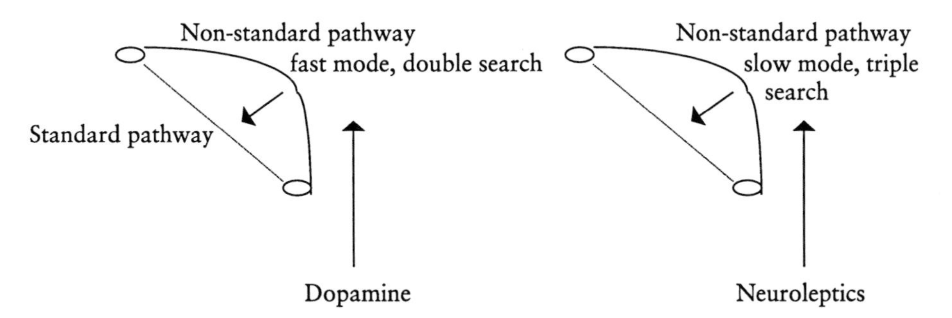

Fig. 119. Situation provided by deficiency or loss of the standard pathways and autocompensation by neuromodulators which generate fast mode and double search. Giving neuroleptics eventually retrieves the standard pathway (if it is only deficient). If this is not possible the therapy abolishes fast mode and double search by reducing the strength of the non-standard pathways

The neuromodulators try to move the pathway back to the standard. If this is not possible, the neuromodulators influence the non-standard pathways for example by reducing their cycEN to fast mode and double search. That means the increased strenght of the non-standard relations provide an increased implicit learning effect. The neuroleptics prevent this auto-compensation and by the low strength of the non-standard pathways slow mode and triple search return.

If this is true patients without neuroleptics should show shorter cycEN! This should be investigated.

Symmetry between the linear pathways

The linear pathways of patients with schizophrenia are not as symmetrical as those of healthy subjects. The high symmetry of the precursor memory sets on each side and the symmetry of the neuromodulatory level guarantee the high symmetry of linear pathways in this group. In some patients with schizophrenia, this symmetry is disturbed either by functional or by structural losses. Is there a preference of the aberrant pathway of one hemisphere? In the above tables, comparing the linear pathways and the cyclical pathways of healty subjects and patients with schizophrenia, the patients are more aberrant in the left-handed tasks (see above). In the scatterplot linEN(a22r) against linEN(a22l) the linear pathway of the a22l lies in most cases *above* the linear pathway of a22r.

Symmetry between the cyclical pathways

There is no clear symmetry between the cyclical pathways. Either an asymmetrical neuromodulatory level or an asymmetrical architecture of the available memory sets may be the reason. If this loss is asymmetrical on both sides, the prolongation of linEN may be accompanied by a prolongation of cycEN.

But how can we explain the fact, that some tasks show an increased linEN and some other tasks an increased cycEN?

Symmetry between the elementary times

The asymmetry of auditory elementary times and the symmetry of visual elementary times are an exciting proof to the involvement of auditory pathways in the symptoms of schizophrenia (auditory hallucinations are more frequent than visual hallucinations in schizophrenia). If one looks into the equations of these patients, one sees that the tasks with increased elementary times are not the tasks with prolonged linear and cyclical pathways. This lack of correlation is confirmed by the above figure. Why do the tasks show either an increased elementary time or a prolonged linear pathway? Because these two facts exclude each other: either the pathway is slowed down or it is interrupted. The slowing down can be seen as the preliminary stage of the interruption.

The interruption of the standard pathway confirms the historical naming of "schizophrenia" by E.Bleuler. If the asymmetry of elementary times is a preliminary stage of the disease this could be harnessed to protect these individuals from

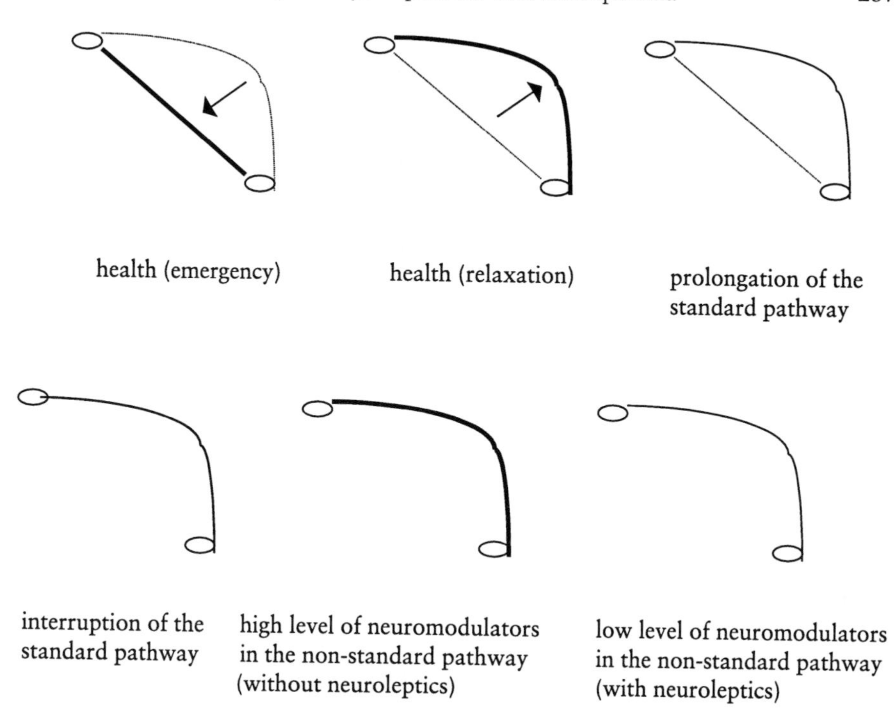

Fig. 120. Deformation of pathways in schizophrenia

becoming schizophrenic. This asymmetry may vanish if the disease progresses and the pathway is interrupted. Then the non-standard pathway has again a symmetrical elementary time.

The changes of linear pathways in repeated tasks

The patient which deviates mostly from the changes in healthy subjects is S17 because the asymmetry between linEN(a22l) and linEN(a22r) is reversed. This is also observable in patient S10. It shows that in some patients the prolonged linear pathways may become shortened in the repeated task and vice versa. Ths can only be the case if the connection is *not interrupted but is deficient* with fluctuating functional losses or gains. This may be the next intermediate state between the asymmetry of elementary times and the irreversible loss of standard pathways. If a patient shows the same prolongation of linEN in all repetitions of tasks, one may assume that he or she has lost its standard pathways (eg patient S02 shows relative constant linEN in the repetition).

Notes

History

The increase of reaction times in patients with schizophrenia is known for a long time. The explanations for this observation have been divergent.

In this work, the auditory tasks of patients with schizophrenia showed some increase of reaction time. But not all patients had this finding and some healthy subjects showed it too (H31A, O16A). This situation is typical for findings in patients with schizophrenia.

Reaction times left of the first peak in patients with prolonged linear pathway

The first observation was made in a22lO16A. There the reaction times left of the first peak have a distance of about 3 elementary times to each other. This would suggest that in each case one area less is used. This gives evidence for a certain variation of linEN but with a clear maximum.

Critical considerations

General consideration for all three groups

The present standard

The present standard design comprises the tasks a11r(100 trials), a22r(200 trials), a11l(100 trials), a22l (200 trials), v11r(100 trials), v22r(200 trials), v11l(100 trials), v22l (200 trials), The results are evaluated by the four programs: FPM31e (x11y), FPM31e(x22y), FPM26f58(x11y), FPM26f58(x22y). Additionally the programs SING106n and SING104r are applied but not used to calculate the elementary times

The results of these programs are combined within the convergence table and the four elementary times are selected by the help of agreements.

With the help of the elementary times and the reaction time distributions the lengths of the linear and the cyclical pathways are calculated.

The results are registered into the eight equations of the eight pathways.

Each subject repeats this design at a second time.

Some subjects are asked to repeat the tasks when an event-related potential is recorded.

Reproducible elementary times are the basis of all

The evaluation programs recognizes the peaks automatically. This is the crucial moment of all further evaluations. If the false peaks are selected, the elementary times cannot be correct. Therefore the recognition of a peak should not depend on one trial.

Special pathways

ET(v11lH30B)=15ms both in chronophoresis and FPM. But in v11lH30B (Fp-con)=40ms. That means there is no place for at least 4 elementary times (2 linET and 2 cycET) to build the minimal possible pathway. This problem may be hit in other patients too. The task v11rH06A showed this low (Fp-con) too.

After 5ET, the distribution of v11rH30A stops. The later peaks have to be explained by cyclical pathways, so we see a mixture of purely linear variants and linear/cyclical mixtures.Because the fastest cyclical pathway takes 4ET (2linET+2cycET), all trials with less than 4 ET have to be generated by pure linear variants (or by variants with shorter constant times).

There are three possible solutions to this problem: either ET=15 is wrong or the constant is smaller or shorter pathways are possible. The next figures show pathways which could fulfil this condition.

There are other reasons why the common linear pathway principle is not fulfilled absolutely. The occurrence of peaks at each elementary time (not only every two elementary times) indicates that linear variations must be present.

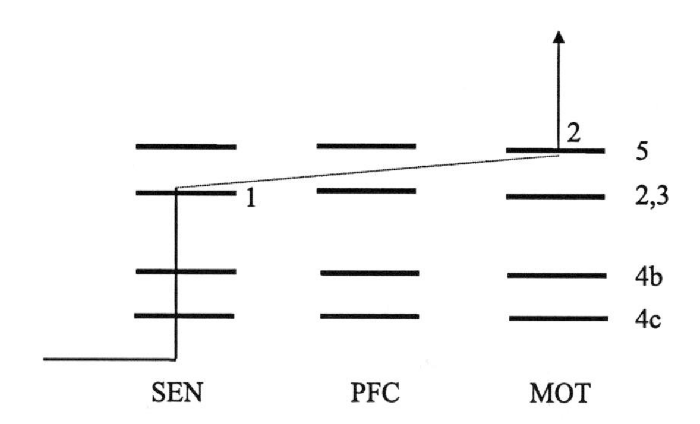

Fig. 121. A pathway using only two elementary times in two areas (purely linear variant)

Such a shortcut is only conceivable in tasks x11y where no decision is necessary. In tasks x22y the pathway has to go through the PFC.

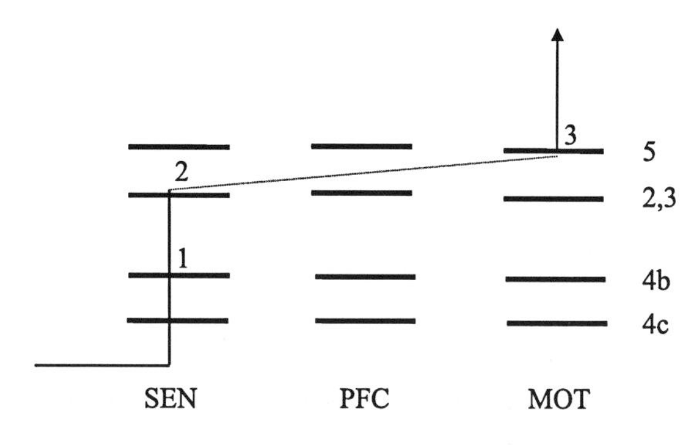

Fig. 122. A pathway using three elementary times in two areas (purely linear variant)

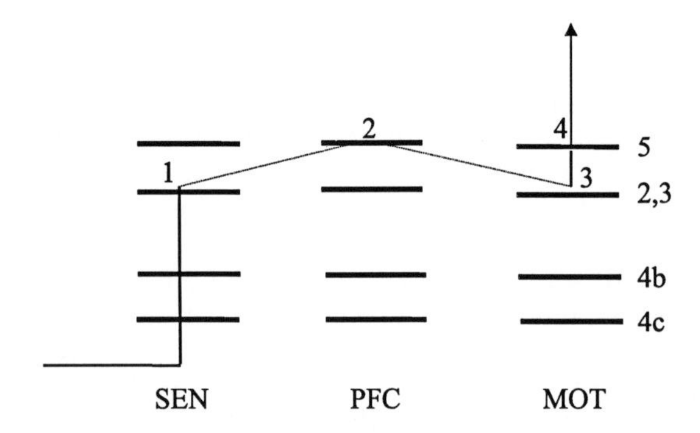

Fig. 123. A pathway using four elementary times in three areas (standard pathway using a combination of linear and cyclical pathway, with one local shortcurt in the sensory area)

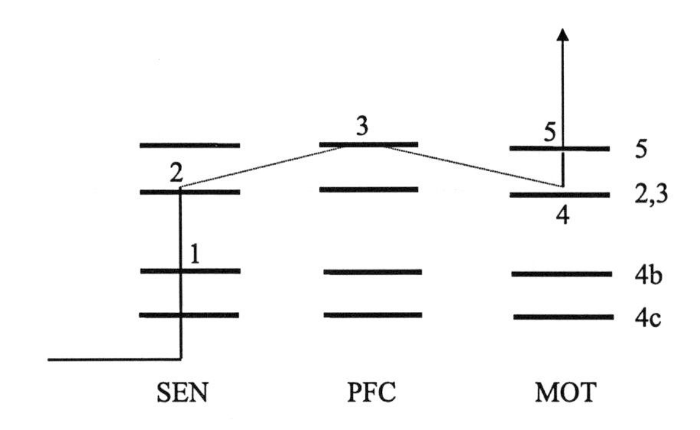

Fig. 124. Standard pathway running through three areas

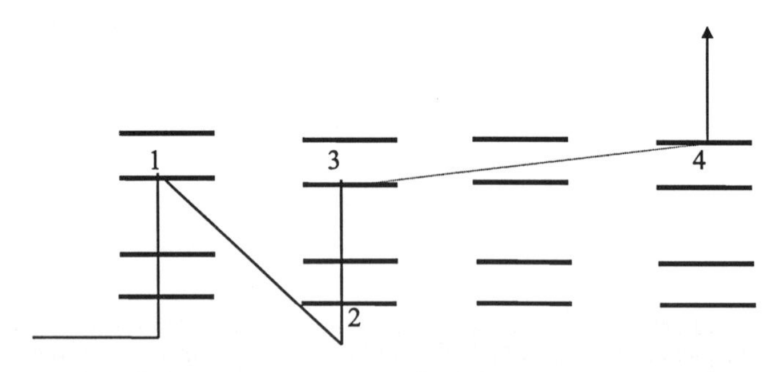

Fig. 125. Purely linear variation using four elementary times in four areas

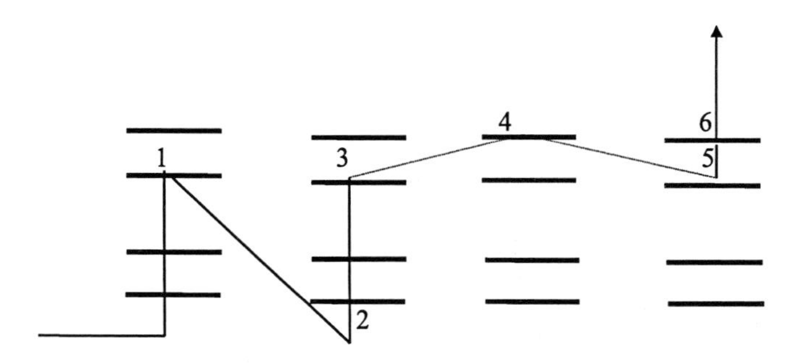

Fig. 126. Standard pathway with two local shortcuts in the sensory areas and minimal cyclical pathway summing up to 6ETs

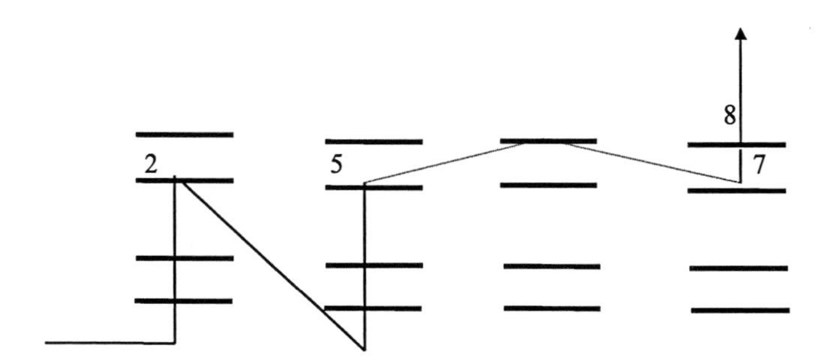

Fig. 127. Standard pathway with 8ETs in four areas without shortcuts but a minimal cyclical pathway of length 2cycET

Competitive stimulus elements in the instruction phase

The insertion of the PFC on every area of the ventral visual pathway remains to be shown. This means, one has a set of stimulus elements and it has to be answered why the searching set contains only *one* stimulus element. Why does the task set not contain *all* these stimulus elements?

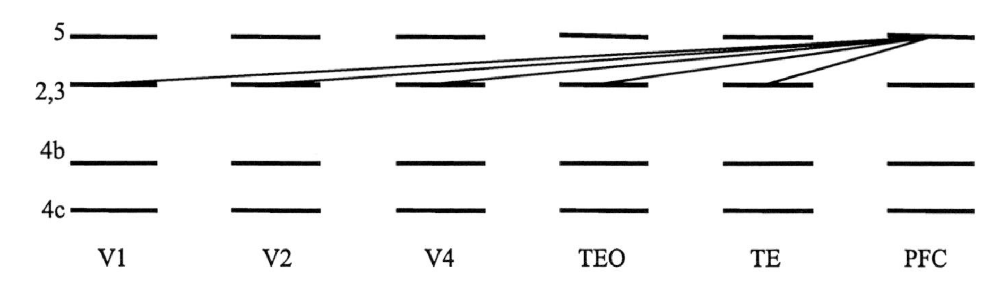

Fig. 128. The auditory sensory areas of the ventral stream. Only a subset of relations is shown. The sequence V1, V2, V4, TEO, and TE may be continued by the hippocampus (declarative memory) and the limbic system (emotional memory)

From all these stimulus elements, different task sets could be constructed with the response element being the same in all cases. These task sets could be active spontaneously in a parallel, asynchronous manner and compete with each other for full activity. The rationale behind these multitude of competing task sets is the fact that they cannot be so extensive as to cover all sensory areas. The result of the competition establishes the length of the linear pathway. And the relative constancy of this length within a certain period implies that the task set which has won the competition has a good chance of winning it again. This is the issue of simultaneous searching sets which can be met in reaction tasks with more than 6 or 7 alternative stimuli. Obviously this principles plays an important role even in tasks of the x11y structure.

Which factors influence the selection of one of the parallel task sets when the stimulus arrives and activates its stimulus element?

Remember: the activation of the task set implies the activation of the top element and the beginning of the alternative activity of either the top element or some of the basic elements of the task set, ending with the activation of the response element.

That task set which succeeds first in hitting the stimulus element activates it first from the interior side in the correct phase relation and wins the competition.

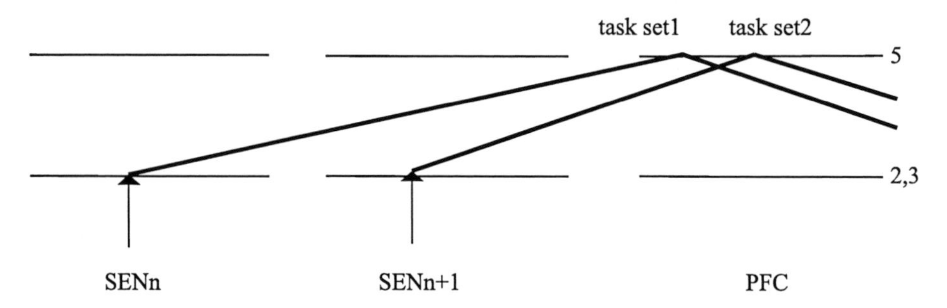

Fig. 129. The task set which succeeds first in hitting the stimulus element activates it first

The early sensory areas show an advantage because the task sets attached to them may start earlier to hit the corresponding stimulus element.

Does the length of different linear pathways (that means the dominant task sets) change in the same manner when replicated. This would support non-random influences, otherwise the selection would be purely accidental. The relative constancy of linEN would be explained by: the task set which wins the competition first becomes dominant and wins all subsequent competitions. The relative, chronologically limited constancy of linEN would therefore be a learning effect.

Another view of the preparing processes states that there is only one large task set at the beginning and the accidental activation of one of the stimulus elements creates the dominant task set. Here, too, the stimulus elements of the early sensory areas exhibit an advantage.

Once selected, the relation between the task set and the stimulus element remains stable for some time. That means the larger searching set of the beginning is reduced to the mature task set which dominates the subsequent trials.

If one extends this principles of competing task sets to different response elements the multitude becomes even larger.

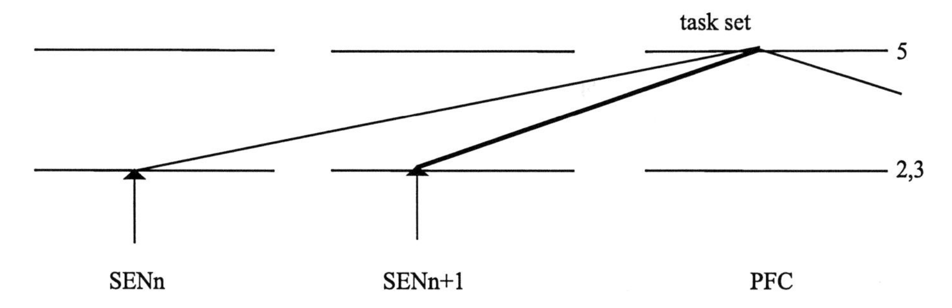

Fig. 130. There is one large task set from which the mature task set is cut out

The healthy subjects which have an extended linear pathway without having the symptoms of schizophrenia prove that this extension alone, at least in the range of the healthy subjects, is not pathological. Perhaps other disorders (eg. an increased contrast) have to be present in order to cause the typical symptoms.

Perhaps the difference between the healthy and the mentally ill lies in the probability of selecting the late sensory areas. Healthy subjects can select late areas and patients with schizophrenia can select early sensory areas. But the probability is different: healthy subjects select early areas more often and patients late areas more often.

The probability theory is contradicted by the fact that the healthy subject H31A replicates its selection of late areas. To answer these questions, observations of a sequence of linEN are necessary.

Longer reaction times at the beginning of a task would support the idea that at first there is a larger searching set with many stimulus elements in different sensory areas – which will be reduced in course of the task.

Attention and the length of the linear pathway (linEN)

Does attention determine the length of the linear part of the stimulus-response pathways?

Subject H22B eg maintained a high attention in series B. The length of the linear pathway was linEN(a11lH22B) = 5.

One could plan series with distractions (hearing in visual tasks and pictures on the screen in auditory tasks) and measure the attention with a self-rating scale (10 items from unattentive to fully attentive).

Common linear pathway principle

If this principle is violated, the distribution of reaction times changes dramatically. The uniformitiy is lost in favor of some widely distributed times.

Alternative writing of the equations

The equations describing the spatio-temporal structure of the stimulus-response pathways could be written:

a11rNNA1 = (12 + 2 (2 + §2))15 + *70*
v22rNNA1 = *50* + (6 + 2 (4+1*4))14 + *70*

That means, the auditory input time is nearly 0, the auditory output time is about 70ms. The visual input time is about 50 ms and the visual output time 70 ms.

Open questions

Why is the linear pathway of the x33y task so long? What happens during this time? A linear progression is not very plausible. What alternative courses are imaginable?

Why do fractional multiples of elementary time arise in FPM26f58? It is the occurrence of these fractional values that impair the NESTLE results of the various elementary times. Are these peaks created by chance or do they convey some unknown meaning?

What are the evidences that indicate the occurence of minimal linear pathway in all pathways of a task. Perhaps longer reaction times posses longer linear pathways? The equations of median pathways propose that these pathways contain the the minimal linear pathway (common linear pathway principle).

What are the elementary times of the middle finger? The x22y tasks contain approximately 100 trials where the middle finger presses the response key. Can these results be used to compute the elementary times of the two middle fingers? Is there a difference to the elementary times of the index fingers?

Do women have more symmetrical elementary times than men (H32A)?

Is the uniformity principle ET(x11y) = ET(x22y) a sign of psychological stability?

Is the asymmetry ET(x11y) $\neq$ ET(x22y) a sign of vulnerability?

The elementary times and the pathways of healthy subjects

The most important findings in this group were:

- The length of the linear pathway being approximately a multiple of 3.
- The preference of certain lengths of the cyclical pathway corresponding to the different search strategies (mode, number of searches).
- The common linear pathway principle, the implicit learning principle, and the symmetry between the hemispheres when the ET, linEN, and cycEN are investigated.

The pathways of patients with monohemispheric brain lesions

The lengths of the linear pathways of subject O16A are astonishing. The length does not depend on the side of the brain lesion and the linEN is a multiple of three. Why does subject O16A exhibit such a long linear pathway at each hemisphere?

An alternative to these prolongation of the linear portions is the prolongation of the minimal cyclical pathway. Trials with mincycEN = 2 are not possible any more. The searching set has difficulties to activate the stimulus element because of the asynchrony between searching set and linear activities or because of the decreased activity exerted by the searching set on the basic elements.

An impaired hearing for example as the consequence of chemotherapy was not examined in these patients.

It remains unsolved why the elementary times of patients with monohemispheric brain lesion are increased in the contralateral tasks. The question whether the linEN/cycEN are increased due to the lesion has not been answered and one unexpected finding namely the symmetrical increase of linEN and cycEN, independent from the lesion, remains unexplained. Future research is necessary to answer these important questions.

The pathways of patients with schizophrenia

The prolongation of the linear pathway (standard and non-standard pathways in patients with schizophrenia)

A very long linear pathway suggests an insufficiency of the neural mechanisms working during the instruction phase. The stimulus elements of the early sensory areas have the advantage of being discovered by the top element at first. This may lead to the stabilization of this connection as part of the task set. If this sequence is disturbed, the stimulus elements of the later sensory areas get their chance.If the interplay between the early stimulus element and the PFC causes no integration of the stimulus element into the task set as in healthy subjects, the variability is increased: many different stimulus elements may be selected, implying many different lengths of linear pathways.

The activation of the (early) stimulus element by the top element of the PFC does not start the task set. This prevents integration of the (early) stimulus element into the task set as is said above but delays the length of the linear pathway. The cause of this non-start could be an insufficiency of the top element or an asynchrony between the top element and the stimulus element or an increased threshold of the task set (respectively the sensorimotor set SMS).

The non-start up opens the door for the later stimulus elements being activated with their additional meanings (and dangers). If the advantage of time cannot be used, then all areas should have the same probability of being read by the prefrontal cortex. But patients with schizophrenia prefer the late sensory areas. Therefore these areas must have an unknown advantage. Perhaps the connections between the early sensory elements and the memory sets are not present any more (either functional or structural). In this case the later sensory elements have to substitute for the earlier elements.

An additional interesting observation can be made. The tasks with prolonged linear pathway have triple search and slow mode. Triple search could suggest an insufficiency of the SMS

The reason for this non start-up cannot be found by the methods of this work. Only an increased linEN or cycEN can be found. The boundary between linEN/cycEN which would be shifted to the right in patients with schizophrenia should be seen in event related potentials too.

The replication of results in patients with schizophrenia

The replication of results is the basis of the whole project. The replication of chronophoresis can be seen for example in S08A and S08B, S15A and S15B, SING106n(x22yS12A) and SING106n(x11yS12B). This supports the notion that factors (keyboard, screen) that may disturb the measurement of elementary times are being compensated.

The independent replication of the results of one series reveals the mistakes of one single evaluation process.

State and trait marker

S13A and S03A show trait marker (asynchrony of ET) and no state marker (prolonged linEN).

The prolonged linEN may be a state marker because it vanishes in some subjects.There is a third phenomenon: the violation of the "common linear pathway principle". The a22yS03A and a22yS13A support the idea that the common linear pathway is not as strongly present as in healthy subjects. Even a multiple of 3 only shows that these reaction times contain that linear pathway while other reaction times contain other linear pathways. This third phenomenon is strongly supported by these two patients.

The asymmetry of elementary times and the prolongation of the linear pathway in patients with schizophrenia

Is there a relation between the asymmetry of elementary times and the prolongation of the linear pathway in these patients? The pathways in both hemispheres run parallel to each other. Perhaps they need to wait until there is coincidence in some late area.

Neuroleptics and the prolongation of the linear pathway

Dopamine increases the contrast in the prefrontal cortex, norepinephrine in the posterior cortex. This means that dopamine prefers the standard pathways and abolishes the non-standard ones (emergency reaction). If the standard pathways are not available any more, an increase of dopamine changes the quality of the non-standard pathways: they become stronger and the danger is that they become uncontrollable (by the standard pathways). Neuroleptics defuse this situation by taking away their power.

It is unclear whether neuroleptics affect the reason of this development: the insufficiency of the standard pathways. In this case they would reverse the prolongation of the linear pathways.

The x33y pathway of patients with schizophrenia

It is important that in the distribution of x33y, only trials with the index finger are present.

The x33y pathways of healthy subjects and patients with schizophrenia should be compared systematically. Perhaps both groups differ in these pathways more than in the x22y pathways.

The most important findings in patients with schizophrenia

The most important findings in patients with schizophrenia are

- The asymmetry of auditory elementary times contrary to the relative symmetry of visual elementary times with no general advantage of one hemisphere. This asymmetry extends to the lengths of the linear and cyclical pathways.
- The prolongation of the auditory (and lesser the visual) linear and cyclical pathways.

Future research

These further studies are proposed:

- Parents of patients with schizophrenia (100 x11y, 200 x22y).
- Patients with schizophrenia (100 x11y, 200 x22y, *300 x33y*).
- Patients with schizophrenia before and after therapy with neuroleptics (100 x11y, 200 x22y).
- Patients with monohemispheric brain lesions (100 x11y, 200 x22y).
- Patients with dementia (100 x11y, 200 x22y).

Part IV

Critical evaluation of the results and the model

Confirmation of elementary times and pathway structure by event-related potentials

The correspondence between reaction time data (ET, linEN) and event-related potentials (latencies) in the a22y pathways of healthy subjects

Method

Design

The following investigation used the VIKING instrument of Nicolet Biomedical. The Department of Neurophysiology at the University Erlangen-Nuremberg and the company itself made it available for us. During the performance of auditory tasks, event-related potentials were recorded and then evaluated.

The design of the ERP study:

a10r (tone only)
a11r (index finger)
a22r (index finger, middle finger)

a10l (tone only)
a11l (index finger)
a22l (index finger, middle finger)

Differences between task and control:

(a11r-a10r)
(a22r – a10r)

(a11l-a10l)
(a22l-a10l)

The rate of the auditory stimulus was 0.5 Hz (this corresponds to an interstimulus time of 2sec). The low tone of 500Hz appeared in 55% of all trials (frequent tone) and required a response by the index finger. The high tone of 1.5kHz appeared in 45% of all trials (rare tone) and required a response by the middle finger. Only the results of the index finger were taken for comparison with the reaction time data where only the index finger had been evaluated. The electrode A1 was attached to the right ear, the electrode A2 to the left ear. Finally two electrodes were placed at a central (Cz) and parietal (Pz) position.

Therefore the activity of the contralateral hemisphere in the task a22r(index finger) can be seen best in the registration F: A2-Cz or F:A2-Pz and for the task a22l (index finger) in the registrations F:A1-Cz or F:A2-Pz. In the data, the upward potentials are negative, the downward potentials positive.

Finally, the reaction time data and the ERP-data are correlated with each other. It is convenient to take these reaction time data which lie closest to the ERP data chronologically. It is known that the length of the linear pathway varies from series to series.Therefore, the second series of reaction times data was used for the comparison because the ERP task is a repetition of reaction tasks ie. a third series.

Principally, the moment of the ERP and that of the task should lie close together. For example a22rH11A with linEN = 9 lies chronologically closer to H11r#2 than a22rH11B with linEN = 14. Therefore I take linEN = 9 to compare it with the event-related potentials (eg. FINPFC).

The comparison between the two data sets has two main objectives. First the elementary time of the reaction time data should be replicable by the ERP data and second the common linear pathway should be seen in the ERP data. The first objective can be reached by taking the latencies as the input of the NESTLE procedure. The second objective can be reached by attaching the latencies to the hypothetical structure of the cortex. The boundary between linear pathway and cyclical pathway would be marked by the transition from 3 step rhythm of the linear pathway to the 2 step rhythm of the searching set.

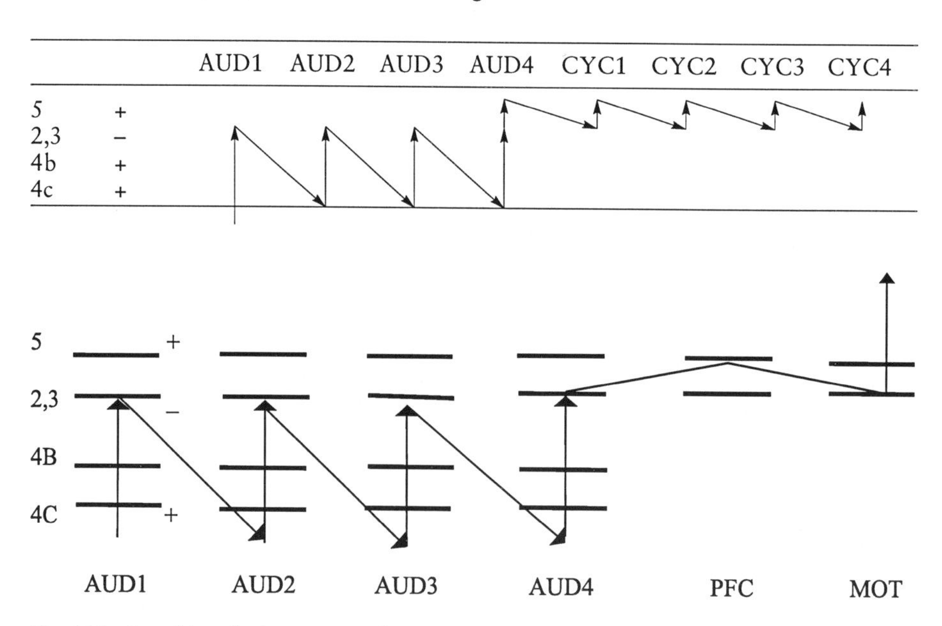

Fig. 131. Attaching the latencies to the hypothetical structure of the linear and cyclical pathway

*Hypothesis: FINPFC = IT + (linEN – 1) * ET*

IT = Input time. This period is nearly 0 in auditory tasks. For this reason it can be omitted in these tasks.

FINPFC = First negativity in PFC, (linEN – 1) is linEN without the relation between layer 2,3 and layer 5 in the motor area.

This first hypothetical relationship between reaction time data and ERP data shall be explained by examples.

Subject H11

(linEN – motor relation)*ET
= (9 – 1) * 16
= 8 * 16
= 128.
This fits well to -P123 in H11r#2

Subject H23

a22lH23A lies closer to H23l#1 than a22lH23B. The latter lies nearly 1 year apart from H23l#1.
linEN(a22lH23A) = 11
linEN(a22lH23B) = 15

(linEN – motor relation)*ET
= (11-1)*14
= 140
FINPFC = -P147 in H23l#1

In both subjects the first positivity lies 1ET after the first negativity:

FIPPFC = FINPFC + 1ET
FIPPFC(a22rH11A) = 128 + 16 = 144. This meets + P(144)
FIPPFC(a22rH23A) = 140 + 14 = 154.

In both cases the FINPFC in the event-related potential is considerably large. This could be an indication that the same linEN occurs more frequently in successive trials (= > relative constancy of linEN).

FINPFC(a22rH23B) = (linEN – 1) * ET
= (12 – 1) * 15
= 11 * 15
= 165

If one adds 1ET, the activity is in layer 5 of the PFC = > 12*15 = 180
In H23B there is a positive peak at P176 (but it is small).

Subject H29

linEN(a22rH29A) = 9
linEN(a22rH29B) = 9
both tasks are chronologically close to H29r#3.

(linEN – motor relation)*ET
= (9 – 1)*12.5
= 100
There lies a negative peak at -P(85) in H29r#3

In the series H29B, the above equation is true, but not in the series H29A. This may be due to the fact that the ERP is similar to H29B but not to H29A because *both* H29B and ERP are replications of H29A. H29B and ERP are performed on similar times, too. They both have the same minlinEN.

The latency tables of x22y pathways

H02r#2	AUD1	A2/C1	CYC2	CYC3	CYC4	CYC5
5 +		78				
2,3 –	26	65–67	91	Artifacts ...		
4b +	13	52–57				
4c +	0	39				
,	a22r-a10r,	F:A2-Pz,				

H02l#1	AUD1	A2/C1	CYC2	CYC3	CYC4			
5 +		70–75	98	126	154			
2,3 –	14–10	56–57	84	112	140	182–180	210–205	252–255
4b +		42–42				168		238–240
4c +	0	28				154–157	196–190	224
,	a22l-a10l,	F:A1-Pz,						

Remarks:

1. There is the impression that linEN has two portions. One is before the cycEN, the other after the cycEN.

$a22rH02A = 70 + (6 + 2(4 + 1*4))13$ $6.8 + 14.5$
$a22lH02A = 70 + (7 + 2(4 + 2*4))14.5$ $7 + 15.2$
$a22rH02B = 70 + (12 + 2(4 + 1*4))13$ $11.5 + 12.7$
$a22lH02B = 70 + (12 + 2(4 + 1*4))14$ $11.9 + 12.8$

H03r#1	AUD1	AUD2	AUD3	A4/C1	CYC2	CYC3	CYC4
5 +							
2,3 –	28–25	84–80	126	168–160	196–200	228	256–260
4b +	14	56	112	154			
4c +	0	42	98	140–140			
a22rH03B = 70 + (14 + 2(4 + 2*4))14,					a22r–a10r,		F:A2-Pz,

Remarks:

1. linEN = 11 + 1

H03l#2	AUD1	AUD2	AUD3	A4/C1	CYC2	CYC3
5 +				150	180	210
2,3 –	15	45–45	90	135	165–160	195–200
4b +			75	120		
4c +	0	30	60	105–100		

a22lH03B = 70 + (11 + 2(4 + 2*4))15, a22l-a10l, F:A1-Pz,

Remarks:
1. linEN = 9 + 1

H05B#5r	AUD1	A2/C1	CYC2	CYC3	CYC4	CYC5	CYC6
5 +		84	112	140	168	196	224
2,3 –	28–35	70	98–95	126–125	154	182	210–200
4b +	14–20	56–60					
4c +	0	42					

a22rH05B = 70 + (9 + 2(4 + 1*4))14, a22r-a10r, F:A2-Pz,

H05B#4l	AUD1	AUD2	AUD3	AUD4	A5/C1	CYC2	CYC3	CYC4
5 +								
2,3 –	13	52–50	91	130–135	169–170	197–195	225	253
4b +		39	78	117	156			
4c +	0	26	65	104–102	143			

a22lH05B = 70 + (9 + 2(4 + 1*4))13, a22l-a10l, F:A1-Cz,

Remarks:
1. The linEN in H05B#4l has the same length as in a22lH05A.
   a22rH05A = 70 + (9 + 2(4 + 1*4))14
   a22lH05A = 70 + (15 + 2(4 + 1*4))12.5

2. It is striking how the program SPRING replicates the elementary times known from the methods of FPM.

H07r#2	AUD1	AUD2	AUD3	AUD4	A5/C1	CYC2	CYC3	CYC4
5 +					238	272–275	306	340–340
2,3 –	34–27	68–75	119–112	170	221–217	255	289	323
4b +			102	153	204			
4c +	0	51–50	85–90	136	187–180			

a22rH07A = 70 + (12 + 2(4 + 2))18, a22r-a10r, F:A2-Pz,

Remarks:
1. The next latencies are: 357, 374–*370*, 391–*400,* 408
2. There is no correlation between 340–340.

H07l#1		AUD1	AUD2	A3/C1	CYC2	CYC3	CYC4		
5	+			128	160	192			
2,3	–	16	64–72	112	144–140	176	208–205	256–253	304–300
4b	+		48	96–102				240	288
4c	+	0	32–40	80				224–222	272–280

a22lH07A = 70 + (13 + 2(4 + 2))16, a22l-a10l, F:A1-Pz,

Remarks:

1. Is it possible that there is another linear pathway after the cyclical pathway? The length of the first linear pathway plus the length of the second linear pathway is 13ET which is the value of the reaction time data.

a22rH07A = 70 + (12 + 2(4 + 2))18	11.9 + 9.1
a22lH07A = 70 + (13 + 2(4 + 2))16	13 + 9.3
a22rH07B = 70 + (12 + 2(4 + 2))16	12.4 + 8.6
a22lH07B = 70 + (12 + 2(4 + 2))15	12 + 10.3

H08r#2		AUD1	AUD2	AUD3	AUD4	A5/C1	CYC2	CYC3	CYC4	
5	+					196	224–230		252	280
2,3	–	14–15	56	98	140	182–175		210	238	266
4b	+		42	84	126–120	168				
4c	+	0	28	70	112	154				

a22r H08 = 70 + (12 + 2(4 + 2))14, a22r-a10l, F:A2-Pz,

H08l#1		AUD1	AUD2	AUD3	A4/C1	CYC2	CYC3	CYC4	CYC5
5	+				188.5	217.5	246.5	275.5–285	304.5
2,3	–	29	72.5	116–125	174–185	203	232–230	261	290
4b	+	14.5	58	101.5	145				
4c	+	0	43.5	87	130,5				

a22l H08 = 70 + (10 + 2(4 + 2))14.5, a22l-a10l, F:A1-Pz,

Remarks:

1. The next latencies in layer2,3 are 319, 348, 377–*380*, etc.

a22r H08 = 70 + (12 + 2(4 + 2))14	11.9 + 8.4
a22l H08 = 70 + (10 + 2(4 + 2))14.5	10.1 + 9

H11Br#2		AUD1	AUD2	AUD3	CYC1	CYC2	CYC3	CYC4	CYC5
5	+							304	336
2,3	–	32	80–80	112–112	160–160	192–195	240–240	288–280	320–315
4b	+	16–10	64		144		224–225	272	
4c	+	0	48	96–90	128–135	176–175	208	256	

a22rH11B = 70 + (14 + 2(4 + 1*4))16, a22l-a10l, F:A1-Pz, F:A1-Cz

Remarks:
1. The next latencies after 320 are: 350–*345*, 380–*375*, 410–*410*, 440–*445* etc.
2. This times the ERP results do better fit to H11B with linEN = 14.

H11Bl#1	AUD1	AUD2	AUD3	CYC1	CYC2	CYC3	CYC4	CYC5
5 +					182	210–210	238	266
2,3 –	28	70	98	126	168–168	196	224	252
4b +	14	56			154			
4c +	0	42	84	112	140–145			

a22lH11B = 70 + (14 + 2(4 + 2*4))14, a22l-a10l, F:A1-Cz,

Remarks:
1. It is unknown where the two shortcuts are!
2. The next layer5 activity is at 294–*295*.
3. This record is impaired by 50Hz artifacts but at some latencies the peaks are so high or low that the artifacts can be neglected.

a22rH11A = 70 + (9 + 2(4 + 2*4))16	8.9 + 18.1
a22lH11A = 70 + (9 + 2(4 + 2*4))14	9.4 + 19.5
a22rH11B = 70 + (14 + 2(4 + 1*4))16	13.6 + 13.1
a22lH11B = 70 + (14 + 2(4 + 2*4))14	13.4 + 20.3

H11Cr#2	AUD1	A2/C1	CYC2	CYC3	CYC4			
5 +								
2,3 –	16–10	64–62	96–97	128	160–155	208–212	256	304–300
4b +		48–40				192	240	288
4c +	0	32–27	80	112–118	144	176	224	272

a22rH11B = 70 + (14 + 2(4 + 1*4))16, a22l-a10l, F:A1-Pz, F:A1-Cz

Remarks:
1. The pathway of a22rH11C is different from the pathway of a22rH11B. The pattern of a22rH11C is *2–3–2–2–2–3–3–*
   whereas the pattern of a22rH11B is *3–3–2–3–2–3–3–*

H11Cl#1	AUD1	AUD2	AUD3	A4/C1	CYC2	CYC3	CYC4	CYC5
5 +				126–125	154	182–188	210	238
2,3 –	14–17	42	70	112–108	140	168–173	196–202	224
4b +				98				
4c +	0	28	56	84–85				

a22lH11B = 70 + (14 + 2(4 + 2*4))14, a22l-a10l, F:A1-Cz,

Remarks:
1. The next multiple is P252–*255*
2. The pathway of a22lH11C is different from the pathway of a22lH11B. The pattern of a22lH11C is *2–2–2–3–2–2–2–*
   whereas the pattern of a22lH11B is *3–3–2–2–3–2–2–*
   There are some areas with similar pathways, some areas where the number of active layers is reduced from 3 to 2 and only one area where the opposite takes place.

H13r#2	AUD1	AUD2	AUD3	A4/C1	CYC2	CYC3	CYC4	CYC5
5 +				154	182	210	238	266
2,3 –	28–30	70–62	98–100	140–140	168–165	196–200	224	252
4b +	14–8	56		126–130				
4c +	0	42–50	84	112				
a22rH13C = 70 + (9 + 2(4 + 1*4))14,					a22l-a10l,		F:A1-Cz,	

Remarks:

1. After 252 comes 280–*280*.
2. The border between linear and cyclical pathway is difficult to draw because the shortcurt uses only 2 layers of one area. Therefore 168 could be a shortcut value or a cyclical value.

H13l#1	AUD1	AUD2	AUD3	AUD4	A5/C1	CYC2	CYC3	CYC4
5 +								
2,3 –	30–28	60–60	90–90	135–130	180	210–210	240	270
4b +	15			120	165–169			
4c +	0	45–40	75–78	105–110	150			
a22lH13C = 70 + (8 + 2(4 + 1*4))15,					a22l-a10l,		F:A1-Cz,	

Remarks:

1. The further cycles are at 300–*310*, 330, 360–*355*, 390–*382*, 420–*430*.
2. The length of the linear pathway is different in ERP and reaction time data.
3. a22rH13A = 70 + (8 + 2(4 + 1*4))14   8 + 15.8
   a22lH13A = 70 + (9 + 2(4 + 1*4))15   9.3 + 13.4
   a22rH13B = 70 + (9 + 2(4 + 1*4))14   9.8 + 12.9
   a22lH13B = 70 + (6 + 2(4 + 1*4))15   6.3 + 14.5
   a22rH13C = 70 + (9 + 2(4 + 1*4))14   8.4 + 13
   a22lH13C = 70 + (8 + 2(4 + 1*4))15   8 + *10.4*

H14r#2	AUD1	AUD2	AUD3	AUD4	A5/C1	CYC2	CYC3	CYC4
5 +					195	225	255–252	285
2,3 –	15	45	90–95	135	180	210	240	270
4b +			75	120	165			
4c +	0	30	60–60	105	150–157			
a22r H13 = 70 + (9 + 2(4 + 1))15,				a22l-a10l,		F:A1-Pz,		

Remarks:

1. The next latencies in layer 2,3 are 300–*305*

H14l#1	AUD1	AUD2	A3/C1	CYC2	CYC3	CYC4	CYC5	CYC6
5 +			112	140	168–170	196	224–230	252
2,3 –	14	56–50	98	126–132	154	182	210–210	238
4b +		42	84–90					
4c +	0	28	70					
a22l H13 = 70 + (9 + 2(4 + 1))14,				a22l-a10l,		F:A1-Pz,		

Remarks:
1. The first shortcut could be in AUD1 or AUD2.
2. The next latencies are 266, 294–*286*

a22r H13 = 70 + (9 + 2(4 + 1))15 9.7 + 7.8
a22l H13 = 70 + (9 + 2(4 + 1))14 9.1 + 8.3

H15r#2		AUD1	AUD2	AUD3	A4/C1	CYC2	CYC3	CYC4	CYC5
5	+				210	252–260	294	336–325	
2,3	–	42–35	84	126–125	189	231–225	273	315–310	357–360
4b	+	21–15			168–178				
4c	+	0	63	105–100	147				
a22rH15 = 70 + (12 + 2(4 + 1*4))22,						a22r-a10r,		F:A2-Pz,	

Remarks:
1. The latencies P178 and P260 do not fit into a pattern with ET = 22ms, therefore ET = 21ms was used the first and only time.

a22rH15 = 70 + (12 + 2(4 + *1**4))22 12.9 + *10.4*
a22lH15 = 70 + (9 + 2(4 + 1*4))22 8.9 + 11.2

H15l#1		AUD1	AUD2	A3/C1	CYC2	CYC3	CYC4	CYC5	CYC6
5	+			176–170	220–210	264	308–310	352–350	396–390
2,3	–	22–20	88–80	154–150	198	242–235	286	330	374
4b	+		66	132					
4c	+	0	44–40	110–116					
a22lH15 = 70 + (9 + 2(4 + 1*4))22,						a22l-a10l,		F:A1-Pz,	

H16#2r		A1/C1	CYC2	CYC3	CYC4	CYC5			
5	+	30	60	90	120	150	180	210	240
2,3	–	15	45–45	75–70	105–100	135–130	165	195–200	225–230
4b	+								
4c	+	0							
a22rH16B = 70 + (7.3 + 2(4 + 2))14.5,						a22r-a10r,		F:A2-Pz,	

Remarks:
1. The latencies support linEN = 5 + 1 = 6 in contrast to linEN = 7.3 in a22rH16B.

H16#1l		AUD1	AUD2	A3/C1	CYC2	CYC3	CYC4	CYC5
5	+							
2,3	–	26–20	65	104–100	130–140	156	182–180	208–210
4b	+	13	52	91				
4c	+	0	39	78–80				
a22lH16B = 70 + (8.4 + 2(4 + 1*4))13,					a22l-a10l,		F:A1-Cz,	

Remarks:
1. The latencies support linEN = 8 + 1 = 9 in contrast to linEN = 8.4 in a22lH16B.
2. The next step involves the precise values of the latencies (by enlargement of the records).

H18r		AUD1	AUD2	AUD3	A4/C1	CYC2	CYC3	CYC4	CYC5
5	+				143	169	195	212	247
2,3	–	26–20	65–60	91–90	130	156	182	208–200	234–230
4b	+	13	52		117–120				
4c	+	0	39–40	78–80	104				

a22rH18A = con + (11.5 + 2(4 + 1*4))13, a22l-a10l, F:A1-Pz,

Remarks:
1. The next cycles are at 260–*260*, 286, 312–*312*

H18l		AUD1	AUD2	AUD3	AUD5	AUD6	AUD7	AUD8
5	+							
2,3	–	24	48–45	84–82	120–125	144–148	180–180	216–215
4b	+	12–15		72	108		168–165	204
4c	+	0	36	60–65	96–95	132–140	156	192–195

a22lH18A = con + (11.8 + 2(4*2))12, a22l-a10l, F:A1-Cz,

H18l								
5	+							
2,3	–	24	48–45	84–82	120–125	156–148	180–180	216–215
4b	+	12–15		72	108	144–140		204
4c	+	0	36	60–65	96–95	132	168–165	192–195

a22lH18A = con + (11.8 + 2(4*2))12, a22l-a10l, F:A1-Cz,

Remarks:
1. There is no visible boundary.

H20#14r		AUD1	AUD2	AUD3	A4/C1	CYC2	CYC3
5	+						
2,3	–	26–20	65	104–100	144	170–160	196–200
4b	+	13	52–55	91	131–130		
4c	+	0	39	78	118		

a22rH20B = 70 + (12 + 2(4 + 2))13* a22r-a10r, F:A2-Pz,

Remarks:
1. Here ET(a22rH20) = 13ms (see the original records) is used in contradiction to the text where ET(a22rH20) = 15 ms is used.
The 3ET rhythm *stops* after 144ms.
The non-appearance of certain periods may have anatomical reasons.
The record R:A1-Pz which shows the cortical activities of the pathway leading to the action of the middlefinger has a 15ms rhythm in H20B.
Is the end of the linear pathway (linEN-1)*ET visible in the ERPs?

H20#12l		AUD1	AUD2	AUD3	AUD4	AUD5
5	+					
2,3	–	26–30	65–70	104–110	144–150	183–180
4b	+	13	52–55	91	131–130	170–165
4c	+	0	39	78	118	157

a22lH20B = 70 + (12 + 2(4 + 1*4))13, a22l-a10l, F:A1-Pz,

Remarks:
1. The transition between 3ET and 2ET is the end of the linear pathway.
2. The 3ET rhythms *stops* after 180ms.

H23r		AUD1	AUD2	AUD3	A4/C1	CYC2	CYC3	CYC4	CYC5
5	+				165–165	195	225	255	285
2,3	–	15	60	105–105	150	180	210	240–232	270
4b	+		45	90	135				
4c	+	0	30–35	75	120				

a22rH23B = 70 + (12 + 2(4 + 1))15, a22l-a10l, F:A1-Pz,

H23l		AUD1	AUD2	AUD3	A4/C1	CYC2	CYC3	CYC4	CYC5
5	+				180	210	240	270	
2,3	–	45–47	90–85	120–115	165–175	195–195	225–220	255–262	300–300
4b	+	15–21	75		150–145				
4c	+	0	60	105	135				285

a22lH23B = 70 + (15 + 2(4 + 2))14, a22l-a10l, F:A1-Pz,

Remarks:
1. I took ET(H23l) = 15 as the mean value of 16,14, and 15.
2. Perhaps the ERP lies chronologically nearer to H23A than to H23B.

3. a22rH23A = 70 + (10 + 2(4 + 1*4))15 10 + 12.3
a22lH23A = 70 + (11 + 2(4 + 2))16 11.1 + 9.7
a22rH23B = 70 + (12 + 2(4 + 1))15 12.7 + 7.2
a22lH23B = 70 + (15 + 2(4 + 2))14 15.9 + 9.3
a22rH23C = 70 + (15 + 2(4 + 1))15 15 + 6.4
a22lH23C = 70 + (17 + 2(4 + 1))15 17 + 6.9

H24r#2		AUD1	AUD2	AUD3	AUD4	A5/C1	CYC2	CYC3	CYC4
5	+					208	240	272	304
2,3	–	32–35	80	112–115	144	192	224	256	288
4b	+	16	64–67			176			
4c	+	0	48	96	128	160–168			

a22r H24 = 70 + (12 + 2(4 + 1*4))16, a22r-a10r, F:A2-Pz,

Remarks:
1. The next latencies in layer 2,3 are 320–*320*, 352, 384–*385*

H24l#1		AUD1	AUD2	AUD3					
5	+								
2,3	–	16.5	49.5–50	82.5	115.5	148.5–150	181.5	214.5	247.5–250
4b	+								
4c	+	0	33	66	99–100	132	165	198–200	231

a22l H24 = 70 + (15 + 2(4 + 2))16.5, a22l-a10l, F:A1-Pz,

Remarks:

1. This subjects uses only two layers in each area. A differentiation between the linear and cyclical portion of the pathway is therefore not possible. The ERP looks very symmetrical.

a22r H24 = 70 + (12 + 2(4 + 1*4))16 12.1 + 11.3
a22l H24 = 70 + (15 + 2(4 + 2))16.5 15.9 + 10.4

H25r#2		AUD1	AUD2	A3/C1	CYC2	CYC3	CYC4	CYC5	CYC6
5	+			126	154	182	210	238	266
2,3	–	28	70	112–105	140	168	196	224	252
4b	+	14	56–50	98					
4c	+	0	42	84					

a22rH25 = 70 + (9 + 2(4 + 2))14, a22l-a10l, F:A1-Pz,

Remarks:

1. The next latencies in layer 2.3 are 280, 308–*310*.

H25l#1		AUD1	A2/C1	CYC2	CYC3	CYC4	CYC5	CYC6	CYC7
5	+		84	112–110	140	168–170	196	224	
2,3	–	28	70–70	98–90	126	154–150	182	210–215	238
4b	+	14	56						
4c	+	0	42–42						

a22lH25 = 70 + (9 + 2(4 + 2))14, a22l-a10l, F:A1-Pz,

Remarks:

1. The latencies after 238 ms are not in synchrony with the latencies before. Perhaps there is one elementary time intermingled.

a22rH25 = 70 + (9 + 2(4 + 2))14 8.4 + 9.3
a22lH25 = 70 + (9 + 2(4 + 2))14 8.7 + 9.7

H26#2		AUD1	AUD2	AUD3	AUD4	A5 / C1	CYC2	CYC3	CYC4
5	+					110	132	154–150	176
2,3	–	11	33–30	55	77	99–102	121	143	165
4b	+								
4c	+	0	22–25	44	66–63	88			

a22rH26B = 70 + (8 + 2(4 + 2*4))11, a22r-a10r, F:A2-Pz,

Remarks:
1. There is no visible boundary between the linear and the cyclical pathway because both of them use two steps.
2. The next values are: 176, 187, 198–*199*. That means the evoked potential uses at last a 3 step structure.

H26l#1		AUD1	AUD2	AUD3	A4 / C1	CYC2	CYC3	CYC4	MOT1
5	+				132	154	176	198	
2,3	-	22	55–52	88	121	143	165–165	187	220
4b	+	11–12	44	77	110–115				209–205
4c	+	0	33	66	99				198

a22lH26B = 70 + (8 + 2(4 + 2*4))11, a22l-a10l, F:A1-Pz,

a22rH26A = 70 + (9 + 2(4 + 2*4))10.5 9.4 + 17.1
a22lH26A = 70 + (9 + 2(4 + 2*4))12.5 8.8 + *13.2*
a22rH26B = 70 + (8 + 2(4 + 2*4))11 8 + 14.7
a22lH26B = 70 + (8 + 2(4 + 2*4))11 8 + 18.2

H27r#2		AUD1	AUD2	AUD3	AUD4	A5/C1	CYC2	CYC3	CYC4
5	+					203–200	232–220	261	290–295
2,3	-	29–30	72.5	101.5–90	145–155	188.5	217.5	246.5	275.5–270
4b	+	14.5	58–65		130.5–125	174			
4c	+	0	43.5	87	116	159.5			

a22rH27 = 70 + (15 + 2(4 + 1*4))14.5, a22l-a10l, F:A1-Pz,

Remarks:
1. The next latencies in layer2,3 are 304.5, 333.5–*332*, 362.5, 391.5–*400*
2. Sometimes the latencies and the expected values do not match very well.

H27l#1		AUD1	AUD2	AUD3	A4/C1	CYC2	CYC3	CYC4	CYC5
5	+				154	182	210	238	266
2,3	-	14–10	56–60	98–110	140	168	196	224	252–245
4b	+		42	84–85	126–130				
4c	+	0	28–25	70	112				

a22lH27 = 70 + (17 + 2(4 + 2))14, a22r-a10r, F:A2-Pz,

Remarks: 294 322–*330*
1. The next latencies are 280, 308, 336, 364, 392, 420, 448–*450*

a22rH27 = 70 + (15 + 2(4 + 1*4))14.5 15.9 + 12.3
a22lH27 = 70 + (17 + 2(4 + 2))14 16.6 + 10.6

H28r#2		AUD1	AUD2	AUD3	AUD4	AUD5	A6 / C1	CYC2	CYC3
5	+						208	234	260–262
2,3	-	13	39	78–75	117–122	156	195–192	221–220	247
4b	+			65	104–100	143	182–180		
4c	+	0	26	52–55	91	130	169		

a22rH28C = 70 + (14 + 2(4 + 1))13, a22r-a10r, F:A2-Pz,

H28r#2		CYC4	CYC5	
5	+	286	312	
2,3	–	273	299–300	351–352
4b	+			338–335
4c	+			325

a22rH28C = 70 + (14 + 2(4 + 1))13, a22r-a10r, F:A2-Pz,

H28l#1		AUD1	A2/C1	CYC2	CYC3	CYC4	CYC5		
5	+								
2,3	–	26–30	65–64	91	130	169–170	208–205	247	273–270
4b	+	13	52		117	156–150	195–190	234	
4c	+	0	39–40	78	104–100	143	182	221–225	260

a22lH28C = 70 + (8 + 2(4 + 2*4))13, a22l-a10l, F:A1-Pz,

H28l#1			
5	+		
2,3	–	299–300	338–335
4b	+		325–320
4c	+	286	312

a22lH28C = 70 + (8 + 2(4 + 2*4))13, a22l-a10l, F:A1-Pz,

Remarks:

1. There is no clear boundary between the linear and the cyclical pathway but after the area AUD2(5ET) and after 19ET

   a22rH28B = 70 + (14 + 2(4 + 1))13 13.8 + 8.3
   a22lH28B = 70 + (15 + 2(4 + 2*4))12.5 14.4 + *14.1*
   a22rH28C = 70 + (14 + 2(4 + *1*))13 13.4 + *9.4*
   a22lH28C = 70 + (8 + 2(4 + 2*4))13 8 + 16.3

H29r#3		AUD1	AUD2	A3/C1	CYC2	CYC3	CYC4	CYC5	CYC6
5	+			112.5	137.5–140	162.5	187.5–185	212.5–218	237.5–240
2,3	–	25	62.5–70	100	125	150	175–173	200–200	225–230
4b	+	12.5	50	87.5					
4c	+	0	37.5–30	75					

a22rH29B = 70 + (9 + 2(4 + 2))12.5, a22l-a10l, F:A1-Cz,

H29l#2		AUD1	AUD2	A3/C1	CYC2	CYC3	CYC4	CYC5	CYC6
5	+			96	120	144–145	168	192–200	
2,3	–	12	48–48	84	108	132	156	180–173	
4b	+		36	72–70					
4c	+	0	24	60					

a22lH29B = 70 + (6 + 2(4 + 1*4))12, a22l-a10l, F:A1-Cz,

Remarks:

1. The first two areas of a22lH29 could take another course.
2. In H29l#2 the layer 5 is shown because of the latency 145 ms.
3. Equations:

a22rH29A = 70 + (9 + 2(4 + 2*4))12.5	9.2 + 17.4
a22lH29A = 70 + (10 + 2(4 + 2*4))12	10.1 + 16.3
a22rH29B = 70 + (9 + 2(4 + 2))12.5	9.2 + 9.8
a22lH29B = 70 + (6 + 2(4 + 1*4))12	6.3 + 11.5

H29r#5		AUD1	AUD2	A3/C1	CYC21	CYC3	CYC4	CYC5	CYC6
5	+			100	125	150	175	200	225–225
2,3	-	12.5	50–54	87.5–95	112.5	137.5	162.5–165	187.5	212.5–205
4b	+		37.5	75					
4c	+	0	25–20	62.5					

a22rH29B = 70 + (9 + 2(4 + 2))12.5, a22l-a10l, F:A1-Cz,

H29l#4		AUD1	AUD2	A3/C1	CYC2	CYC3	CYC4	CYC5	CYC6
5	+			84–85	108	132	156	180	204–200
2,3	-	12	36–37	72	96	120–122	144	168–170	192
4b	+			60					
4c	+	0	24	48					

a22lH29B = 70 + (6 + 2(4 + 1*4))12, a22l-a10l, F:A1-Pz,

Remarks:

1. There are two latencies P145 and N155 which do not fit into the theoretical pattern.

H31r#2		AUD1	AUD2	AUD3	AUD4	AUD5	A6/C1	CYC2	CYC3
5	+						289	323	
2,3	-	34	68–60	119–125	170–167	221	272–275	306	340–343
4b	+	17		102	153	204	255		
4c	+	0	51–43	85–90	136–142	187	238–235		

a22rH31A = 70 + (18 + 2(4 + 1*4))17, a22r-a10r, F:A2-Pz,

H31r#2			
5	+		
2,3	-	391–390	
4b	+	374	425–420
4c	+	357–360	408

a22rH31A = 70 + (18 + 2(4 + 1*4))17, a22r-a10r, F:A2-Pz,

Remarks:

1. I have taken ET(aNNrH31A) = 17.
2. The difference is 58 (12 correlations)

H31l#1		AUD1	AUD2	AUD3	AUD4	AUD5	AUD6	A7/C1	CYC2
5	+							279–275	310
2,3	–	15.5	62–70	93	139.5–135	170.5–170	217–225	263.5	294.5
4b	+		46.5		124		201.5–210	248	
4c	+	0	31–35	77.5	108.5–112	155	186	232.5	
,		a22l-a10l,		F:A1-Pz,					

H31l#1					
5	+				
2,3	–	325.5–325	372–373	418.5	449.5–440
4b	+		356.5–357	403–407	
4c	+		341	387.5	434
,		a22l-a10l,	F:A1-Pz,		

Remarks:
1. The latencies of H31l#1 are correlated with ET = 17 in the discussion to show the ambiguity of the procedure.

a22rH31A = 70 + (18 + 2(4 + *1**4))17   17.7 + *11.5*
a22lH31A = 70 + (26 + 2(4 + 2))15.5   26.1 + 10.5
a22rH31B = 70 + (18 + 2(4 + *1**4))16   18.6 + 11.9
a22lH31B = 70 + (20 + 2(4 + 2*4))15   20 + 15.4

The event-related potentials of the x11y pathways

The objective of this correlation is to show the relative constancy of the boundary between the linear and the cyclical pathway in most trials of a task. This is the only case when the boundary between the linear and the cyclical part can be seen in the evoked potential. This relative constancy would support the equational description of the stimulus-response pathway.

If a latency is correlated with an integer multiple of the elementary time, the difference between the latency and the multiple of elementary time should be less than half the elementary time: (lat – n*ET) < ET/2. Otherwise there is no sure correspondency between these two values.

Latency tables of ERP(a11y)

H11Br		AUD1	AUD2	AUD3	CYC1	CYC2	CYC3	CYC4	CYC5
5	+								
2,3	–	16	48–50	80	112–115	160–162	192–190	240–240	
4b	+					144		224	
4c	+	0	32	64–70	96	128	176	208	
a11rH11B = 70 + (3 + 2(2 + §2))16,					a22r-a10r,		F:A2-Pz,	a11rH11B = 216	

Remarks:
1. If the two step rhythm prevails from the beginning, experience shows that one area as the minimal number of sensory areas should be taken as the linear pathway.

H11Bl		AUD1	AUD2	AUD3	A4/C1	CYC2	CYC3	CYC4	CYC5
5	+								
2,3	–	14–8	56–60	98–100	140–140	168–170	196	224	252–255
4b	+		42	84	126–125				
4c	+	0	28	70–70	112	154–155	182	210	238

a11lH11B = 70 + (5 + 2(2 + §2))14, a22l-a10l, F:A1-Pz, a22lH11B = 224

a11rH11A = 70 + (3 + 2(2 + §2))16	3 + 5.2
a11lH11A = 70 + (5 + 2(2 + §2))14	4.8 + 5
a11rH11B = 70 + (3 + 2(2 + §2))16	3.6 + 5.5
a11lH11B = 70 + (5 + 2(2 + §2))14	5.5 + 5.5

H11Cr		A1/C1	CYC2	CYC3	CYC4	CYC5	CYC6	
5	+							
2,3	–	16–8	48–42	80–75	112	144–138	176–178	224–220
4b	+							208–208
4c	+	0	32–28	64–67	96	128	160–155	192

a11rH11B = 70 + (3 + 2(2 + §2))16, a22r-a10r, F:A2-Pz, a11rH11B = 216

Remarks:
1. There is the same boundary as in H11Br. Additionally the first four areas use the same pathway as in a11rH11B.

H11Cl		AUD1	AUD2	AUD3	AUD4	A5/C1	CYC2	CYC3	
5	+								
2,3	–	28–32	56–50	98–100	140–145	182–185	210–203	238–235	280–285
4b	+	14		84	126	168			266–268
4c	+	0	42	70–71	112–115	154–153	196–195	224–220	252

a11lH11B = 70 + (5 + 2(2 + §2))14, a22l-a10l, F:A1-Pz, a22lH11B = 224

Remarks:
1. The transition between triple and double steps occurs after the fifth area. The first four areas use nearly the same pathway as in a11lH11B (with the exception of the first two areas but their pathway is very similar and eventually identical).
2. If one takes the transition from 3 layer activity to 2 layer activity as the boundary between linear and cyclical pathway a11lH11C deviates from the reaction data length of 5linET very much. It may be considered that the latencies represent the *unspecific spreading of activity* in the layers 2,3,4b, and 4c.
3. It remains the question whether the searching cycles of the memory sets leave some marks on the event related potential?

H14r		A1/C1	CYC2	CYC3	CYC4	CYC5			
5	+								
2,3	–	30–27	60	90–88	120	150	190	220–225	250
4b	+	15							
4c	+	0	45–42	75	105	135	175–175	205	235

a11rH13 = 70 + (8 + 2(2 + §2))15, a22l-a10l, F:A1-Pz, a11rH13 = 253

Remarks:

1. In this case it is difficult to fix the moment when the linear pathway end and the cyclical one starts. It can be at each area with only two active layers (see the next table for example). *This is a very important observation which is valid for every correlation: the start of the cyclical pathway is always ambiguous because the last area of the linear pathway could only have two active layers.*

H14r		A1/C1	CYC2	CYC3	CYC4	CYC5			
5	+	45–42	75	105	135	175–175	205	235	275
2,3	–	30–27	60	90–88	120	150	190	220–225	250
4b	+	15							
4c	+	0							

a11rH13 = 70 + (8 + 2(2 + §2))15, a22l-a10l, F:A1-Pz, a11rH13 = 253

H14l		AUD1	A2/C1	CYC2	CYC3	CYC4	CYC5		
5	+		84	112	140	168–165	196		
2,3	–	28–25	70–65	98	126–128	154	182	210–210	252–250
4b	+	14–13	56						238
4c	+	0	42–47						224–227

a11l H13 = 70 + (8 + 2(2 + §2))14, a22l-a10l, F:A1-Pz, a11lH13 = 250

Remarks:

1. In this case the length of the linear pathway could be (5 + 2) + 1 = 8 if the last area of the linear pathway has only two active layers and the motor area takes 1 step.

a11rH13 = 70 + (8 + 2(2 + §2))15 7.7 + 4.6
a11l H13 = 70 + (8 + 2(2 + §2))14 8.4 + 4.5

H20r		AUD1	A2/C1	CYC2	CYC3			
5	+		78	104	130–128			
2,3	–	26–30	65–72	91	117	156–155	182	208
4b	+	13	52			143		
4c	+	0	39			130	169–175	195

a11rH20B = 70 + (5 + 2(2 + §2))15, a22r-a10r, F:A2-Pz, a11rH20C = 219

H20l		A1/C1	CYC2	CYC3	CYC4	CYC5			
5	+	39	65–65	91	117	143	169	195–202	221
2,3	–	26–30	52	78–85	104	130	156–150	182	208
4b	+	13							
4c	+	0							

a11lH20B = 70 + (6 + 2(2 + §2))13, a22l-a10l, F:A1-Pz, a11lH20C = 230

a11rH20A = 70 + (5 + 2(2 + §2))15	5 + 5
a11lH20A = 70 + (6 + 2(2 + §2))13	6.1 + 5.2
a11rH20B = 70 + (5 + 2(2 + §2))15	5 + 5.3
a11lH20B = 70 + (6 + 2(2 + §2))13	5.7 + 5.2

H23r		A1/C1	CYC2	CYC3	CYC4	CYC5		
5	+	30–25	60–65	90	120	150–145	180	
2,3	–	15	45–50	75	105–100	135	165–163	210
4b	+							
4c	+	0						195–195

a11rH23C = 70 + (3 + 2(2 + §2))15, a22r-a10r, F:A2-Pz, a11rH23B = 202

Remarks:

1. The boundary is taken after the first area because the pathway uses only two layers in each area.

H23l		AUD1	AUD2	A3/C1	CYC2	CYC3		
5	+							
2,3	–	15–6	60–60	105	135	175	220–221	250–245
4b	+		45	90			205	
4c	+	0	30–36	75	120	150–155	190	235–235

a11lH23C = 70 + (6 + 2(2 + §2))15, a22l-a10l, F:A1-Pz, a11lH23B = 242

a11rH23A = 70 + (3 + 2(2 + §2))15	3.3 + 5
a11lH23A = 70 + (4 + 2(2 + §2))16	4.3 + 4.7
a11rH23B = 70 + (3 + 2(2 + 1))15	3 + 4.2
a11lH23B = 70 + (4 + 2(2 + *1*))14	4.4 + *4.2*
a11rH23C = 70 + (3 + 2(2 + §2))15	3.7 + 5.1
a11lH23C = 70 + (6 + 2(2 + §2))15	6.3 + 5.1

H25r		AUD1	AUD2	A3/C1	CYC2	CYC3	
5	+						
2,3	–	28	70–72	112–115	140	168–175	210–210
4b	+	14–15	56–50	98			196
4c	+	0	42	84	126	154–155	182

a11rH25 = 70 + (3 + 2(2 + §2))14, a22r-a10r, F:A2-Pz, a11rH25 = 192

H25l		AUD1	A2/C1	CYC2			
5	+						
2,3	–	28	70	98–95	140	182	224–220
4b	+	14	56–50		126–130	168	210
4c	+	0	42	84	112	154	196

a11lH25 = 70 + (3 + 2(2 + §2))14, a22l-a10l, F:A1-Pz, a11lH25 = 181

a11rH25 = 70 + (3 + 2(2 + §2))14 3.7 + 5
a11lH25 = 70 + (3 + 2(2 + §2))14 2.6 + 5.3

H27r		AUD1	AUD2	AUD3	AUD4	AUD5	A6/C1	CYC2	CYC3
5	+								
2,3	–	29–33	58	87–85	116–112	145–150	188.5	217.5–221	246.5
4b	+	14.5–10					174		
4c	+	0	43.5	72.5–65	101.5–102	130.5–130	159.5	203–202	232–238

a11rH27 = 70 + (15 + 2(2 + 2*2))14.5, a22r-a10r, F:A2-Pz, a11rH27 = 391

Remarks:
1. The next steps are 261, 275.5–275
2. There are two possible boundaries after the first area and after the sixth area. The reaction time data support the latter one.

70

H27l		AUD1	AUD2	A3/C1	CYC2	CYC3	CYC4	CYC5	CYC6
5	+								
2,3	–	28–30	70	112–112	140	168–170	196–190	224	252–260
4b	+	14–18	56–60	98					
4c	+	0	42	84–80	126	154–155	182–180	210	238

a11lH27 = 70 + (12 + 2(2 + 2*2))14, a22l-a10l, F:A1-Pz, a11lH27 = 355

Remarks:
1. There is a mistake in 56–60 (false sign) which cannot be explained.

a11rH27 = 70 + (15 + 2(2 + 2*2))14.5 15 + 7.6
a11lH27 = 70 + (12 + 2(2 + 2*2))14 12.3 + 8.1

H29#3r		AUD1	AUD2	AUD3	A4 / C1	CYC2	CYC3	CYC4	CYC5
5	+				137.5	162.5–162	187.5–182		
2,3	–	12.5–12	50–50	87.5–90	125–122	150	175–175	200–195	237.5
4b	+		37.5	75	112.5				225–220
4c	+	0	25–20	62.5–66	100–105				212.5

a11rH29B = 70 + (4 + 2(2 + §2))12.5, a22r-a10r, F:A2-Pz, a11rH29B = 181

H29#2l		AUD1	AUD2	AUD3	A4 / C1	CYC2	CYC3	CYC4	CYC5
5	+				108	132			
2,3	–	12–12	36	60	96	120–115	144	180–175	216
4b	+				84			168	204–202
4c	+	0	24–25	48	72–78			156–155	192

a11lH29B = 70 + (4 + 2(2 + §2))12, a22l-a10l, F:A1-Pz, a11lH29B = 184

H29#5r		AUD1	A2/C1	CYC2	CYC3	CYC4			
5	+								
2,3	–	12.5	50–55	75	100	125–123	162.5–165	187.5	212.5–215
4b	+		37.5				150		
4c	+	0	25	62.2	87.5–90	112.5	137.5–140	175	200–200

a11rH29B = 70 + (4 + 2(2 + §2))12.5, a22r-a10r, F:A2-Pz, a11rH29B = 181

H29#4l		A1/C1	CYC2	CYC3	CYC4	CYC5			
5	+								
2,3	–	24–23	48	72	96	120–120	156–150	180	216
4b	+	12					144		204
4c	+	0	36	60	84	108	132	168	192–195

a11lH29B = 70 + (4 + 2(2 + §2))12, a22l-a10l, F:A1-Pz, a11lH29B = 184

Remarks:

1. Eventually a second area with two active layers is added to give the four steps of the reaction time data.

a11rH29A = 70 + (4 + 2(2 + §2))12.5	4 + 5.7
a11lH29A = 70 + (6 + 2(2 + §2))12	6.3 + 4.8
a11rH29B = 70 + (4 + 2(2 + §2))12.5	4:4 + 4.5
a11lH29B = 70 + (4 + 2(2 + §2))12	4.3 + 5.3

Discussion

The assignment of latencies to the theoretical lattice derived from elementary times and anatomical considerations can be ambiguous. The table below shows an example of another assignment than that taken above.

H25l#1		AUD1	AUD2	AUD3					
5	+								
2,3	–	28	70–70	98–90	140–135	182	210–215	252–252	294
4b	+	14	56		126	168–170		238	280–280
4c	+	0	42–42	84	112–110	154	196	224	266

a22lH25 = 70 + (9 + 2(4 + 2))14, a22l-a10l, F:A1-Pz,

Remarks:

1. There is no visible border between the linear pathway and the cyclical pathway.

Now comes an example where I incorrectly used the latencies of H31*l*#1 and the ET(a22*r*H31A).

H31l#1		AUD1	AUD2	AUD3	AUD4	AUD5	AUD6	AUD7	AUD8
5	+								
2,3	–	17	68–70	102	136–135	170–170	221–225	255	289
4b	+		51				204–210		
4c	+	0	34–35	85	119–112	153–150	187	238	272–275

a22rH31A = 70 + (18 + 2(4 + 1*4))17, a22r-a10r, F:A2-Pz,

H31l#1		AUD9	AUD10	AUD11	
5	+				
2,3	–	323–325	374–373	425	440–440
4b	+		357–357	408–407	
4c	+	306	340	391	

a22rH31A = 70 + (18 + 2(4 + 1*4))17, a22r-a10r, F:A2-Pz,

Remarks:

1. I have taken ET(xNNyH31A) = 17 because the multiples of this give a better fit to the ERP latencies.
2. There is no visible border between linear and cyclical pathway in the ERP.
3. The 440 latency does not fit to the theoretical pattern, perhaps a layer2,3 to layer2,3 connection takes place.

The following two tables show the ambiguity of the correlation ET-ERP too. Instead of ET = 17, ET = 16 is used. The sum of the differences between the multiples of ET and the latencies may be a measure for the correlation. Difference = 69 (12 correlations).

H31r#2		AUD1	AUD2	AUD3	AUD4	AUD5	AUD6	AUD7	AUD8
5	+								
2,3	–	32	64–60	96	128–125	160–167	192	224	272–275
4b	+	16							256
4c	+	0	48–43	80–90	112	144–142	176	208	240–235

a22rH31A = 70 + (18 + 2(4 + 1*4))17, a22r-a10r, F:A2-Pz,

H31r#2		AUD9	AUD10		
5	+				
2,3	–	304	336–343	384–390	
4b	+			368	416–420
4c	+	288	320–325	352–360	400

a22rH31A = 70 + (18 + 2(4 + 1*4))17, a22r-a10r, F:A2-Pz,

If I accept one mistake (240–253), the cyclical pathway is visible:

H07l#1		AUD1	AUD2	A3/C1	CYC2	CYC3	CYC4	CYC5	CYC6
5	+			128	160	192	224–222	256	288–280
2,3	–	16–13	64–72	112	144–140	176	208–205	240–253	272
4b	+		48	96–102					
4c	+	0	32–40	80					

a22lH07A = 70 + (13 + 2(4 + 2))16, a22l-a10l, F:A1-Pz,

H07l#1		CYC7	CYC8
5	+	320–317	352
2,3	–	304–300	336
4b	+		
4c	+		

a22lH07A = 70 + (13 + 2(4 + 2))16, a22l-a10l, F:A1-Pz,

Is it possible that there is another linear pathway *after* the cyclical pathway?

H07l#1		AUD1	AUD2	A3/C1	CYC2	CYC3	CYC4		
5	+								
2,3	–	16	64–72	112	144–140	176	208–205	256–253	304–300
4b	+		48	96–102				240	288
4c	+	0	32–40	80	128	160	192	224–222	272–280

a22lH07A = 70 + (13 + 2(4 + 2))16, a22l-a10l, F:A1-Pz,

Finally, ERP(a11lH11B) and ERP(a11lH11C) showed no correlation between the latencies of the contralateral hemisphere and multiples of ET, but a clear correlation between the latencies of the homolateral hemisphere and multiples of ET.

Statistical correlations between reaction time data (ET, linEN) and ERP data (ET, break)

Correlation between the reaction time data (linEN) and the ERP data (break) of the a22y pathway

The reaction tasks accompanied by event-related potentials are a series of their own. Two different series have similar linear pathways. This is shown for the linear pathway of the task a22r.

Now the length of the linear pathway defined by the break in the latency tables is compared with the length of the linear pathway defined by reaction time data.

The 1 has to be added because the linear pathway consists of the first linear part and one motor relation.

Now the same procedure is applied to the left-handed tasks.

The left-handed tasks change more than the right-handed ones. The linear pathway can be prolonged or reduced.

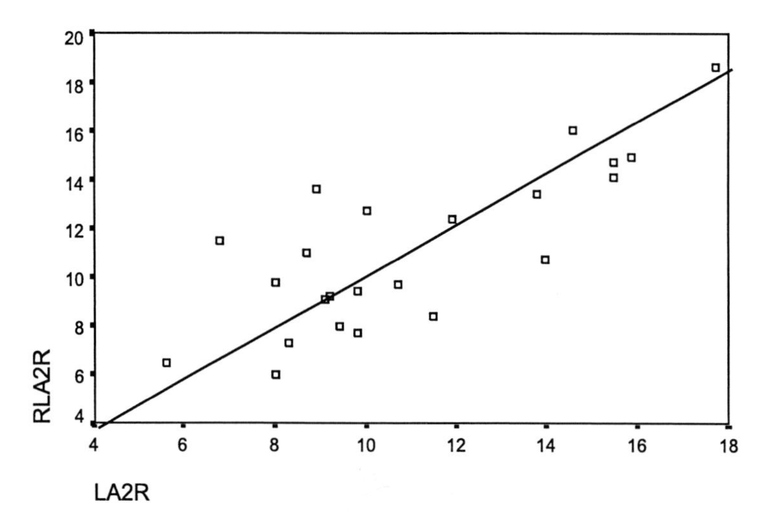

Fig. 132. Correlation between LA2R = linEN(a22rSEA) and RLA2R = linEN(a22rSEB) with SEA = Series A and SEB = Series B. The Pearson correlation coefficient is $r = 0.787$, the significance level is $p < 0.01$

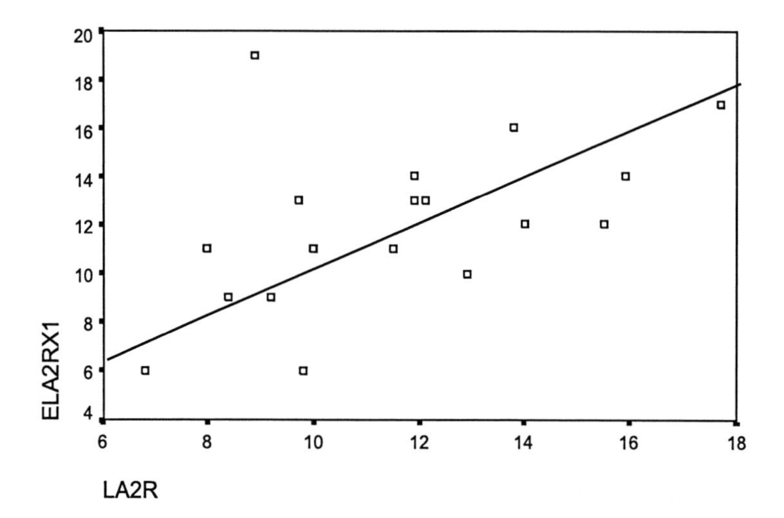

Fig. 133. Correlation between LA2R = linEN(a22rSEA) and ELA2RX1 = Break (ERP (a22rSEB)) + 1 with SEA = Series A and SEB = Series B. The Pearson correlation coefficient is $r = 0.504$, the significance level is $p < 0.05$

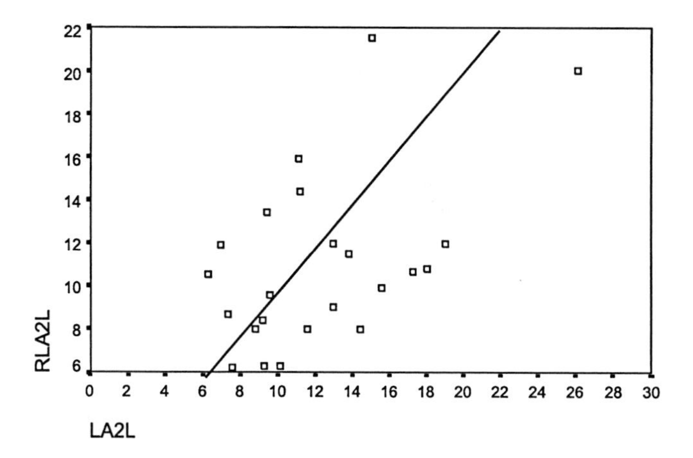

Fig. 134. Correlation between LA2L = linEN(a22lSEA) and RLA2L = linEN(a22lSEB) with SEA = Series A and SEB = Series B. The Pearson correlation coefficient is r = 0.514, the significance level is p< 0.05

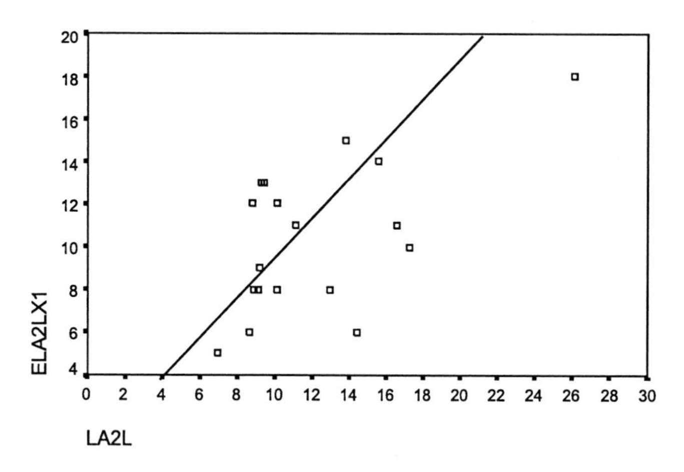

Fig. 135. Correlation between LA2L = linEN(a22lSEA) and ELA2LX1 = Break(ERP (a22lSEB)) + 1 with SEA = Series A and SEB = Series B. The Pearson correlation coefficient is r = 0.579, the significance level is p< 0.05

The 1 has to be added because the linear pathway consists of the first linear part and one motor relation.

This relation is similar to the comparison between linEN(a22lSEA) and linEN (a22lSEB). In both cases, the variability is higher than in the right-handed counterparts. The length of the linear pathway may be reduced or increased.

Correlation between the elementary time from the reaction time data and the elementary time from the ERP data of the a22y pathway

Computing the elementary times from the ERP latencies with SPRING16:

Table 36. aNNyAB contains the elementary times from reaction time data, ABy#2 contains the elementary times from the ERP data (the variable y implies a set of tasks)

	aNNrAB	ABr#1	aNNlAB	ABl#2
H05B	14	14.5	13	12.5

Computing the elementary times with ERPET15. This program applies the NESTLE procedure to the latencies of the ERP data. Because of the symmetry between ET(aNNr) and ET(aNNl) in healthy subjects, this principle is used to select these values which occur at both sides. Addionally the values near 20 are omitted. Most values are cycle times CT = 2ET.

Table 37. The column ERP(a22y) contains the results of the NESTLE procedure (within the program ERPET15), the column eaETy contains the elementary time calculated from the ERP data, the column aETy the elementary time calculated from reaction time data

Subject	ERP(a22r)	ERP(a22l)	eaETr	eaETl	aETr	aETl
H02	33 /artifacts	26	16.5/artifac	13	13	14.5
H03	20,26	20,25	13	12.5	14	15
H05B	20, 32	17, 34	16	17	14	13
H07	25,37,45	14,20,35	18.5	17.5	18	16
H08	25,29	32,41,47	14.5	16	14	14.5
H11B	28	21,24,29, 35,42	14	14.5	16	14
H11C	30	17,21	15	17	16	14
H13	20,28	21,26,28	14	13.5	14	15
H14	16,31,50	19,42	16	19	15	14
H15	25,32,45	39	22.5	19.5	22	22
H16	25,33	20,26,35	12.5, 16.5	13, 17.5	14.5	13
H18	14,20,45	20,26,35	14	13	13	12
H20	20,25	26	12.5	13	13	13
H23	21,33	22,37	16.5	18.5	15	14
H24	32,35	17,25,50	16.5	17	16	16.5
H25	13,17,26, 50	22,30,36	13, 17	15, 18	14	14
H26	15,25	23,41	12.5	11.5	11	11
H27	22,30	22,30,41	15	15	14.5	14
H28	20,25	15,25,30,33	12,5	12.5	13	13
H29B	24,35	18,25,48	12	12.5	12.5	12
H29C	19	12,17,29,41	19	17	12.5	12
H31	15,30	17,35	15	17	17	17

Here, the same effect can be seen as in the comparison between the length of the left-and right-handed linear pathways, which differed in left-handed tasks more than

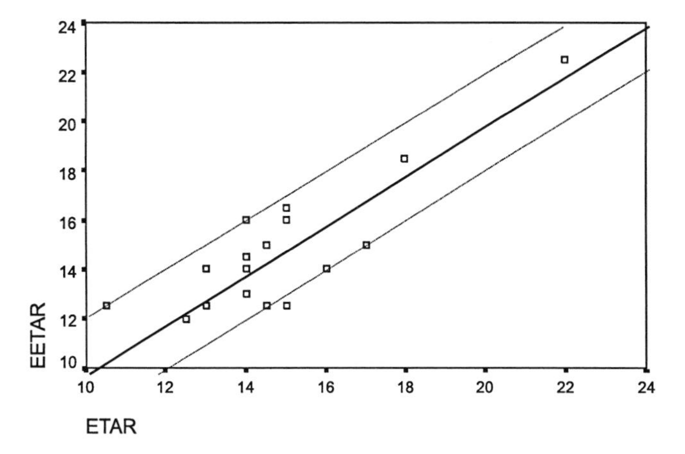

Fig. 136. Relation between ETAR = elementary time derived from reaction time data and EETAR = elementary time derived from ERP data. Each point represents these two elementary times in the right-handed tasks of one subject. The broken lines represent the measurement uncertainty which is ± 2ms. The Pearson correlation coefficient is r = 0.847, the significance level is p< 0.01

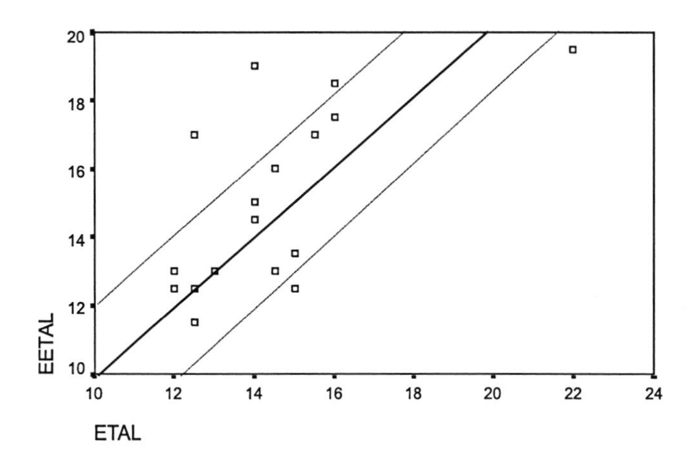

Fig. 137. Relation between ETAL = elemenary time derived from reaction time data and EETAL = elementary time derived from ERP data. Each point represents these two elementary times in the left-handed tasks of one subject. The broken lines represent the measurement uncertainty which is ± 2ms. The Pearson correlation coefficient is r = 0.648, the significance level is p< 0.01

in right-handed tasks. The reason for this observation is unknown. Perhaps some subjects use the left hemisphere in left-handed tasks. But the subjects which deviate in linear pathways and those which deviate in elementary times are different. These preliminary results have also to be replicated in a larger sample.

Correlation between the reaction time data and the ERP data of the a11y pathway

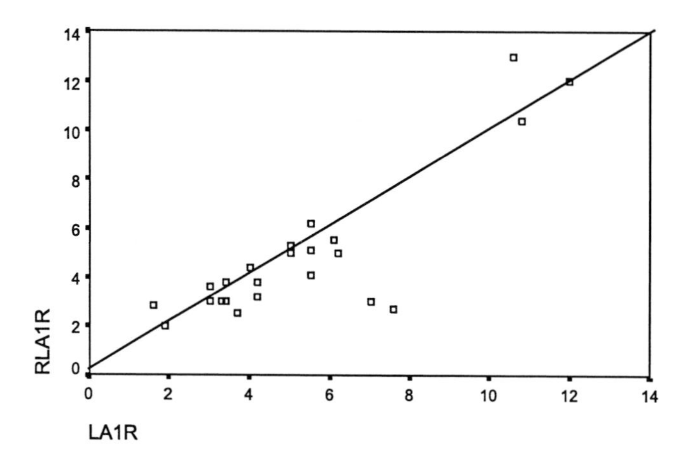

Fig. 138. Correlation between LA1R = linEN(a11rSEA) and RLA1R = linEN(a11rSEB) with SEA = Series A and SEB = Series B. The Pearson correlation coefficient is $r = 0.861$, the significance level is $p < 0.01$

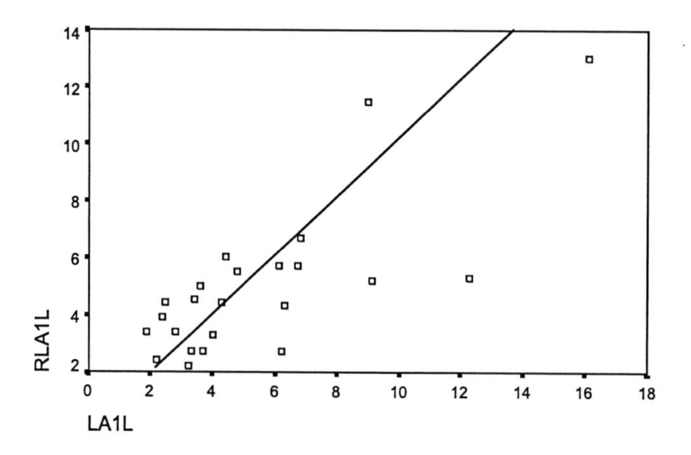

Fig. 139. Correlation between LA1L = linEN(a11lSEA) and RLA1L = linEN(a11lSEB) with SEA = Series A and SEB = Series B. The Pearson correlation coefficient is $r = 0.766$, the significance level is $p < 0.01$

The a11r pathway shows a greater variability of the linear part compared with the a11l pathway. This greater variability of the linear part in repeated tasks had already been found in the a22r pathway compared with the a22l pathway.

Because only 7 subjects have repeated the ERP(a11y), at the present time no statistical calculations are possible. The latency tables of these subjects are summarized at the end of this chapter.

The replication of event related potentials

Can the latencies been replicated? A study which tests the findings by replication of ERP(aNNy) is under way.

The side differences of event related potentials during reaction tasks in single subjects

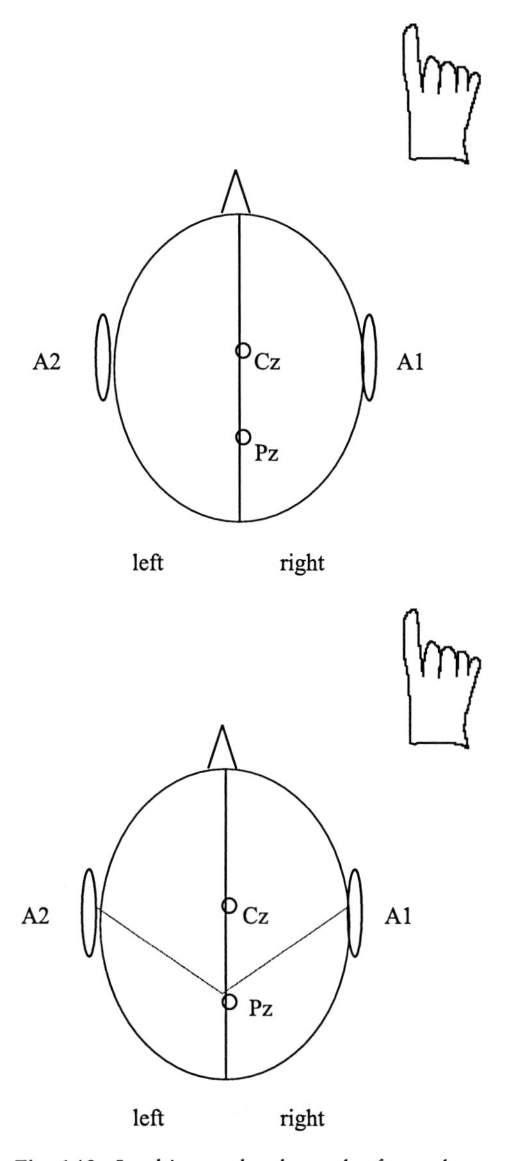

Fig. 140. Looking at the electrodes from above

If one compares the two event related potentials F:A1-Pz and F:A2-Pz one can see different periods for the subject H11C:

Some reaction time data of H11A and H11B

a11rH11A = 70 + (3 + 2(2 + §2))16	3 + 5.2
a11lH11A = 70 + (5 + 2(2 + §2))14	4.8 + 5
a11rH11B = 70 + (3 + 2(2 + §2))16	3.6 + 5.5
a11lH11B = 70 + (5 + 2(2 + §2))14	5.5 + 5.5

Task (a11l-a10l)H11C

Period 1 between 0 and 71 ms (this could be the linear pathway)
The two sides show nearly the same potential

Period 2 between 71 and 153 ms (this could be the cyclical pathway)
F: A2-Pz (homolateral) is more negative than F:A1-Pz (contralateral)

Period 3 between 153 ms and reaction at 220 ms (output time between cortex and reaction)
F:A1-Pz is more negative than F:A2-Pz

What is the cause of this asymmetry of potentials? The homolateral cortex shows more activity in layer 2,3 (negative) than the contralateral cortex. Is there any causal relation with the prefrontal cortex searching on the layer 2,3?

Correlations between RT data and ERP data of H11

a11lH11B = 70 + (5 + 2(2 + §2))14 5.5 + 5.5

↓ ↓

(6–1)*14 = 70	6*14 = 84ms	con = 70
71ms	(153–71)ms = 82ms	(220–153)ms = 67ms

In the linear pathway, 1 "motor" ET has to be subtracted.

Task (a11r-a10r)H11C

The points of intersection (contact) are: 28ms, 120ms, 290ms
These points are dependent from the kind of matching between the two ERPs.

Period 1 between 0 and 28 ms
F:A2-Pz(contralateral) is more negative than F:A1-Pz (homolateral).

Period 2 between 28 and 120 ms
F:A2-Pz(contralateral) is more negative than F:A1-Pz (homolateral).

Correlations between RT data and ERP data of H11

a11rH11B = 70 + (3 + 2(2 + §2))16 3.6 + 5.5

↓ ↓

(3–1)*16 = 32 5.5*16 = 88ms
28 ms (120–28)ms = 92ms

In the linear pathway, 1 "motor" ET has to be subtracted.

Can these results been replicated in another subject?

In (a11r–a10r)H23, the potential F:A1-Pz and the potential F:A2-Pz are separated at 50 ms. 50/15 = 3.3, linEN = 3.7

Discussion

The NESTLE procedure applied to the ERP latencies frequently produces a second result at 20ms (±2ms)

This is the 50Hz artifact. The recording of event-related potentials was rendered more difficult by these artifacts which are due to the 50Hz-frequency of the alternating current in the laboratory. The NESTLE procedure can find these periods which have a length of 1000ms / 50 = 20ms.

The relation between the break in ERP(a22rSEB) and linEN(a22rSEA) in 20 healthy subjects

In this relation, some subjects deviate from the linear correspondence between the two variables (see figure above). There are some cases in which the ERP derived linear pathway (ELA2R) is shorter than the linear pathway computed by reaction time data (LA2R). Do these subjects have something in common?

16 = H16A, 15 = H15, 19 = H20A, 26 = H27, 3 = H03A

H03A uses triple search
H15 uses triple search in a22y
H16 uses triple search in a22y
H20A uses triple search in a22y
H27 uses triple search in a22r

On the contrary there are the subjects which lie near the correlation line:

24 = H25, 28 = H29A, 5 = H05A, 22 = H23, 7 = H07A, 8 = H08, 27 = H28B

H05A uses triple search in a22r
H07A uses double search in a22r
H08 uses double search in a22r
H23B uses double search in a22r (not H23A)
H25 uses double search in a22r
H28B uses double search in a22r
H29B uses double search in a22r (not H29A)

This is the confirmation that subjects with double search have a minimal cyclical pathway mincycEN = 2.

But the subjects with triple search have a longer minimal cyclical pathway because the third searching set has to coincide with the second searching set even in the minimal pathway. Therefore the minimal cyclical pathway of these subjects is mincycEN = 4 (or 6). That means that the equations of all the subjects with triple search have to be corrected by reducing the linear pathway by 2ET and increasing the cyclical pathway by 2ET (the minimal cyclical pathway needs these additional 2ET). With this correction, the correspondence between the break of ERP(a22rSEB) and linEN(a22rSEA) is much better.

There are subjects which deviate in the other direction with the ERP derived linear pathway (ELA2R) being longer than the linear pathway computed by reaction time data (LA2R). Do these subjects, too, have something in common? The subject 11 = H11A does not have any break in the range predicted by the reaction time data. There is a late second break where the two layer working proceeds to a three layer working. This subject has no primary three layer working.

There are three reasons why I have not generally reduced the linear pathway by 2ET and increased the cyclical pathway by the same amount in the equation systems:

- there are some subjects with fast mode and triple search who have approximately the same linear pathway in the reaction time data and the ERP data. If one changes all linear pathways in subjects with triple search, the results of these subjects would deteriorate.
- the correlation between linEN(SEA) and linEN(SEB) shows deviations of linear pathway length in the repeated task. This variability may be observable in ERP tasks too.
- the probability of the minimal pathways (see Chapter 42) is p(mincycEN = 2) = 3.1% and p(mincycEN = 4) = 5.1% for slow mode plus triple search x22y tasks and p(mincycEN) = 2) = 6.3% for x22y tasks with fast mode plus triple search.

The conclusion therefore is that in some subjects (with triple search) the reaction time distribution does not show a first peak in the expected place at mincycEN = 2 but at mincycEN = 4. This implies an additional reduction of the linear pathway of 2ET and a prolongation of the cyclical pathway of the same amount. To take the decision where the first peak may be localized, ERP data may be helpful. It is not convenient to change the linear pathways in all subjects with triple search. In this work the linear pathways have not been changed at all after knowing the ERP data but the necessity to reduce the linear pathway may be assumed in some subjects.

The correspondence of reaction time data with event-related potentials in patients with schizophrenia

Until now only five patients with schizophrenia have taken part in this study. Therefore, only the presentation of single cases can be offered so far. It is interesting whether a prolonged linear pathway of some of these patients corresponds with a break (= transition point) shifted to the right. In the latency tables of these patients, the *largest gap* after a 3step area is taken to be the transition point from 3 step working to 2 step working. A cortical area is represented in the latency tables by a column.

Latency tables

S03r#2		AUD1	AUD2	AUD3	...				
5	+								
2,3	–	28	70	112	140–135	182	224	252–255	294
4b	+	14	56	98		168	210		280
4c	+	0	42	84–85	126	154	196–195	238	266

a22rS03A = 70 + (23 + 2(4 + 2*4))14, a22r-a10r, F:A2-Pz,

S03r#2			
5	+		
2,3	–	336	378–375
4b	+	322–325	364
4c	+	308	350

a22rS03A = 70 + (23 + 2(4 + 2*4))14, a22r-a10r, F:A2-Pz,

Remarks:

1. The ERP shows more than 27 steps without any clear boundary between linear and cyclical pathway. There are only two short gaps.

17

S03l#1		AUD1	AUD2	AUD3	...				
5	+								
2,3	–	24	60	84–85	120	144–140	168–170	204–202	228
4b	+	12	48–45		108			192–190	
4c	+	0	36	72	96	132	156–153	180	216

a22lS03A = 70 + (24 + 2(4 + 2*4))12, a22l-a10l, F:A1-Pz,

29

S03l#1							
5	+						
2,3	–	252	276–277	312	348	372	396–395
4b	+			300	336–340		
4c	+	240–237	264	288	324	360	384

a22lS03A = 70 + (24 + 2(4 + 2*4))12, a22l-a10l, F:A1-Pz,

Remarks:

1. There is no clear boundary between the linear and the cyclical pathway. There are two possible breaks the first after 17ET and the second after 29ET.

a22rS03A = 70 + (*23* + 2(4 + 2*4))14   22.5 + 17.4
a22lS03A = 70 + (*24* + 2(4 + 2*4))12   23.8 + 23.9

17

S05l#1		AUD1	AUD2	AUD3	AUD4	AUD5	AUD6	A7/C1	CYC2
5	+							234–240	260
2,3	–	26–20	65–65	91	117–117	156–160	182	221–215	247
4b	+	13	52–45			143		208	
4c	+	0	39	78–80	104	130–138	169	195–190	

a22lS05A = 70 + (18 + 2(4 + 1*4))13, a22l-a10l, F:A1-Pz,

S05l#1		CYC3	CYC4	CYC5			
5	+	286	312–308	338			
2,3	–	272–273	299	325–330	351	377–375	416–417
4b	+				351–350		403–400
4c	+				338	364	390

a22lS05A = 70 + (18 + 2(4 + 1*4))13, a22l-a10l, F:A1-Pz,

a22rS05A = 70 + (15 + 2(4 + 2*4))12.5   15.6 + 22.6
a22lS05A = 70 + (*18* + 2(4 + 1*4))13   18.4 + 11.2

12

S05r#2		AUD1	AUD2	AUD3	A4/C1	CYC2	CYC3	CYC4	CYC5
5	+				150	175	200	225–230	250
2,3	–	25	62.5–65	100	137.5	162.5–165	187.5	212.5	237.5
4b	+	12.5–15	50	87.5	125				
4c	+	0	37.5	75	112.5–110				

a22rS05A = 70 + (15 + 2(4 + 2*4))12.5, a22r-a10r, F:A2-Pz,

S05r#2		CYC6	CYC7	CYC8	
5	+	287.5	312.5	337.5	
2,3	–	275–280	300	325	387.5
4b	+				375
4c	+				350–350

a22rS05A = 70 + (15 + 2(4 + 2*4))12.5, a22r-a10r, F:A2-Pz,

15

S12r#2		AUD1	AUD2	AUD3	...		A6/C1	CYC2	CYC3
5	+								
2,3	–	13.5–15	54–55	81	121.5–120	162	202.5–200	229.5	256.5–260
4b	+		40.5–35		108	148.5	189		
4c	+	0	27	67.5	94.5–95	135–135	175.5	216–215	243

a22rS12B = 70 + (11 + 2(4 + 1*4))13.5, a22r-a10r, F:A2-Pz,

S12r#2					
5	+				
2,3	-	297–300	337.5–345	378–380	418.5–422
4b	+	283.5–280	324–320	364.5	405–405
4c	+	270	310.5	351–355	391.5

a22rS12B = 70 + (11 + 2(4 + 1*4))13.5, a22r-a10r, F:A2-Pz,

Remarks:
1. The largest gap after a 3step area comes after 15ET.

							14		
S12l#1		AUD1	AUD2	AUD3	...		A6/C1	CYC2	CYC3
5	+								
2,3	-	26–30	52–58	78	104	143–150	182–180	208	234–240
4b	+	13				130–132	169		
4c	+	0	39	65	91	117	156	195–202	221–225

a22lS12B = 70 + (15 + 2(4 + 1*4))13, a22l-a10l, F:A1-Pz,

S12l#1		CYC4	CYC5	CYC6	CYC7	CYC8
5	+					
2,3	-	260	299–295	338–335	377–380	416–420
4b	+		286	325	364	403
4c	+	247–252	273	312–310	351	390–395

a22lS12B = 70 + (15 + 2(4 + 1*4))13, a22l-a10l, F:A1-Pz,

Remarks:
1. There are two large gaps after a 3step area, the first at 2ET and the second at 14ET.

a22rS12A = 70 + (9 + 2(4 + 2))15	9.7 + 9.8
a22lS12A = 70 + (11 + 2(4 + 2*4))13.5	11.3 + 16.3
a22rS12B = 70 + (11 + 2(4 + 1*4))13.5	11.3 + 11.5
a22lS12B = 70 + (15 + 2(4 + 1*4))13	14.5 + 12.8

					10				
S16Ar		AUD1	AUD2	AUD3	A4/C1	CYC2	CYC3	CYC4	CYC5
5	+				165	195	225	255	295
2,3	-	15–20	60–55	105–105	150	180–172	210	240	270–275
4b	+		45	90	135–135				
4c	+	0	30–32	75–75	120				

a22r = 70 + (6 + (4 + 2*4))15, a22r-a10r, F:A2-Pz,

This pathway has a clear boundary between the linear and cyclical portion.

10

S16Al		AUD1	AUD2	AUD3	...	A5/C1	CYC2		
5	+								
2,3	–	15–15	45	75–80	105	150–155	180	210–205	255
4b	+					135			240–240
4c	+	0	30–32	60	90	120–125	165–163	195	225
a22l = 70 + (17 + (4 + 2*4))15,						a22l-a10l,	F:A1-Pz,		

Remarks:
1. The first gap does not come after a 3step area.
2. The next steps are: 270, 285, 300–*305*. That means linEN > 20.

Discussion

Because of the small number of patients in this chapter no definitiv results can be presented. All findings have to be replicated. The break between 3step and 2step working is difficult to find. Perhaps the linear pathway uses 2step working in the form of shortcuts.

Models of the xNNy pathways

Memory sets and set systems

The structure of memory sets

The elements of the proposed model are neuron sets. An element may be weakly or strongly active. If an element is active (weakly or strongly) it may activate a neighboring element (weakly or strongly). This lasts one elementary time.

An element is only activated strongly if it is activated by two or more elements which are synchronous to each other. Activations may be asymmetrical or symmetrical. Symmetrical activation means an element activates its neighboring element after one elementary time and the neighboring element reactivates the first element after another elementary time.

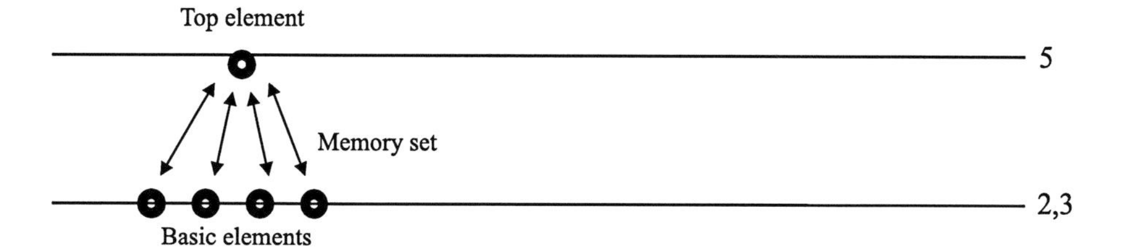

Fig. 141. Memory sets are pattern generators by symmetrical relations between their top and their basic elements. The top element activates one of its basic elements (by chance), then this basic element activates its top element etc..These oscillations may be spontaneous, asynchronous, and slow or stimulated, synchronous, and fast

The term "memory set" is used as the generic term for all these sets (task set, sensorimotor set etc.).The task set (TS) is a special memory set. The basic elements of this memory set are called task elements. These task elements do not have any real relationship to each other, therefore the task set can search on them randomly. If there would be any relationship between the task elements, this relationship would influence the searching sequence on the task elements. The sensorimotor set (SMS)

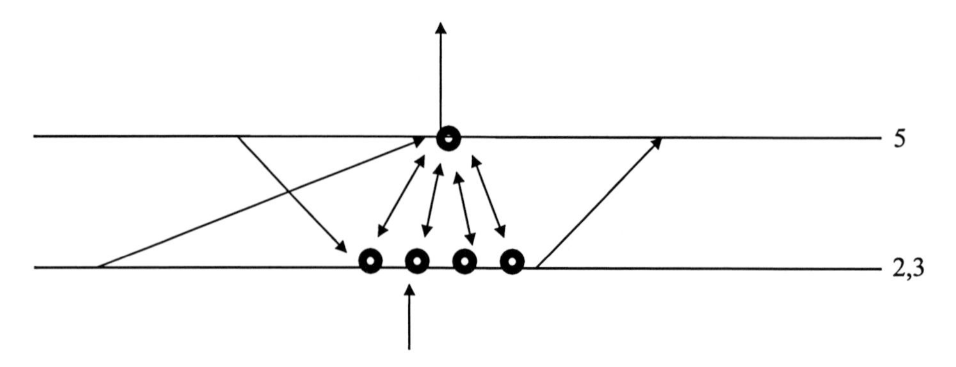

Fig. 142. Memory sets may have asymetrical inputs and outputs

is another specialized memory set. It combines a stimulus element with a response element. In other words: a sensorimotor set consists of one sensory and one motor element. Another name for the sensorimotor set (SMS) would be stimulus-response-set.

The spontaneous activity of memory sets

Spontaneous, asynchronous, and slow activity of memory sets

Here the top element weakly activates its basic elements one after the other. The sequence of basic elements is accidental. This may be referred to as the top element scans its basic elements or it searches on its basic elements. All the different spontaneously active memory sets are asynchronous to each other.

Such a spontaneous activity preserves the structural integrity (synapses and neurons) of the memory set. A decrease of spontaneous activity deteriorates the structure, some elements may break away and the memory set desintegrates.It is possible that the frequency of the spontaneously active memory sets represents its importance within the whole structure (= > personality)? That means there would exist many different spontaneous frequencies.

Consequences of the slow spontaneous asynchronous activity of task sets

The spontaneity of the task set has some important consequences for the performance of a task. The subject's response is faster because the top-down relation of the first searching cycle does not appear in the reaction time because it is simultaneous to the afferent weak activation of the stimulus element. This activation of the stimulus element is considered as part of the linear pathway. The last bottom-up relation of the first search is simultaneous to the (2,3 –>5) motor relation. Therefore only (n-2) cycET are really observable as the length of the cyclical pathway.

Slow spontaneous asynchronous activity of sensorimotor sets

Sensorimotor sets are special memory sets with two basic elements, a stimulus element and a response element. Sensorimotor sets may also be spontaneously active. This spontaneity is the prerequisite of staying alive (housekeeping activity). Such an asynchronous, weakly active sensorimotor set cannot elicit any action. The asynchronicity prevents from interfering with the activation of the stimulus element by the combined effort of the task set and the input. The weakness prevents from activating the response element.

Fast stimulated synchronous activity

After the arrival of the stimulus, the stimulus element is activated simultaneously by the task set and by the input. This causes a strong activation of the stimulus element and a subsequent strong activation of the sensorimotor set.

There is a frequency for the top element of a memory set ($f_{top} = 1000 / (2 * ET)$). Different from this is the frequency of its basic elements. This depends on the number of elements of a memory set: $f_{basic} = 1000 / ((2 * ET) * n)$

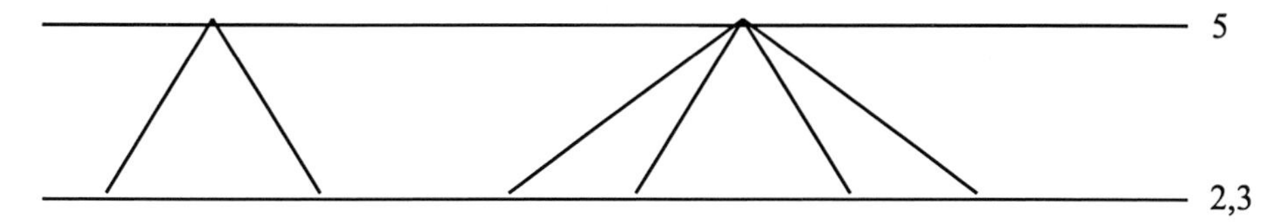

Fig. 143. A four element memory set stimulates its basal elements more rarely than a two element memory set

Memory sets in the cerebellum

In the cerebellum, the Purkinje cells represent the top elements and the granular cells the basic elements of memory sets. The climbing fiber increases the activity of the top elements, the mossy fiber of the basic elements. The output comes from the the top elements. The stellar and basket cells produce the lateral inhibition between the top elements, the Golgi II cells between the basic elements. Dentate cells fire preferentially at the onset of movements that are triggered by mental associations with either visual or auditory stimuli.These results suggest that the dentate helps initiate movements triggered by stimuli which are mentally associated with the movement. In tasks where movements were triggered by light, the order of activity was dentate, motor cortex, interpositus, muscles. Lesions of the dentate produce a slight delay in the reaction time of movement triggered by light or sound (Zigmond 981,982).The motor memory sets (= motor pattern generators) of the cerebellum are tonically active even in the abscence of movement. They exert a lasting effect on other motor pattern generators in the vestibular and reticular nuclei and the motor cortex (via the thalamus).

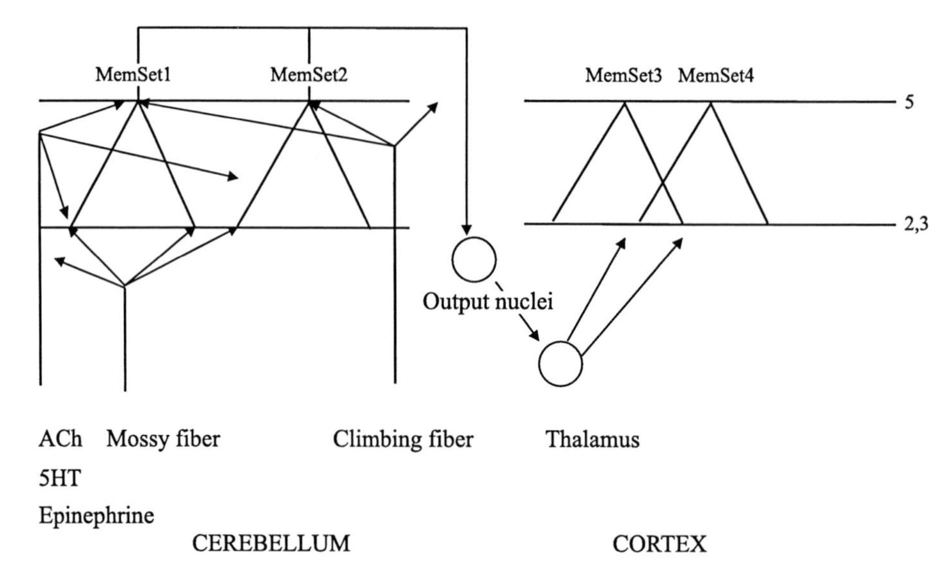

Fig. 144. The motor memory sets (= motor pattern generators) of the cerebellum are tonically active even in the absence of movement. They exert a tonic effect on other motor pattern generators in the vestibular and reticular nuclei and the motor cortex (via the thalamus)

Set systems

A set system is a set of memory sets which interact to perform a common task. The memory sets participating in a set system show a strong, synchronous, and fast activity. The memory sets of a set system are all synchronous to each other, i.e. display all the same frequency and phase. If all memory sets had different frequencies and phases it would be too difficult to interact.

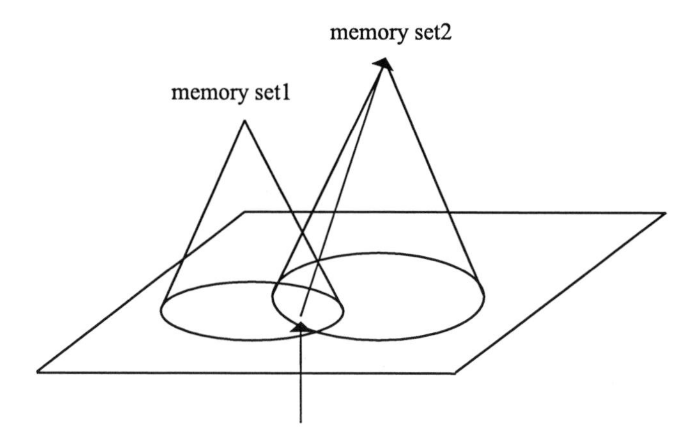

Fig. 145. Memory set1 scans one of its basic elements which gets an input. As a consequence, this basic element is strongly activated and may activate the memory set2

If the set system is not performing a task, each memory set of a set system is weakly and asynchronously oscillating between its top and its basic elements. The arrival of a stimulus gives rise to the synchronisation and strong activity of the participating memory sets.

Between the memory sets exists a lateral inhibition. But there is cooperation, too: memory sets may increase the frequency of other memory sets which have common basic elements with them.

There arises an interesting question, if two sensorimotor subsets oscillate within an oscillating task set. Are the three memory sets synchronous or asynchronous to each other?

The order of activation and cancellation of memory sets within a set system

In a reaction time task, the spontaneously oscillating task set is the first set to be activated because of its low threshold. If one of the task elements becomes strongly activated, additional memory sets attached to these task elements are activated. This occurs by *spatio-temporal addition* at the top elements of these memory sets. These memory sets are activated in the order of their threshold level: the higher the threshold, the later their activation. Examples of such later memory sets are the sensorimotor sets. The spontaneously oscillating task set ist the earliest memory set, the later memory sets are only temporarily active. If a later memory set is activated, one may say that the earlier task sets have *instructed* this later memory set. This process means a recruitment of additional memory sets above the same task elements or additional task elements (examples of this process are the additional memory sets which are produced during the search process of x22y). A strongly activated memory set is represented by an increased number of firing neurons in the elements of this memory set.

The order of decline is determined by the threshold once again: the higher the threshold, the earlier the *cancellation* of the memory set. The system of memory sets is reduced by inhibiting influences. The top elements (and with them the memory sets) with the highest thresholds return first to the resting state. The inhibition set can reduce the activity of the task elements (not to zero) and therefore switch off the memory sets with high thresholds.

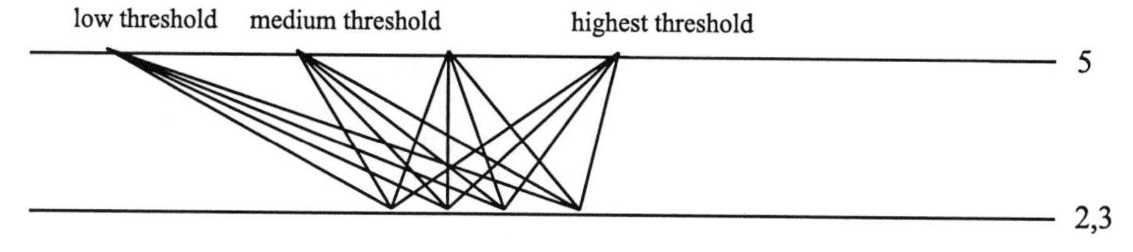

Fig. 146. Four memory sets with different thresholds

If there are many memory sets using the same basic elements, two (or more) of them may activate one common basic element by chance. As a consequence this element is strongly activated and may activate another memory set. These parallel oscillations are shown in the figure below.

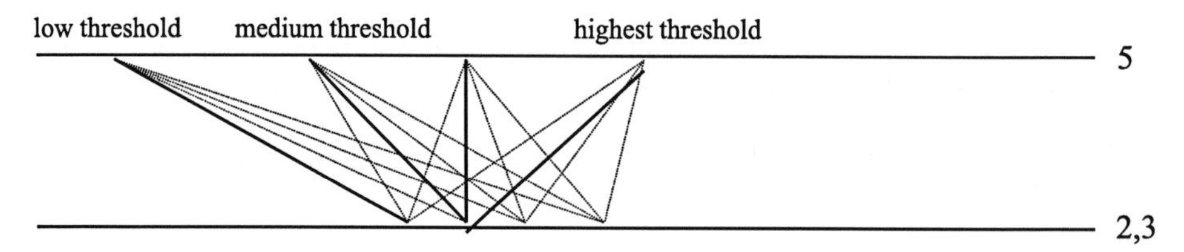

Fig. 147. Set system with memory sets having different thresholds. The present oscillations are drawn boldly

The memory sets with a higher threshold are activated in two ways. The active sets meet with each other or an afferent, for example, the target stimulus element.

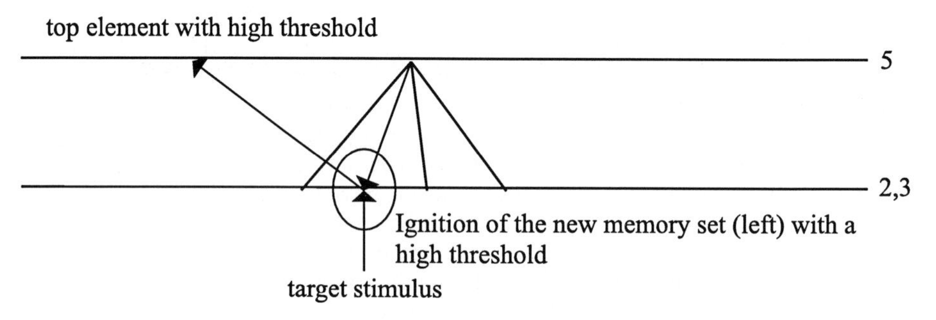

Fig. 148. The memory sets with a higher threshold are activated in two ways. The active sets interact with an afferent (e.g. the target stimulus element) or interact with each other

The Set System of x11y

The set system of x11y contains two memory sets, a task set and a sensorimotor set. Omitting the inhibition set, one has the following structure for the x11y set system:

1. The task set spontaneously searches for the stimulus.
2. The coincidence of the task set oscillation and the afferent stimulus generates a strong activity in the stimulus element. This stimulates the top element of the sensorimotor set. The sensorimotor set oscillates on the stimulus element and the response element.
3. If the sensorimotor set finds the response element, this element causes the motor output.

There exist some important variations:

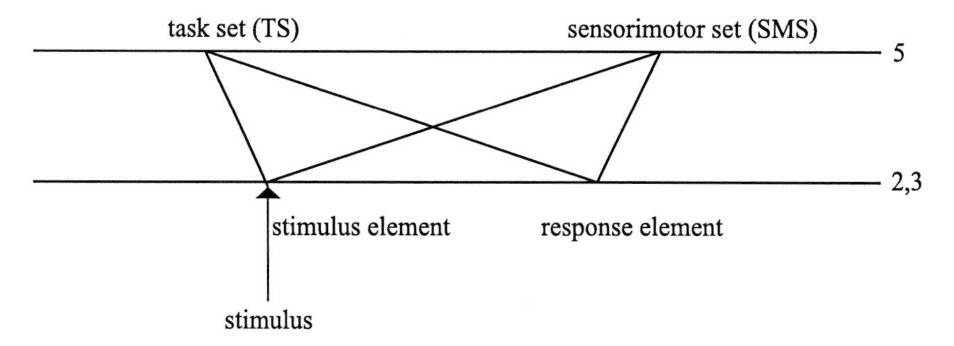

Fig. 149. The set system of x11y

1. The oscillating sensorimotor set activates the response element so much that the motor output is possible without any additional activation by the task set (double search). Alternatively the sensorimotor set cannot activate the response element alone and needs another set to do this (the task set, for example, in triple search).
2. The sensorimotor set does not oscillate at all but activates the response element right after the activation of its top element (fast mode). A search of the sensorimotor set for the response element needs a longer time (slow mode).

The structure of the x11y set system should be compatible with the structure of the x22y set system: the sensorimotor set of x11y is included in every higher task set.

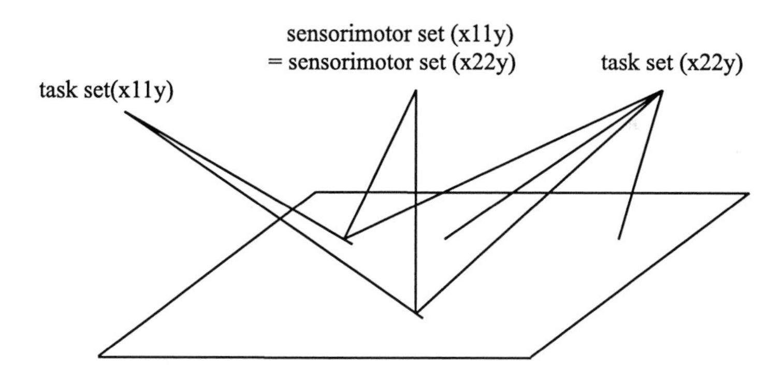

Fig. 150. The sensorimotor set of x11y is the same as in x22y. This is the cause of the implicit learning axiom: If a sensorimotor set uses the fast mode in x11y, it should use the fast mode too in x22y

The minimal pathway of x11y

How long does the *minimal search* in x11y take? I have defined the minimal pathway of x11y as a truly linear pathway without any search.

In this figure, the third set which causes the triple search is drawn apart. It is not known whether this third searching set is identical with the task set or whether it is a "second task set" built in the instruction phase.

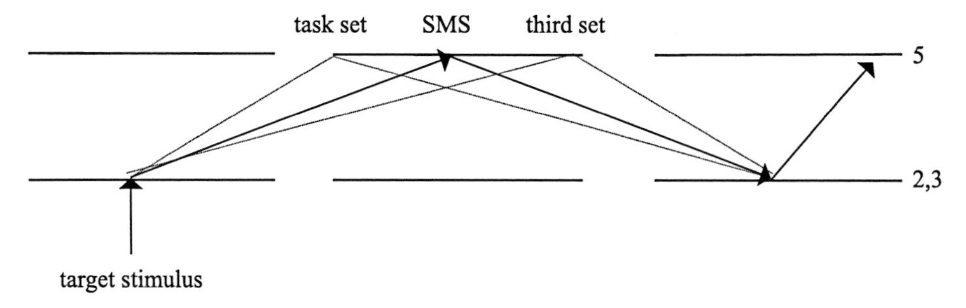

Fig. 151. Minimal pathway of x11y with sensory and motor linear part and a minimal cyclical pathway of length 2ET (thick lines)

A minimal linear pathway is conceivable by the accidental coincidence of the spontaneously oscillating task set and the target stimulus in layer 2,3 of the sensory area. One has to distinguish the following different constructions of the x11y pathway:

- fast mode, double search
  If the SMS is learned very well, it is activated instantly and the minimal cyclical pathway (mincycEN) lasts only 2ET.

- slow mode, double search
  in this case, the shortest pathway is the same as in fast mode, double search. Only the probability is lower than in fast mode, double search (because there could be extra cycles scanning the stimulus element before scanning the response element).

- fast mode, triple search
  here too the shortest pathway has the length 2ET but the probability is lower than in fast mode, double search because the third searching set has to coincide with the sensorimotor set in order to activate the response element.

- slow mode, triple search
  in this case the shortest pathway of length 2ET is still more unprobable than in the cases above because both the sensorimotor set and the third searching are scanning by chance and have to meet in the response element in order to activate it.

The probability that the task set activates the stimulus element when the target stimulus arrives is *1/number of task elements*. The cases which are not minimal may vary up to several searching cycles. More variation comes from the length of the linear pathway (linEN). The minimal length of the linear pathway (minlinEN) which was used to formulate the equations of the median pathway is considered as *the most frequent* linear time. This prominence of minlinEN is proven by the latency tables of the event-related potentials.

In the case of x11y the probability of a zero search of the task set for the stimulus element is 1/number of task elements = 1/2.

In the fast mode, double search variant the response element is activated with a probability of 1 only 2ET later.

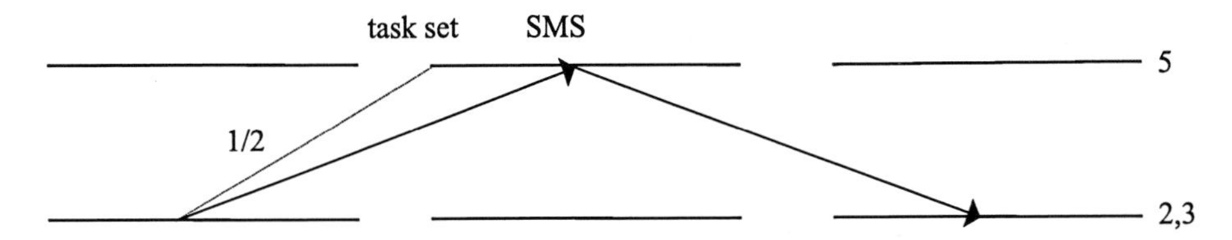

Fig. 152. Probability of the minimal x11y pathway in fast mode, double search: p(mincyc EN = 2) = 1/2*1

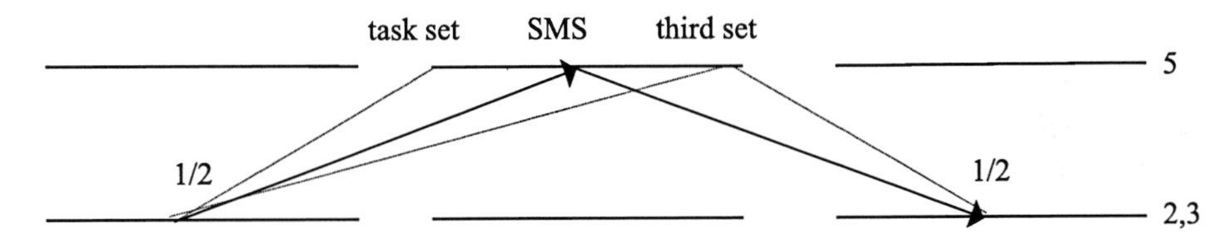

Fig. 153. Probability of the minimal x11y pathway in fast mode, triple search: p(mincyc EN = 2) = 1/2*1/2

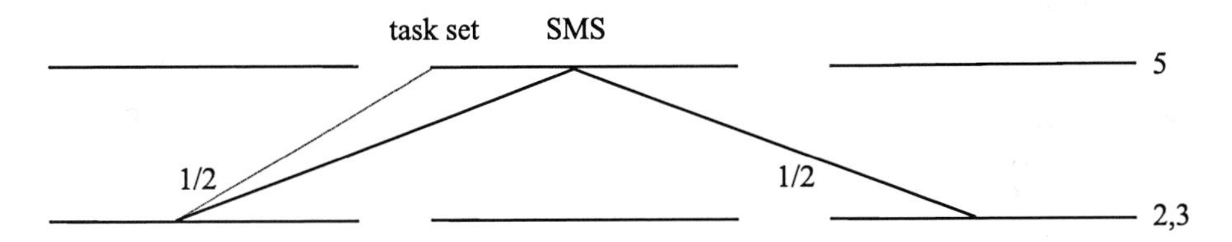

Fig. 154. Probability of the minimal x11y pathway in slow mode, double search: p (mincycEN = 2) = 1/2*1/2

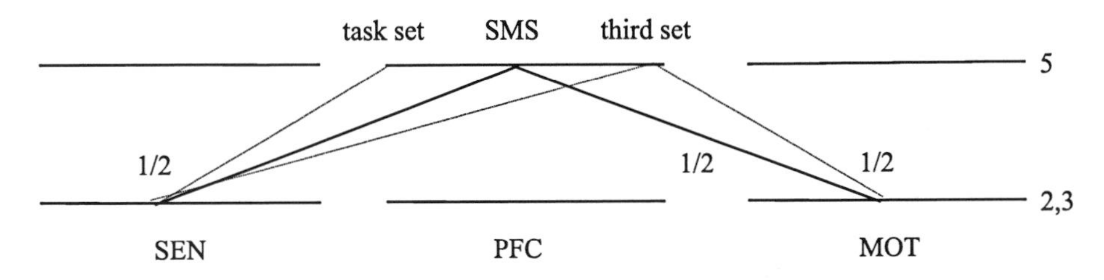

Fig. 155. Probability of the minimal x11y pathway in slow mode, triple search: p(mincyc EN = 2) = 1/2*(1/2 *1/2) = 1/8 = 4/32

The probability of the minimal x11y pathway in slow mode, triple search is p(mincycEN = 4) = 3*(1/2*(1/2 *1/2)*(1/2*1/2)) + (1/2*1/2*1/2*1/2) = 3/32 + 1/16 = *5/32*.

These numbers have to be considered in determining the equations of median x11y pathway.

The set system of x22y

The arriving target stimulus can meet the spontaneously oscillating task set. In the task x22y, the average number of searching cycles of the task set until it finds the stimulus element is (4 + 1)/2 = 2.5 cycles if the wrong choices are remembered, and 4 cycles if this is not the case.

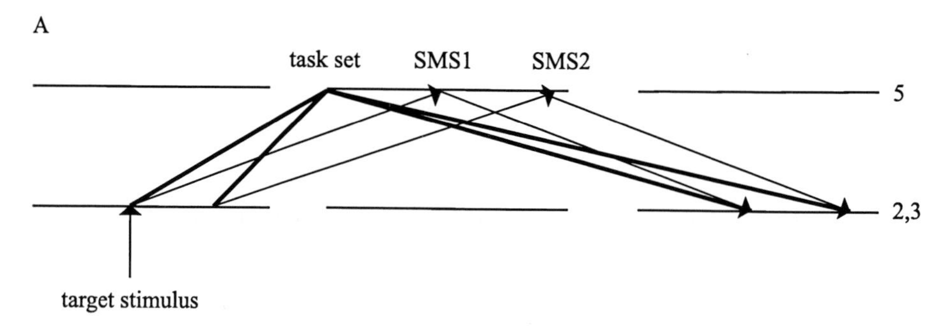

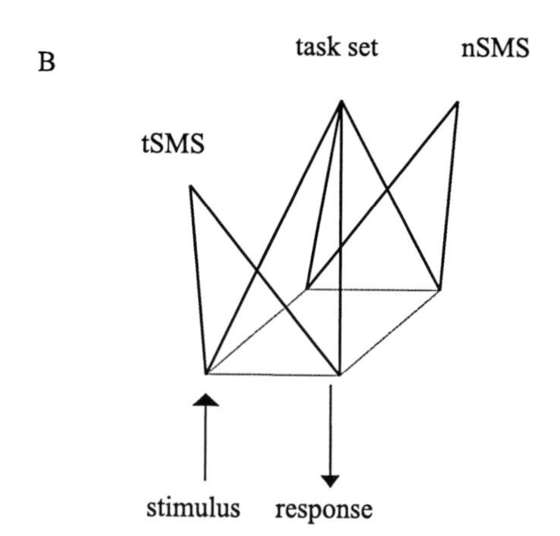

Fig. 156. Set system of x22y. (A) (tSMS = target sensorimotor set, nSMS = nontarget sensorimotor set). (B) An equivalent figure of the set system (A) is the respresentation shown in (B)

The sequence is the same as in the task x11y: at first only the task set is active.

After finding the stimulus element, another memory set (the sensorimotor set) with a higher threshold is activated.

The minimal pathway of x22y

The structure of the fastest trials of x22y

Which is the structure of the fastest trials of x22y? First comes the linear part of the x22y pathway, then the cyclical part of this pathway. The shortest linear pathway which uses the prefrontal cortex is:

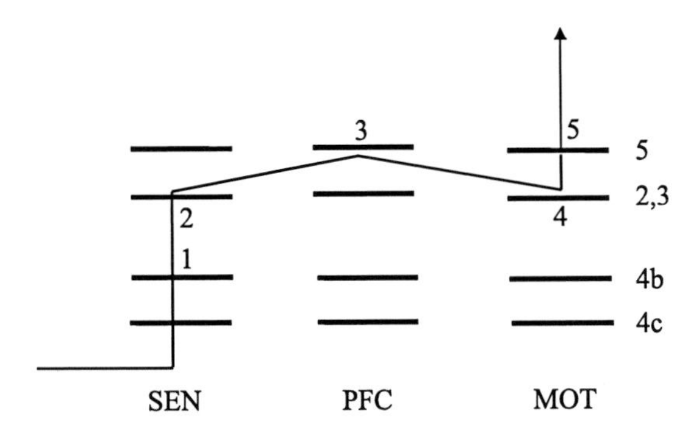

Fig. 157. Fastest cortical pathway of x22y. The above pathway uses five steps. The linear pathway lasts 3ETs, the additional search would last minimally 1 cycle time (1CT = 2ET). That means the minimal time is 3 linET + 2 cycET = 5ET

These considerations are valid for the *fast mode*. In this mode there may be no search at all.

The task set is highly oscillating *before* the stimulus arrives. That means in minimal pathways there is no search at all for the stimulus and the sensorimotor sets are so strong that the arriving target stimulus may activate its coordinated response element 2ETs later:

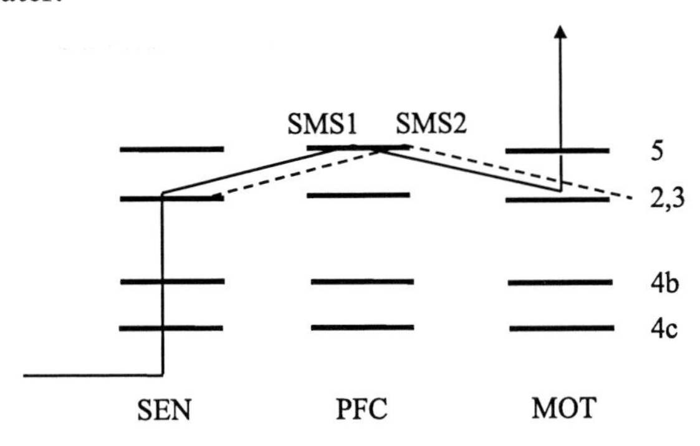

Fig. 158. Fastest cortical pathway of x22y.The task set has been omitted. Its basic elements are the four task elements and its top element lies at the same level as the top sensorimotor elements

In this fast mode, the minimal pathways (first peak pathways in the reaction time distribution) last only 5 linET (like the x11y trial). An example of this mode is subject H01 (see his results).

The accidental coincidence of the spontaneously oscillating task set and the target stimulus

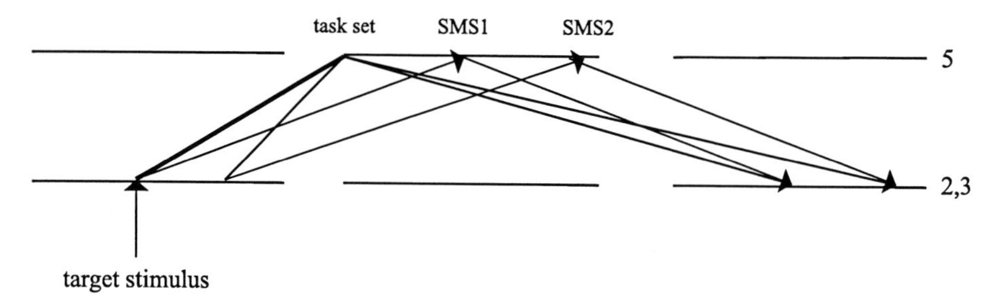

Fig. 159. The set system of the task x22y. The facultative third set causing the slow mode has been omitted

One hypothesis how a minimal pathway (first peak) with 5 ET is conceivable is the accidental coincidence of the oscillating task set and the target stimulus in layer 2,3 of the sensory area. In the fast mode the SMS1 is implicitely learned so well that it activates the response element 2ETs later. This is only possible in highly learned strong SMS and appears especially in the task v22r. The average number of task set oscillations is 4 cycles until the target stimulus is found.

Variations of the minimal pathway of x22y

Strictly speaking the above statements are only valid for the fast mode, double search variants.

The other variants (slow mode and/or triple search) show a much lesser probability of this minimal pathway with mincycEN = 2. In the case of x22y the probability of a zero search of the task set for the stimulus element is 1/number of task elements = 1/4. In the fast mode, double search variant the response element is activated with a probability of 1 only 2ET later.

The probability of the minimal pathway p(mincycEN = 2) for the different variants is as follows.

Probability of the minimal x22y pathway in fast mode, double search:
p(mincycEN = 2) = 1/4*1 = 64/256

Probability of the minimal x22y pathway in fast mode, triple search:
p(mincycEN = 2) = 1/4*1*1/4 = 16/256

Probability of the minimal x22y pathway in slow mode, double search:
p(mincycEN = 2) = 1/4*1/2 = 32/256

Probability of the minimal x22y pathway in slow mode, triple search:
p(mincycEN = 2) = 1/4*(1/2 *1/4) = 1/32 = 8/256

Probability of the minimal x22y pathway in slow mode, triple search:

$$p(\text{mincycEN} = 4) = 7*(1/4*(1/2*1/4)*(1/2*1/4)) + 3*(1/4*1/4*(1/2*1/4) = 13/256$$

The probability that mincycEN = 4 is determined by the number of processes which lead to a coincidence at 4ET. If one draws a tree of possible processes, one gets the above probability of 13/256.

The low probability of mincycEN = 2 in slow mode, triple search is the reason why the length of the linear pathway in the event-related potentials of subjects using this variant is shorter than the reaction time data (with mincycEN = 2) suggest. In the reaction time distribution of these subjects the break between the linear and the cyclical pathway does not lie mincycEN = 2ET left of the first peak but for example 4ET left of the first peak (see above).

Discussion

Previous models of the x11y pathway

There are many hypotheses to explain the distribution of x11y:

- *Variation of linear pathway*, the numbers (x11y-con)/n with n = 5,7,8,10,11,13,15 lay in the near of the plots in the chronophoresis. Which is the cause of these numbers?
- *Some kind of search mechanism.*
- *Some kind of inhibition* of the previous task, the present task has to wait until the previous task is inhibited. The question is, whether unknown tasks are inhibited or the same task (previous trial).
- *Perhaps another task has to be completed* before the new task may be attacked
- *"Next train phenomenon".*

I have doubted my previous view of the x11y pathways because of the trials with very long reaction times. There is no linear pathway which can represent these very long times. The trials with a high reaction time can only be explained by cycles. Another argument are the peak latencies of the ERP(x11y).

The previous conception for the x11y pathway proposed the variation of the number of areas

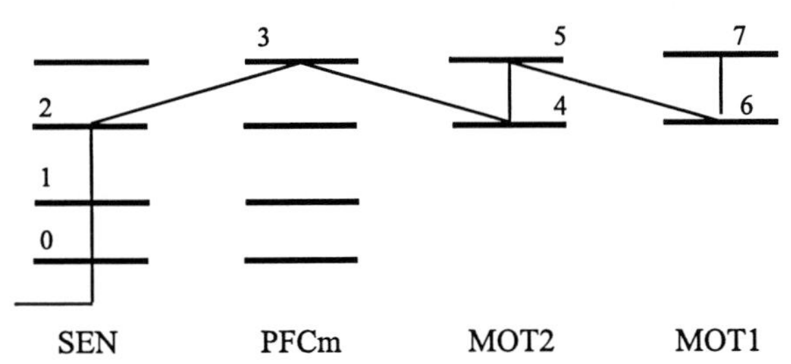

Fig. 160. The 5ET pathway and the 7ET pathway with one sensor area and one or two motor areas

1. The 5ET pathway and the 7ET pathway with one sensor area and one or two motor areas.

Does PFCm operate only on SEN2 and MOT2?

2. The 8ET and 10ET pathways (version1) with two sensor areas and one or two motor areas

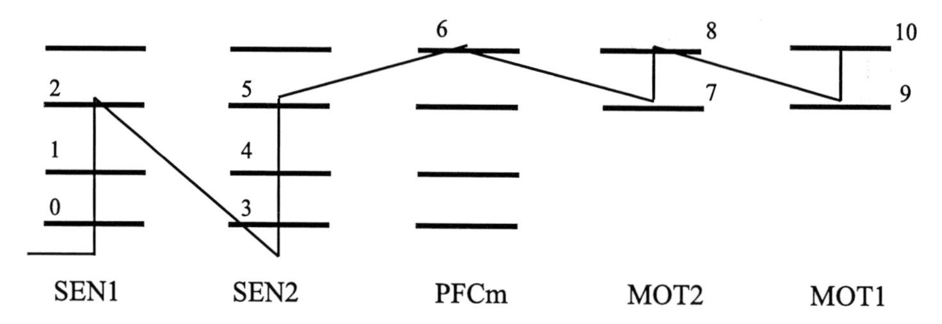

Fig. 161. The 8ET and 10ET pathways (version1) with two sensor areas and one or two motor areas

3. The 11ET pathway and the 13ET pathway with three sensor areas, the PFC area and one or two motor areas:

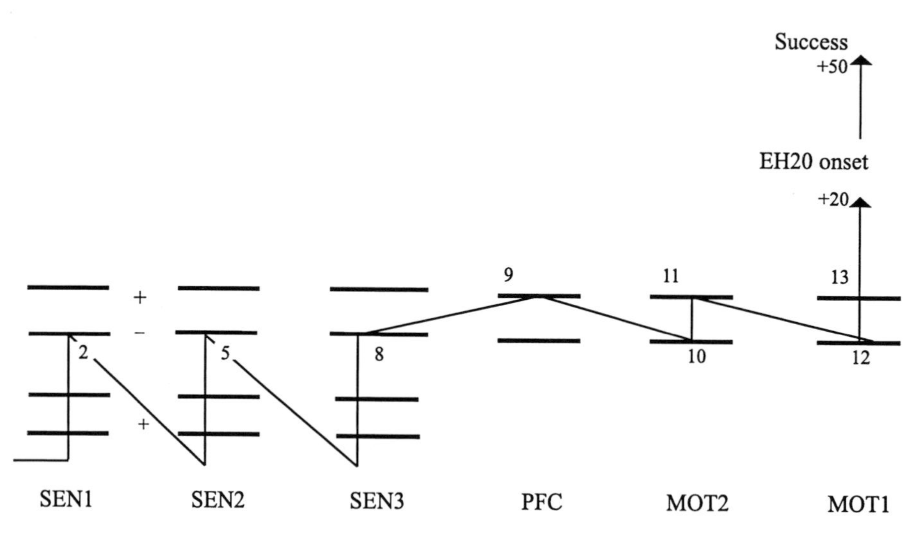

Fig. 162. A purely linear 11ET or 13ET pathway with three sensory areas, the PFC area and one or two motor areas

Now an alternative model of the a11lH23 pathway:

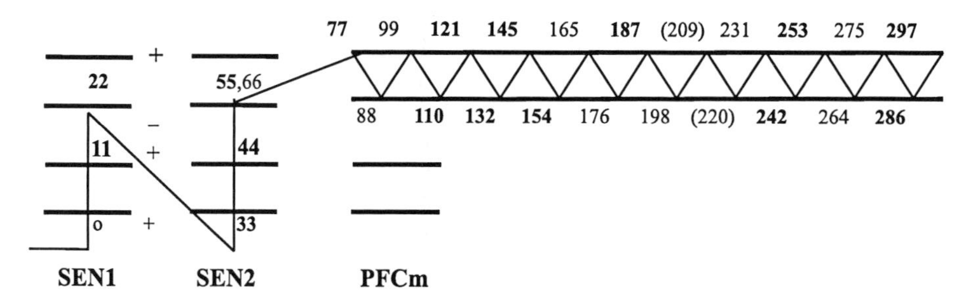

Fig. 163. The model of the a11lH23 pathway with two sensory areas and cycles within the PFC. The hypothetical latencies are multiples of linET = 11ms

This last model shows a better fit to the latencies of (a11il-a10l)H23 if one introduces a delay of one linET. The linET = 11ms for a11lH23 has been taken from SINGLE and FPM results.

The bold latencies correspond to latencies in the ERP; the latencies printed in brackets contradict latencies of ERP, and the other latencies do not appear in the ERP.

What could be the purpose of these cycles? Are they similar to the x22y searching cycles? Is there a random process, too?

Conclusive structure of the x11y pathway

The x11y pathway:

- There is always a linear portion of the x11y pathway.
- Sometimes there is an inhibitory portion of the x11y pathway (if a competitive task is present).
- Mostly there is a search portion of the x11y pathway (except the minimal x11y pathways).

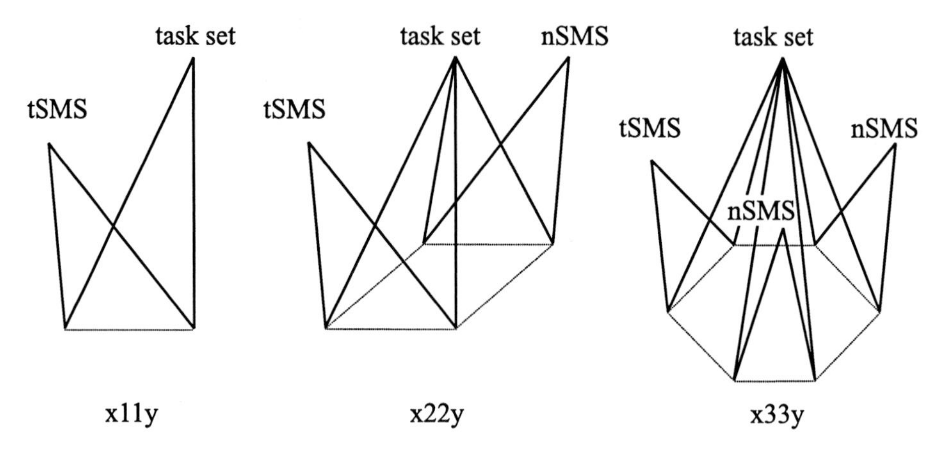

Fig. 164. Models of the xNNy pathways. A third searching set which causes the triple search has been omitted. tSMS = target sensorimotor set, nSMS = non-target sensorimotor set

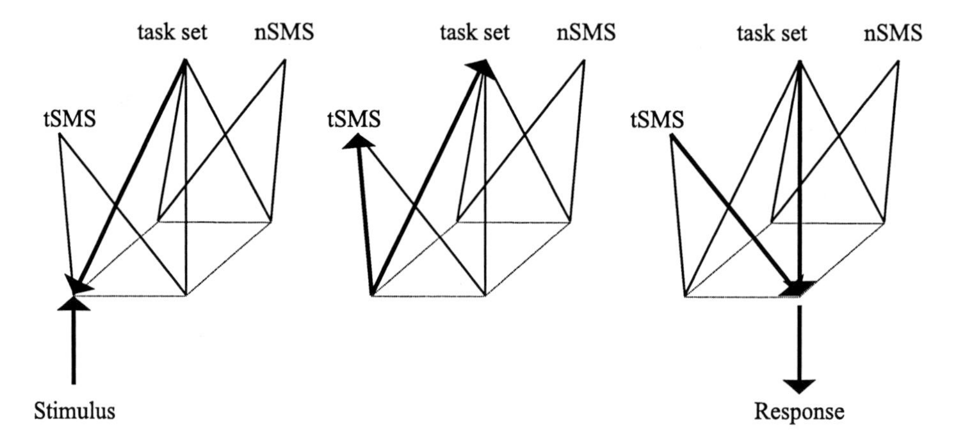

Fig. 165. The above sequence shows the minimal pathway for x22y (fast mode, triple search). As the third memory set, the task set has been used. Because of the spontaneity of the task set, the first search can be simultaneous to the arrival of the stimulus. Then comes the tSMS and eventually the third set (represented here by the task set)

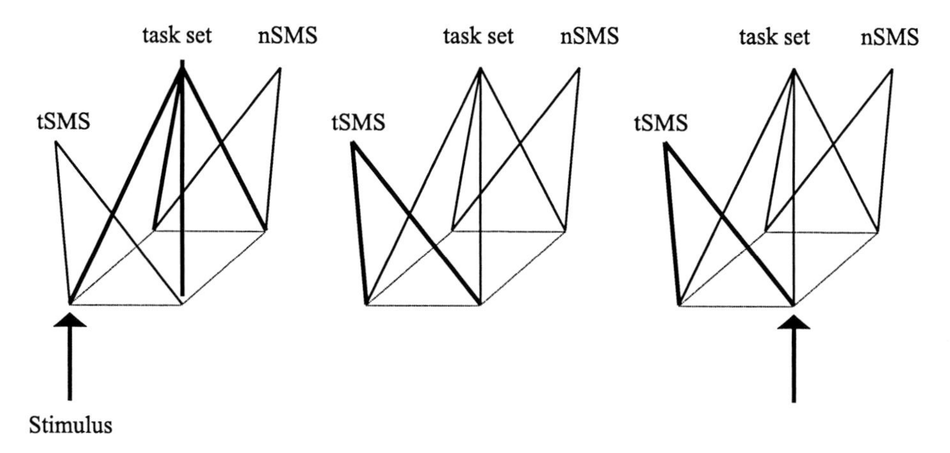

Fig. 166. The progression of the x22y set system. At first, only the task set searches for its stimulus element.Then the sensorimotor set is activated. At last, the third set (not drawn) supports the search for the response element. The basis elements are supposed to be part of the layer 2,3 and the top elements of layer 5 in the cortex cerebri

The Simulation of Set Systems

This chapter deals with the simulation of set systems and its contribution to the validation and understanding of the model. Knowing the spatiotemporal structures of the stimulus- response pathways, it is possible to emulate them in a computer program. This simuluation produces the same distributions as the subjects (if the equations of the program and the subject are identical).

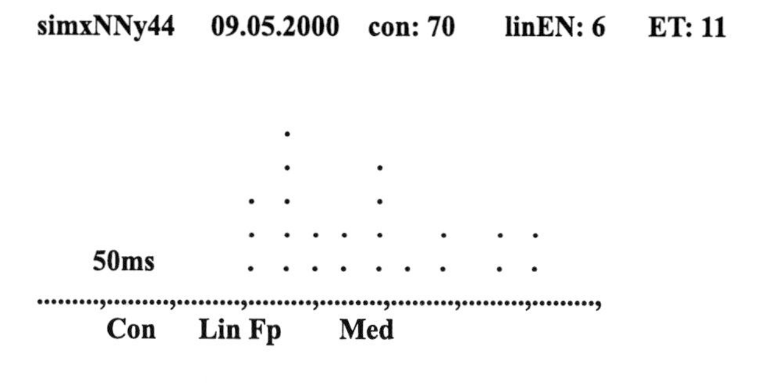

Fig. 167. Output of the simulation program SIMxNNy44/45. The program produces the distribution, the first peak (Fp), the median reaction time (Med), the internal meancycEN, and the externally observable (med-lin)/ET.Only the peaks of the x11y distribution are simulated, the intermediate values are produced by a small random factor which has to be added (± z). A comma means 50ms on the x-axis. The times of Con, Lin etc. are given in millisecond, too

Time	tSMS	tSET	nSMS	tSTIM	tRESP	nSTIM	nRESP
-1	tSMS 0	**tSET 1** (spon)	nSMS 0	tSTIM 0	tRESP 0	nSTIM 0	nRESP 0
0	tSMS 0	tSET 0	nSMS 0	**tSTIM 1** (aff)	tRESP 0	**nSTIM 1** (scan)	nRESP 0
1	tSMS 0	**tSET 1** (rev)	nSMS 0	tSTIM 0	tRESP 0	nSTIM 0	nRESP 0
2	tSMS 0	tSET 0	nSMS 0	**tSTIM 2** (aff+scan)	tRESP 0	nSTIM 0	nRESP 0
3	**tSMS 2** (rev)	**tSET 1** (rev)	nSMS 0	tSTIM 0	tRESP 0	nSTIM 0	nRESP 0
4	tSMS 0	tSET 0	nSMS 0	**tSTIM 1** (aff)	**tRESP 2** (scan)	nSTIM 0	**nRESP 1** (scan)
5	**tSMS 2** (rev)	**tSET 1** (rev)	nSMS 0	tSTIM 0	tRESP 0	nSTIM 0	nRESP 0
6	tSMS 0	tSET 0	nSMS 0	tSTIM 1	tRESP 3	nSTIM 0	nRESP 0

Fig. 168. The set system of the task (x22y) with successive interaction between three memory sets and four basic elements: the task set tSET with its four basic elements, the target stimulus element tSTIM, the target response element tRESP, the non-target stimulus element nSTIM and the non-target response element nRESP. The target elements are scanned by the target sensorimotor set tSMS and the non-target elements by the non-target sensorimotor set nSMS. The first facilitation of tSTIM by the stimulus is time = 0 because of the spontaneity of the task set. In minimal pathways it is this moment where tSTIM may be facilitated both by the task set and the stimulus. An activation of tRESP (represented by value = 3) results in a motor efferent from the system

The internal structure of simulation programs

The value of the con = input time + output time is constant. The length of the linear pathway (linEN) and the size of the elementary time (ET) are determined before running the program. The precondition for the simulation of a pathway is the precise measurement of the elementary time and the length of the linear pathway. Furthermore, the mode and the number of searches should be known.

Before the program runs, further parameters have to be entered: x11y or x22y, fast mode or slow mode, and double search or triple search. Now the program produces the cyclical pathway by using the following strategy:

1. Spontaneous random search of the task set on its basic elements
2. Efferent facilitation of the stimulus element
3. The task set finds and activates the stimulus element
4. The active stimulus element activates its sensorimotor set
5. Random search of the sensorimotor set on its basic elements for the response element
6. The sensorimotor set finds and activates the response element

The number of searching cycles is registered and the subsequent output is generated.

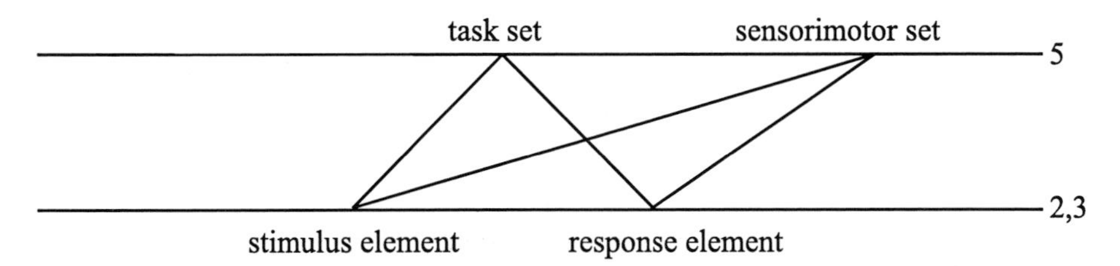

Fig. 169. The connections between the two memory sets and the two basic elements in the task x11y

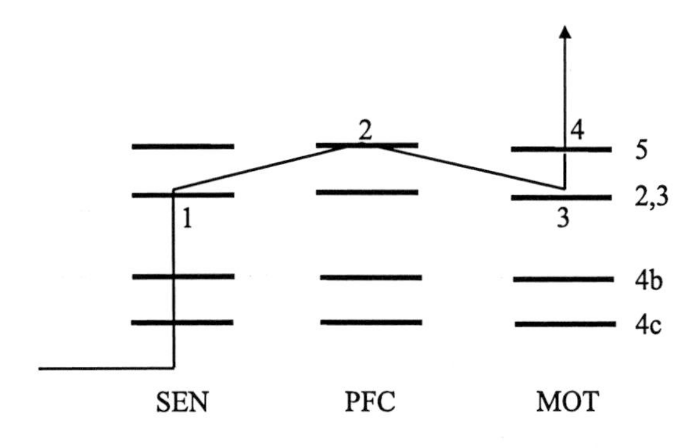

Fig. 170. A pathway with a linear portion of 2 ET (skipping the layer 4b of SEN) and a cyclical portion of 2ET. This cyclical portion may consist of a searching process which needs 5 cycET in v11rH01A (see above)

The program allows the observation of its "interior" by watching the memory sets working (with fictive values of the simulated elements because their real values are not known yet).

The possible representation of the memory sets (of the x11y task) within the cerbral cortex are shown in the last two figures. The last figure shows the minimal pathway of the task x11y.

In the case of a minimal cyclical pathway of 2ET, both elementary times are used by the sensorimotor set. The task set finds (scans) the stimulus element at the moment of its facilitation by the stimulus. How can the first peak (Fp) be as early as 2ET after the linear path? When the target stimulus arrives in layer 2,3, the task set is *just activating* the target stimulus element. As a consequence, the target SMS can become active and the central portion of the pathway can be transversed in 2 ET.

Simulation of xNNyH01A and xNNyH32A with different strategies

The following simulations used the linEN and the ET of subject H01A and H32A. Very different strategies were used.

Table 38. Simulation programs for special tasks

Program	Task	Goal	Notes	Simulated mean cycET	Predicted mean cycET
SIMxNNy11	v22rH01A	target stimulus	without working memory	7.84	4 cycles = 8 cycET
SIMxNNy12	v22rH01A	target response	without working memory, random TRESP. Task set must not find TRESP = > 4 elements + 2 elements	11,8	12 cycET = (8 cycET + 4 cycET)
SIMxNNy13	v22rH01A	target response	without working memory, immediate transition from TSMS to TRESP	10.1	10 cycET (8cycET for the task set and 2 cycET for TSMS)
SIMxNNy14	v22rH01A	target response	as 13, task set has to detect the target response	16.3	18 cycET (8 cycET + 2 cycET + 8 cycET)
SIMxNNy15	v22rH01A	target response	random task set + random SMS + random task set	22.9	20 cycET (8 cycET + 4 cycET + 8 cycET)
SIMxNNy16	v11rH01A	target stimulus	2 elements, random search	4.1	4 cycET
SIMxNNy17	v11rH01A	target response	random task set, random SMS, no detection of target response by task set	7.75	8 cycET (4 cycET + 4 cycET)

Table 38. Continued

Program	Task	Goal	Notes	Simulated mean cycET	Predicted mean cycET
SIMxNNy18	v11rH01A	target response	as 17 but immediate transition from TSMS to TRESP	5.9	6 cycET (4 cycET + 2 cycET)
SIMxNNy19	v11rH01A	target response	random task set, immediate tranistion TSMS-TRESP, random detection of target response by task set	8.1	10 cycET = (4 cycET + 2 cycET + 4 cycET)
SIMxNNy20	v11rH01A	target response	random task set, random SMS, random detection of TRESP by task set	11.9	12 cycET = (4 cycET + 4 cycET + 4 cycET)
SIMxNNy23 (21–23)	v22rH01A	target stimulus	with flop set	5	5 cycET = (4 + 1)/2 cycles
SIMxNNy24	v22rH01A	target response	flop set, immediate transitions from TSMS to TRESP	7	7 cycET = 5 cycET + 2 cycET
	v22rH01A	target response	flop set, random transition from TSMS to TRESP		9 cycET = 5 cycET + 4 cycET
	v22rH01A	target response	flop set, flop set transition from TSPS to TRESP		8 cycET = 5 cycET + 3 cycET (2.5 cycles + 1.5 cycles)
	v22rH01A	target response	flop set, immediate transition, flop set in detecting the TRESP		12 cycET = 5 cycET + 2 cycET + 5 cycET
SIMxNNy25	v22lH01A	target response	flop set, immediate TSMS		7 cycET = 5 cycET + 2 cycET
SIMxNNy26	v11lH01A	target response	flop set, immediate TSMS n = 2 = > (2 + 1)/2 cycles = 3 cycET		5 cycET = 3 cycET + 2 cycET
SIMx11y20	a11rH32A	target response	2cycCT + 1cycCT		3 cycles = 6 cycET
SIMx11y21	a11rH32A	target response	2cycCT + 2cycCT	8cycET	8cycET
SIMxNNy40	a11rH32A	"	1.5cycCT + 1.5 cycCT		
SIMxNNy42	a11rH32A	"	2cycCT + 2cycCT	6	
SIMxNNy43	a11rH32A	"			

The random search in n elements takes n cycles, on average, with 1 cycle = 2 cycET. The number of task elements in x22y is always 4, the number of elements of a sensorimotor set is always 2.

SIMxNNy42 and SIMxNNy43 are the most important simulations because I changed the internal enumeration of cycles. The program starts counting at n = 2. That means, it subtracts 2 cycET from the internal number because these 2 cycET cannot be seen externally. One must not count the descending relation of the first searching cycle and the ascending relation of the last searching cycle because these activities are simultaneous to the arriving stimulus respectively the activation of the layer 5 of the motor area.

Comparison of simulation results with reaction time data

Simulation of tasks xNNyH01A

The simulation of v11lH01A by SIMxNNy26 produces a computed median value of 260 ms with a real median value of 257ms. The distribution produced by SIMxNNy25 is very similar to the real distribution of v22lH01A. It is simple to apply the program SIMxNNy25 to the tasks x11y. One has to take rand1 = int(2*....) Then only two elements are considered. SIMxNNy36 simulates a11rH01A SIMxNNy38 simulates a22rH01A with a (4 cycles + 1 cycle + 4 cycles) structure. The result of the simulation is very good. A 4–1–4 system needs only the present sets, no additonal ones.

The task a22rH01A can be simulated by SIMxNNy38: The equation of a22rH01A is

a22rH01A = 70 + (15 + 2(4 + 2*4))12 14.6 + 19.3

The natural distribution follows.

H01A a22rH01A1+a22rH01A2 FPM31e 10.05.2000 start: 150

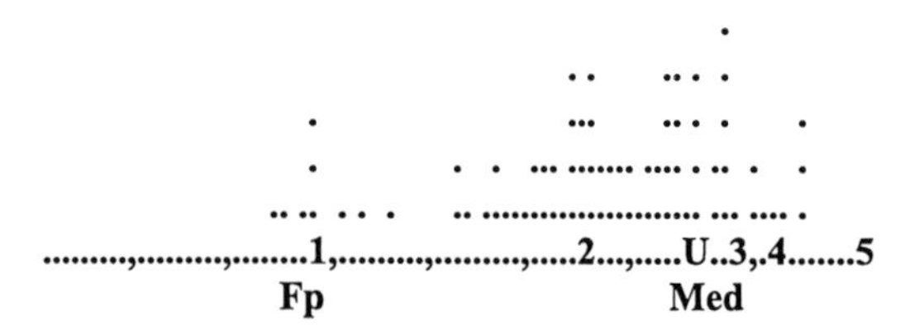

Median= 474.5 Fp= 280 linET= 12 (ChPh+FPM)
Median-Fp= 194.5 cycEN = (Median-Fp)/linET + 2cycET= 18.2
Fp-con= 210 linEN = (Fp-con)/linET - 2cycET= 15.5
a22rHO1A = con + (15.5 +2(4+2*4)) 12

Fig. 171. Real distribution of the task a22rH01A. The distribution begins at 150 ms

The parameters of task a22rH01A (linEN = 15, ET = 12, mode = s, search = 3) are now given to the simulation program SIMx22yC which produces the distribution shown below.

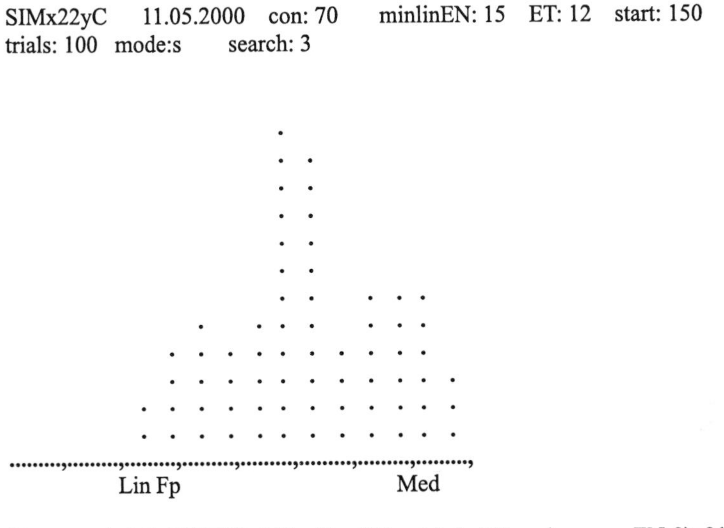

Lin=con+(minlinEN*ET): 250 Fp: 270 Med: 465 (meancycEN-2): 21.2
(mediancycEN-2)=(Med-Lin)/linET: 17.9 (Fp-Lin)/linET: 1.7

Fig. 172. Simulated distribution of a22rH01A by the program SIMx22yC. The program produces a similar median value and a similar meancycEN as the natural task

Simulation of task a11rH32A

Attempt to simulate a11rH32A by SIMxNNy35 with median = (6 linET + 2 cycles + 1 cycle) = (6 linET + 4cycET + 2cycET).

The program uses the structure a11rH32A = (con + 6linET + 2cycCT + 2cycCT) = (con + 6linET + 4cycET + 4cycET) = (con + 6linET + 8cycET) / *internal mean*

The external meancycEN is obtained by subtracting 2 cycEN because these are parallel to other relations (see above):

external meancycEN = 6.3 in the computer simulation
external meancycEN = 6 in theory

Different from these mean values are the median values of searching cycles.The *median* lies left of the *mean*. This can be explained by the shift to the left in the distribution of reaction times.

external median cycEN = (median – lin)/ET
= *5.8* in the computer simulation

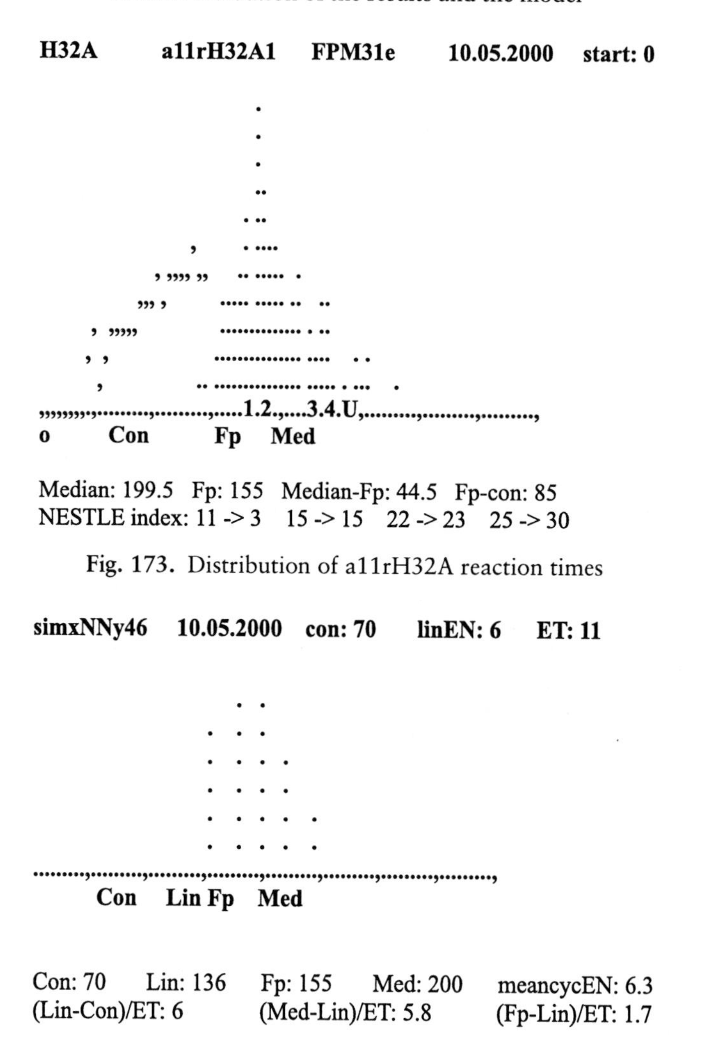

Fig. 173. Distribution of a11rH32A reaction times

Fig. 174. Distribution generated by SIMxNNy45 with the values of a11rH32A: con = 70, linEN = 6, ET = 11, slow mode, double search. Double search means that the response element (tRESP) is activated when scanned only by the sensorimotor set (tSMS)

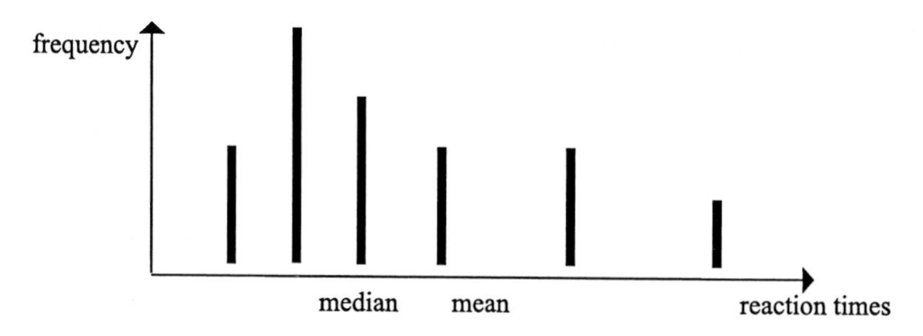

Fig. 175. The median lies left of the mean value in the reaction time distribution

Simulation of task a11lS21A

The simulations of a11lS21A by SIMxNNy43 show that the shape of the distribution varies considerably if one takes 100 trials. Only 1000 trials produce a stable shape of the distribution.

For this reason, it is difficult to draw conclusions from the shape of the distribution to the mechanisms which produce it if the number of trials is 100.

Anyhow, the distribution of a11lS21A is different from the distribution produced by SIMxNNy43.

The spontaneity of the task set has important consequences for the structure of the pathways

The spontaneity of the task set has some important consequences: the subject can react faster because the seize of the first searching cycle of the task set does not appear in the reaction time (at the time account) because it is simultaneous to the arrival of the afferent activation in the stimulus element. The arrival at the stimulus element is counted to the linear path. Only the return of the first searching cycle is added to the cyclical portion of the pathway (cycEN).

In order to get the whole number of searching cycles, I had to take sum = sum + (n + **1**) in the simulating program SIMxNNy.

The return of the last searching cycle runs simultaneously to the first relation of the SMS. That means the searching time between the target stimulus and the target response is

(cycle number – 1) + SMS = (n*cycET – 2 cycET) + 2 cycET = n*cycET
with n = number of task elements

The reaction time therefore contains RT = con + m*linET + (n–1) cycles + SMS

Testing the evaluation programs (FPM31e, FPM26f58, SINGLE106n, SINGLE104r) with artificial data produced by using an artificial elementary time

Two artificial elementary times

The programm ARTIFIC1 takes two different elementary times and generates reaction times by the mathematical expression:

reactiontime(n) = 70 + (6 * 14) + (INT(10 * random1) * *14* + INT(10 * random2) * *23*

The programm ARTIFIC2 takes two different elementary times and generates reaction times by the mathematical expression:

reactiontime(n) = 70 + (6 * 14) + (INT(6 * random1) * *14* + INT(6 * random2) * *23*

The difference between the two version is the range in which the multiples of elementary times vary. In ARTIFIC1 the range is 10, in ARTIFIC2 the range is 6.

The product of these two programs are 100 artificial reaction times such as

c:\workmemo*a11rARB*

```
302 274 210 247 242 191 270 223 293 200 311 200 200
260 237 293 256 316 191 200 242 270 177 210 200 223
242 339 214 270 237 205 210 223 283 302 177 233 219
302 191 247 214 214 274 274 228 191 325 237 224 182
224 339 224 177 219 302 270 246 251 154 223 200 316
339 223 224 260 293 311 283 247 168 237 214 265 200
316 247 214 246 200 177 177 200 191 260 297 205 293
219 311 339 196 214 325 154 228 270
```

Fig. 176. Hundred artificial reaction times produced by the program *ARTIFIC2* (with ET1 = 14ms and ET2 = 23ms). In this version the multiples of ET vary from 1 to 6

These artificial reaction times can now be submitted to the evaluation programs with the question whether these programs may find the hidden elementary times.For SING106n the true ARB data were used, for FPM31f the data a11rARC were used, which have been renamed into a111ARC, v11rARC, and v11lARC (because the program demands all 1-stimulus-tasks to function).

c:\workmemo*a11rARC*

```
293 297 260 270 210 247 214 269 214 214 247 205 269
311 196 311 246 283 311 242 247 247 260 168 210 311
339 302 247 293 288 247 279 214 288 177 269 242 246
274 247 242 270 283 223 242 283 316 154 279 269 223
283 274 168 246 182 154 265 256 311 316 269 265 325
200 274 311 191 260 283 214 233 219 228 269 325 200
219 288 191 242 223 269 297 224 325 325 311 283 293
316 237 297 191 288 274 228 302 242
```

Fig. 177. Hundred artificial reaction times produced by the program *ARTIFIC2* (with ET1 = 14ms and ET2 = 23ms). In this version the multiples of ET vary from 1 to 6

The NESTLE procedure finds elementary times at 15ms and at 24ms. The inaccuracy is 1ms in both cases. Anyhow, the program is capable of distinguishing two elementary times.This could be important in patients with monohemispheric temporal brain lesions.These simulations can be used to improve the evaluation programs. If the covariation of the two elementary times is reduced that means they vary more and more together, the two elementary times are merged by the evaluation programs into one mean elementary time (see the results of SINGL95f).

S Y N O P S I S OF RELATIVE MINIMA OF x11y

arb sing106n.bas 11.05.2000

	a11r	a111	v11r	v111
6.5 .				
7 .				
7.5 .				
8 .				
8.5 .				
9 .				
9.5 .				
10 .				
10.5 .				
11 .				
11.5 .				
12 .	1111	2222	3333	4444
12.5 .	1111	2222	3333	4444
13 .				
13.5 .				
14 .	**1111**	**2222**	**3333**	**4444**
14.5 .				
15 .	1111	2222	3333	4444
15.5 .				
16 .				
16.5 .				
17 .				
17.5 .				
18 .				
18.5 .				
19 .				
19.5 .				
20 .				
20.5 .				
21 .				
21.5 .	1111	2222	3333	4444
22 .	1111	2222	3333	4444
22.5 .	1111	2222	3333	4444
23 .	**1111**	**2222**	**3333**	**4444**
23.5 .	1111	2222	3333	4444
24 .	1111	2222	3333	4444
24.5 .	1111	2222	3333	4444
25 .				
25.5 .				
26 .				
26.5 .				
27 .				

Fig. 178. Results of the evaluation of the artificial data by the program SINGLE. The elementary times are found (14 and 23) but the inaccuracy is rather large: ± 1.5ms for ET = 23ms

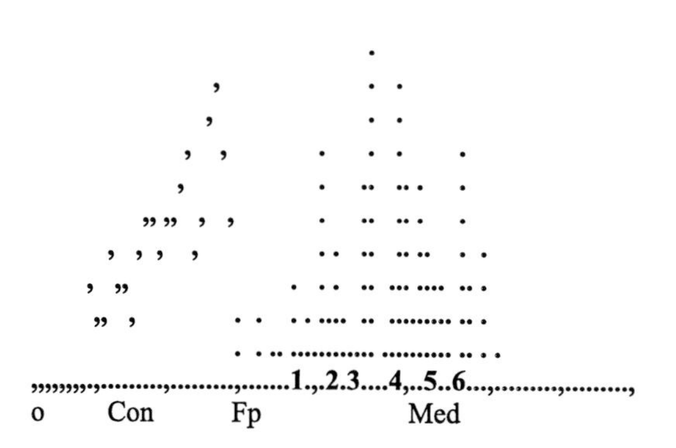

Fig. 179. Results of the program FPM31f when evaluating the artificial data of a11rARC

One artificial elementary time

In the case that only one elementary time is used, the evaluation programs find this without doubt. There are two programs which are used to find the elementary time: SINGLE and FPM. First the results of SINGLE are shown then the results of FPM.

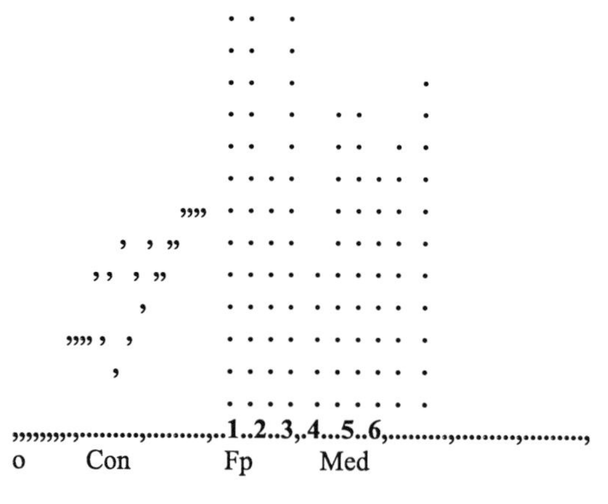

Fig. 180. Results of the program FPM31f when evaluating the artificial data of a11rARD (with an artifical ET = 16ms)

```
Made by ARTIFAC3(one ET=16)

S Y N O P S I S  OF RELATIVE MINIMA OF x11y
ard        sing106n.bas          11.05.2000
      a11r        a11l        v11r        v11l
6.5  .
7    .
7.5  .
8    .
8.5  .
9    .
9.5  .
10   .
10.5 .
11   .
11.5 .
12   .
12.5 .
13   .
13.5 .
14   .
14.5 .111111     222222      333333      444444
15   . 111111    222222      333333      444444
15.5 .111111     222222      333333      444444
16   . 111111    222222      333333      444444
16.5 .111111     222222      333333      444444
17   . 111111    222222      333333      444444
17.5 .111111     222222      333333      444444
18   . 111111    222222      333333      444444
18.5 .
19   .
19.5 .
20   .
20.5 .
21   .
21.5 .
22   .
22.5 .
23   .
23.5 .
24   .
```

Fig. 181. Results of the evaluation of the artificial data by the program SINGLE. The elementary time is found (ET = 16) with an inaccuracy of 0.5 ms

The artificial reaction times are produced by the version ARTIFAC3(with ET = 16ms) by the mathematical expression:

reactiontime(n) = 70 + (6 * 16) + (INT(10 * random) * *16*

The NESTLE procedure finds this elementary time at 16ms. The inaccuracy is 0ms.

Testing other evaluation programs (SPRING, ERPET) with artificial reaction time data produced by using an artificial elementary time

The procedure can be applied to these programs which calculate the elementary time from event-related potentials (latencies). If one uses test latencies which are multiples of 14ms, the program ERPET finds this artificial elementary time from the artificial latencies and produces the figure below.

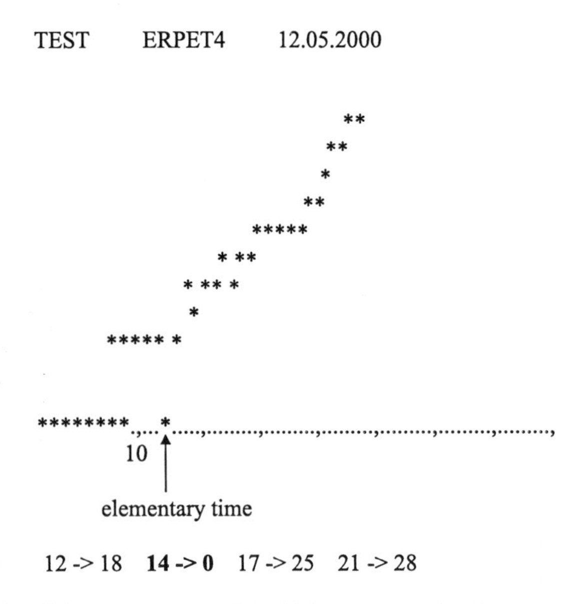

Fig. 182. Results of the program ERPET which computes the elementary time from latencies of an event-related potential when fed with artificial latencies produced by multplicating an artificial elementary time with integer numbers

Conclusion

The evaluation programs used within these work are capable of extracting the elementary time from reaction time data and latencies of even-related potentials with an inaccuracy of 1-2ms. The simulation program is capable of producing the distribution and the median value of natureal reaction time data when fed with the parameters of these subjects. This may be due to the similar construction of the natural and the model processes.

Discussion

Unsolved problems

The accuracy of elementary time

The mean elementary time in healthy subjects is 14.5ms. The range of inaccurate measurement is ± 2ms. This means a possible error of 13.8%. This value is too large. In order to reduce it, better technical devices should be applied with more precise evaluation programs processing the data. The validity of the elementary times has to be tested by different evaluation programs (FPM31e, FPM26f58, SING106n, SING104r) using different reaction time data, by replicating each elementary time in a second series and by the help of event-related potentials replicating the elementary times obtained by the evaluation of the reaction time data.The danger of measuring the elementary time without replication or ERP is shown in the following example.

Some subjects, especially patients, could not be measured a second time because of the unwillingness of the patient. In these patients erraneous elementary times can be obtained.

Table 39. Convergence table of subject O16A which has been examined only once

O16A	FPM31e (x11y)	FPM31 (x22y)	FPM26f (x11y)	FPM26f (x22y)	Chrono-phoresis	MFM	Result
aNNr	10.5,12	13,17	height>1		10.,14.5	14.2	
aNNl	13	23	10,13		13	12.8	
vNNr	15,21	15	height>1		14.5	11.1,19	
vNNl	?	?	10,19,21		11	10,7	

The distances between these reaction times, lying left of the first peak, have helped to find an error of the elementary time in a22rO16A. The first hypothesis was ET(aNNrO16A) = 16.5.

But the distance between the reaction times left of the first peak was 3ET = 40ms with ET = 13.3. The Median-FirstPeak-Method gives 14.2 ms and the chronophoresis P14.5.

How is it possible to avoid such mistakes in the future? Beside the replication are the results of the *chronophoresis and the MFM additional arguments* for an elementary time. The false elementary time was a consequence of FPM26f58 with height>1. *Such distributions should not be accepted for future evaluations.*

The accuracy of the length of the linear pathway (and the cyclical pathway)

The length of the linear pathway is an important value because it determines the length of the cyclical pathway (together with the median reaction time). Here, the same principles are valid. The length of the linear pathway should be replicated in a second series and should correspond to the break between the three layer working and the two layer working in the latency tables. Nevertheless, the variability of this value seems to be larger than that of the elementary time. In some cases there is no visible break in the latency tables because the linear pathway may use a two layer working, too. It was surprising that the ERP derived length caused a correction of the method to compute it from the reaction time data: in subjects with slow mode and triple search, the transition between the two pathways does not lie 2ET left of the first peak but 4ET (or more) left of it.

The accuracy of the event-related potentials

The records of the event-related potentials are distorted by a 50 Hz artifact in some subjects. Thus one has to expect artificial latencies caused by the 50 Hz artifact as the results of ERPET prove. These 50Hz latencies can falsify the break in the latency tables.

The accuracy of the statistical results

The number of subjects is large enough to give evidence to the findings but it is necessary to replicate all statistical evaluations with new samples. The findings of this work are so important that every effort should be made to replicate them. Because of the small number of cases some interesting questions could not be answered:

Do women have more symmetrical elementary times than men (H32A)?
Is the uniformity principle ET(x11y) = ET(x22y) a sign of psychological stability?
Is the asymmetry ET(x11y) ≠ ET(x22y) a sign of vulnerability?

The findings stimulate new questions

The neural basis of the elementary time

This problem has not been adressed within this work. Simply stated, the elementary time is the time between two neurons. When neuron A generates an action potential at its axon hillock this potential has to propagate along its axon to the boutons. There the voltage gated Ca + + channels have to be activated, Ca + + enters the presynaptic bouton, the vesicles filled with transmitter are transported to the presynaptic cell membrane, merge with the membrane and free the transmitter molecules. These diffundate across the synaptic cleft to the receptors at the postsynaptic membrane, where they activate the transmitter gated ionotropic receptors. This generates excitatory postsynaptic potentials which travel along the dendrites, amplified by a few voltage gated Na + channels until they reach the axon hillock of the postsynaptic cell. There a large number of voltage gated Na + channels produce the new action potential.

The elements of the pathways investigated in this work are sets of neurons. The number of active neurons within an set of neurons may also influence the time between two elements. It is not known which of these processes are disturbed by brain lesions.

Histological questions

Is there any histological evidence for the connections between layer 5 of the PFC and layer 2,3 of the posterior brain? The location of the *fast mode memory sets* built by implicit learning may be another one than in the PFC (eg. basal ganglia).

What is the significance of the second break in some latency tables?

Some subjects show a second, reverse transition from two layer working to three layer working in the latency tables? This finding could hint to a stimulus dependent three layer working at the time of the response. The precise start of this activity relative to the reaction time data has to be investigated.

Cued visual reaction tasks

If one gives a visual cue 100 ms prior to the visual stimulus, the reaction time is decreased.

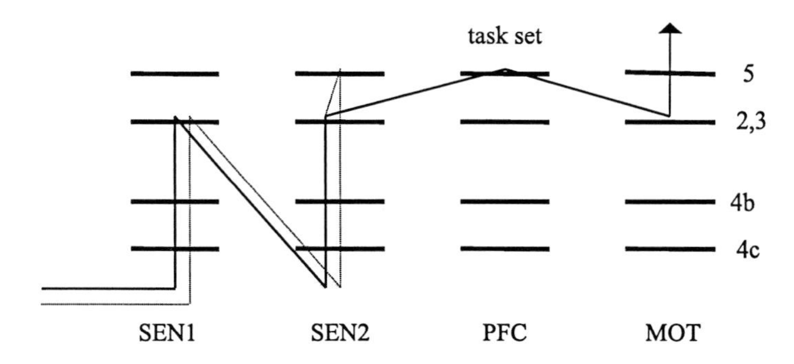

Fig. 183. Hypothetical pathway of a cued visual reaction tasks. The broken line marks the pathway of the visual cue

The visual cue associates the visual stimulus and the task set may search for the visual stimulus *before* this arrives. By that way, the search for the target stimulus may be over when the target stimulus really arrives and the sensorimotor set may be activated at once. The average time to search for the target stimulus is 2 CT. Is this prediction correct?

Inhibition of return

If the time between cue and stimulus is more than 500ms and less than 1500ms, a new phenomenon takes place: the reaction time is increased. The pathway of the cue can activate the task set of the visual stimulus but not the sensorimotor set. Maybe there is an oscillation between the top element and the stimulus element which leads to some exhaustion of both elements and the arriving visual stimulus cannot use the exhausted task set.

Delayed response

A delayed-response task consists of several trials, each several seconds long. At the beginning of each trial, a cue is presented briefly, instructing the subject which response to make. But the response must be witheld until the end of the trial, when a signal to respond is given.

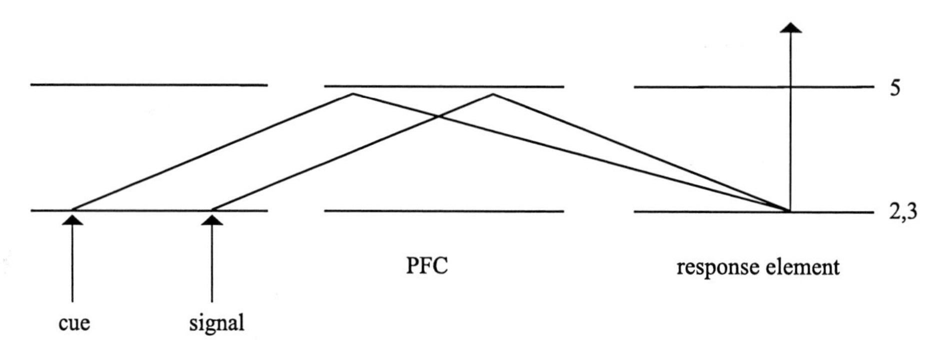

Fig. 184. Hypothetical pathway of a delayed response task

The cue-response-set is memorized by the prefrontal cortex which has been proven by single unit recording in animals. The signal-response set should have a similar location.

Which memory sets are used by a x22y delayed-response task?

Processing steps:

1. CUE1 appears
2. The TASK SET scans CUE1.
3. The SMS1 becomes active and scans RESP1
4. RESP1 is not executed but delayed.

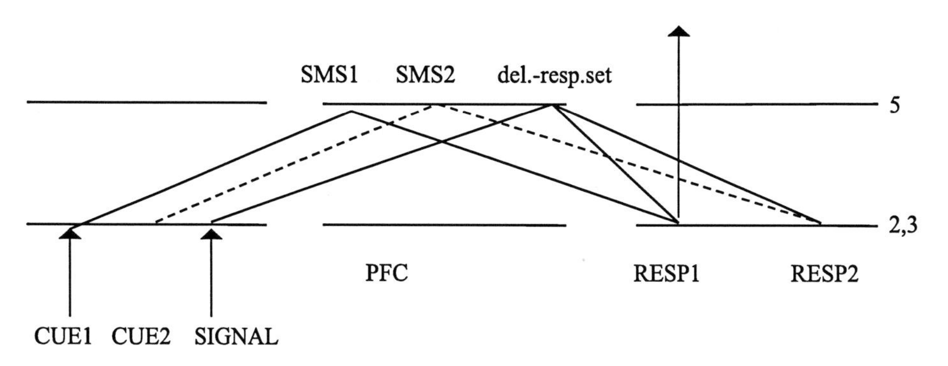

Fig. 185. Memory sets of the x22y delayed response task. The task set is not shown

5. SIGNAL appears, is scanned by the TASK SET. The delayed-response set becomes active, scans and finds RESP1.
6. RESP1 is executed.

Are there motor programs?

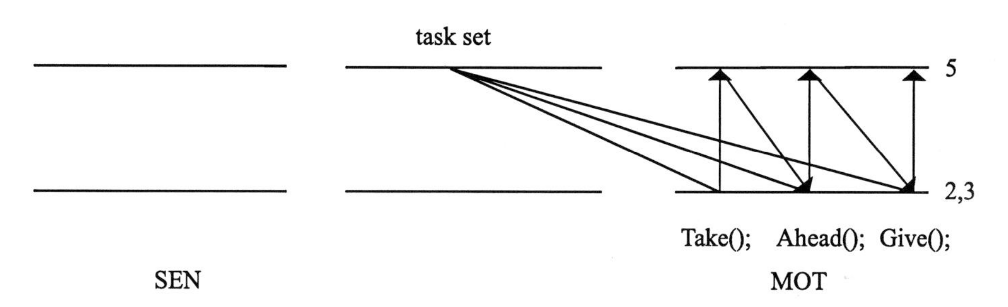

Fig. 186. Hypothetical pathway of a motor program

The single actions may be connected hierarchically and within the motor area. The relations between two response in the motor area may be called response-response-sets.

Comparison between the auditory and the visual elementary time in a subject

Which elementary time is longer, the auditory or the visual one? This question can be answered by a scatterplot with each point representing one subject.

Epilogue

Errors in the present version

In such a work, any number of errors, logical as well as grammatical, occur. I would like to thank the reader for his forbearance for this mistakes. Any comments and criticisms are welcome. All the equations and the models should be tested by other researchers.

The pathways of mind evade the body-mind problem

The "Pathways of Mind" do not put the question who or what is moving along the pathways, only the physical parameters are investigated. Nevertheless, the neuromechanics of cortical information processing reveals structures and processes of great beauty. It is a privilege to possess such structures and let them work for us.

Computers and programs as the adequate tools to investigate the brain

This work could not be done without computers and the special programs which were written to investigate the Pathways of Mind.

Programs:
MERKP27 tests the subject
FPM31e computes the elementary time from the reaction time data
FPM26f58 „
SING106n „
SING104r „
EASY14a prints the equations from the elementary times, the first peaks and the medians.
ERPET15 computes the elementary times from latencies
PROBE12 tests the completeness of the series
CHANGE3 transforms an result file into result.sps
ARTIFIC1,2 generates artificial reaction times.
SIMx22yC simulates the cyclical pathway

Discussion of references

The cortical areas used by stimulus-response pathways in humans (PET, fMRI, rCBF, NIRS)

Delayed cued finger movement task

1.0.1 Richter et al. (1997) measured activity in the human primary motor cortex, the premotor cortex and the supplementary cortex during a delayed cued finger movement task. All three areas were active during movement preparation and movemement execution. Activity in the primary motor cortex was considerable weaker during preparation than execution, in the premotor cortex and the supplementary cortex the activity was of similar intensity during both periods.

v11

v22, v21, both visual fields, right hand, (v22-v0), (v21-v0), (v22-v21)

Kawashima et al. (1996) published a PET study in man about the functional anatomy of GO/ NO-GO discrimination and response selection. In the response selection task, subjects were instructed to flex their thumb or index of the right hand immediately after the LED turned on red or green, respectively. In the GO/ NO-GO task, subjects were asked to flex their thumb immediately after the LED turned on red, however they were asked not to move their fingers when the LED turned on green. In the control task, subjects were asked to look at the LED without any movements.

The authors subtracted the control task from the response selection task, the control task from the GO/ NO-GO task and the response selection task form the GO/ NO-GO task. *They found eight areas showing a significant activation in GO/ NO-GO minus control and also in GO/ NO-GO minus response selection.* Four of this fields were in the left hemisphere (two in the frontal lobe: one in the precentral gyrus and one in the middle frontal gyrus, another two fields in the left hemisphere were located in the posterior part of the insula cortex and in the superior occipital cortex). In the right hemisphere these subtraction yielded fields of activation only in the frontal lobe. Two of them were located in the superior frontal gyrus, one in the superior frontal sulcus, another in the medial aspect of the superior frontal gyrus. Another two fields in the right frontal lobe were located in the surface of the middle frontal gyrus and in the inferior frontal gyrus.

In their study eleven fields showed significant activation not only in response selection minus control but also in response selection minus GO/ NO-GO. Here the fields of activation were mainly located in the left hemisphere. In the frontal lobe of the left hemisphere, two fields were located in the superior frontal gyrus and the cortex lining the posterior lip of the precentral sulcus. One field was located in the superior temporal sulcus, another field in the anterior part of the cingulate gyrus. The insula was activated bilaterally. A field in the left thalamus and a field in the left hippocampus were also active.

Their most striking findings were that fields in the prefrontal cortex of the right hemisphere were specifically active in relation to the GO/ NO-GO task and the acitivation of the anterior cingulate gyrus in response selection tasks.

O'Sullivan et al. (1994) showed activation of the same field during a two alternative forced choice task of somatosensory stimuli. The studies of Paus et al. (1993), Deiber et al. (1991) and Frith et al. (1991) indicate that the cingulate cortex is functionally heterogeneous and that *one of the important roles of the anterior cingulate is response selection* (Kawashima et al.).

Godefroy and Rousseaux (1995) assessed the two-choice response task (CRT) and the simple response task (SRT) in patients with frontal or posterior brain damage and normal subjects. They gave a warning stimulus, told the subjects to use the prefered hand and varied the stimulus probabilities. They used the relative judgement theory of Link and Heath (1975) and the work of Green et al. (1983) to analyse their data. In this theory a random walk process to decision is developed and the decision and non-decision components of latencies are separated using a linear regression model with CRT as a dependend variable an the distance traversed by the decision process as an independent variable. Using this approach, the *slope* of the regression is the time required for the average step towards the correct boundary and the *intercept* value corresponds to the duration of non-decision components. *Their data support the prominent role of the left hemisphere in the visual decision process* suggested by the results from Dee and Van Allen (1973). *The results suggest, that the binary decision process is spared in patients with focal lesions and that their longer choice reaction time is mainly related to slowing the perceptual and motor stages.*

In a second study Godefroy and Rousseax (1997) assessed novel and previously practiced decisions in normal subjects and patients with frontal or posterior brain damage using a unimanual two-choice response time test. Patients with frontal damage had a dramatic impairment on novel decision, whereas practiced decision was normal. *They suggest that the prefrontal cortex* is critical for the ability to create internal referents that are determined from instructions and *is required to associate the current stimulus with the appropriate response.*

Klingberg and Roland (1997) measured the interference between two concurrent tasks (an auditory and a visual go/ no-go task) as an increased reaction time during simultaneous performance compared to when each task is performed alone. With positron emission tomography they measured the cortical activation as fields with significant increase in regional cerebral blood flow during single task performance. The two tasks activated overlapping parts of the cortex and interfered significantly during dual task performance.

Sasaki et al. (1996) review their MEG studies on the human frontal association cortex. The no-go potential was first found at go/ no-go reaction-time hand movement tasks with discrimination between different colour light stimuli in the prefrontal cortex of monkeys. In humans it's current dipoles could be localized by use of MEG *in the dorsolateral part of the frontal association cortex in both hemispheres. The function for no-go decision and subsequent suppressor action was thus substantiated in the human frontal cortex.*

Schluter et al. (1998) used transcranial magnetic stimulation (TMS) to disrupt the processing in the human premotor cortex while the subjects carried out visual choice reaction tasks and simple reaction tasks. They were able to delay responses in the contralateral hand by stimulating the premotor cortex. *They were also able to delay responses with the ipsilateral hand while stimulating over the left premotor cortex, but not while stimulating over the right premotor cortex or either sensorimotor cortex.* They concluded that the premotor cortex is important for selecting movements after a visual cue and that the *left hemisphere is dominant for the rapid selection of action.*

Pratt et al. (1997) used visual choice reaction tasks to determine whether inhibition of return can be best characterized as an attentional or a motor phenomenon.

Their results are consistent with the attentional, not the motor, explanation of inhibition of return.

Manoach et al. (1997) investigated whether a nonspatial working memory (WM) task would activate dorsolateral prefrontal cortex (DLPFC) and whether activation would be correlated with WM load. Using functional magnetic resonance imaging they measured regional brain signal changes in 12 normal subjects peforming a continous performance, choice reaction time task that requires WM. A high WM load condition was compared with a non-WM choice reaction time control condition and a low WM load condition. Significant changes in signal intensity occurred in the DLPFC, frontal motor regions and the intraparietel sulcus (IPS) in both comparisons. *These findings support the role of DLPFC and IPS in working memory required by choice reaction time tasks and suggest that signal changes in DLPFC correlate with working memory load.*

Taylor and McCloskey (1996) presented visual stimuli at random in one of two different locations to normal human subjects in a choice-reaction time task. When the stimulus appeared in one of the locations, subjects made a motor response. When sthe stimulus appeared in the other location, subjects made a different motor response. Large and small stimuli were presented in either location. In some trials, the small stimulus was followed 50 ms later by the large stimulus. The small stimulus was then "masked" by the large stimulus and could not be perceived. Despite this subjects were able to select and execute the appropriate motor response. *That means the subjects could react to the masked stimuli without subjective awareness.*

v33

Paus et al. (1993) obtained task-specifif changes in a Positron Emission Tomography Study in a manual task where one of three response keys had to be pressed according to a particular visual stimulus shown. The stimuli (simple geometric forms such as a cross or a circle) were presented for 200 ms inside an empty circle that was displayed permanently in the center of the screen. The three response keys (the second, third or fourth finger of the right hand) were arranged in a row.

They used an overpracticed and a reversal version of this task. In the baseline scan, the subjects were not required to execute any responses other than fixating the center of the screen.

They carried out three types of subtractions:

reversal minus baseline subtraction (activation foci in the left central sulcus, left anterior cingulate /caudal, right intraparietal sulcus /rostral, right thalamus or corpus callosum, left substantia nigra),

overpracticed minus baseline subtraction (activation foci in the left postcentral gyrus, left central sulcus, left anterior cingulate /caudal, right calcarine sulcus, left lingual gyrus, left substantia nigra, left cingulus or corpus callosum)

and reversal minus overpracticed subtraction (activation foci in the left inferior frontal sulcus, the left anterior cingulate /caudal, right intraparietal sulcus, left intraparietal sulcus and left putamen).

These tasks are analog to v33, both visual fields, right hand , subtraction (v33-v0)

v44

v55

a11

Remy et al. (1994) investigated the cerebral blood flow with positron emission tomography induced in 10 healthy subjects by two different tasks: a repeated flexion-extension of all fingers and a repeated flexion-extension of the middle finger. The all-finger movement only activated the primary sensorimotor cortex (SM) and the supplementary motor area (SMA) contralateral to the movement. However the SMA activation was only observed when the movement was triggered by an *auditory cue* but not when it was self-paced. The middle finger movement, performed during self-paced conditions, induced a much more complex pattern of activation than the all-finger movement, characterized by a high degree of SM and SMA activation contralateral to the side of the movement, as well as a slight ipsilateral activation of these areas. The authors suggest that this pattern of cortical activation may reflect the process of individuating finger movement or the early stages of motor learning of this unusual and technically difficult movement.

a22

Holocomb et al. (1996) and Holocomb and Lahti (1997) generated an rCBF comparison between schizophrenic patients and matched healthy control subjects while both groups are performing a practiced auditory recognition task. The task involves discriminating between two auditory tones. The sensorimotor control task involves repetitive tone and motor stimulation, there also is a rest condition task. Image analysis resulted in two subtracted group images: sensorimotor control task minus rest and auditory discrimination task minus sensorimotor control task. *For the decision task, healthy control subjects activated the anterior cingulate, right insula, and right middle frontal cortex.*

The schizophrenic subjects had considerable more activation in the sensorimotor control task, in the decision task they showed very little incremental increase in activation except for a mild flow increase in right frontal cortex. One might speculate that schizophrenic persons need already "full effort" to perform the easiest of tasks.

Pugh et al. (1996) did a fMRI investigation with 25 adult subjects who discriminated speech tokens ([ba]/[da]) or made pitch judgements on tone stimuli (rising/falling) under both binaural and dichotic listening conditions. Under the dichotic conditions activation within the *posterior (parietal) attention system* and at primary processing sites in the superior temporal and inferior frontal regions was increased. The cingulate gyrus within the *anterior attention system* was not influenced by this manipulation.

a33

a44

(a44, both ears, right hand, subtraction a44-a11). Deiber et al. (1991) measured in normal subjects with positron emission tomography while these had to move a joystick on hearing a tone. In one task they had to select between four possible directions (conditional task). In the control task they always pushed the joystick forwards (fixed task).

In one subtraction they compared the Fixed task with the Conditional task

When selection of a movement in the conditional task was made, significant increases in regional cerebral blood flow were found in the visual association areas of the *left superior parietal cortex.* Because the fixed task compares to the a11 task, the subtraction can be written (a44-a11). These two tasks differ only in the left superior parietal cortex significantly. The authors think, that the activity of this area is related to the process by which movements are selected and not because the movements were spatial. (see also Roland PE. 1993, pages 249, 250)

Other cognitive tasks

In a PET study of the Tower of London task, Baker et al. (1996) observed activation in distributed network of cortical areas incorporating prefrontal, cingulate, premotor, parietal and occipital cortices. Activation in corresponding areas has been observed in visuospatial working memory task with the exception of the rostral prefrontal cortex. The authors therefore suggest that this area may be identified with the executive components of planning comprising response selection and evaluation.

Osmon and Suchy (1998) devised reaction time measures of three basic aspects of mental set by using the Milwaukee Card Sorting Test (a slight variation of the Wisconsin Card Sorting Test). *Forming the mental set* – the ability to collect together the necessary perceptual and conceptual information (posterior cortex processing) in working memory (dorsolateral frontal region) for the purpose of developing a mental set or intentional goal that is used to guide subsequent behavior . *Switching mental set* – the ability to respond to external social and emotional contingecies for the purpose of recognizing when the current mental set is no longer appropriate and a switch in set is needed (Orbtial frontal region). *Maintaining mental set* – the ability to integrate internal perceptual, conceptual, memorial information with situationally relevant motor stereotypes for the purpose of gating out distracting stimuli and maintaining the current mental set (medial frontal region, including the supplementary motor area, and the cingulate gyrus motor areas).

Osmon suggests that the task, requiring the participant to utilize a rule, must be kept in working memory (dorsolateral prefrontal cortex) to respond quickly to a stimulus.

Volz et al (1997) studied brain activation during the Wisconsin Card Sorting Test (WCST) by using functional MRI. In healthy volunteers WCST stimulation resulted in a right lateralized frontal activation. In 13 chronich schizophrenics on stable neuroleptic medication, a lack of activation in the right prefrontal cortex and – as a trend – an increased left temporal activity was noted compared to controls. The suggest a reduced ability of schizophrenics to coordinate cerebral function.

Posner and Raichle (1994) When subjects were asked to look for animal names as they read a series of nouns, strong PET activation appeared in the anterior cingulate gyrus. Moreover, the activation became stronger as the number of targets in the list of nouns increased. In addition, there is a reduction in such activation as the task becomes automated with practice. This made the authors think that the area served as an attention system.

Villringer et al. (1997) performed simultaneously near-infrared spectroscopy (NIRS) and positron emissions tomography (PET) in five healthy subjects during rest and during performance of a calculation task and a Stroop task. A statistically significant correlation between changes in CBF and changes in [total-Hb] was found.

The cortical areas used by the stimulus-response pathways in animals (PET, lesion studies, single or multi unit studies)

PET

Tsujimoto et al. (1997) mapped regions of monkey brain involved in go/ no-go reaction time tasks with positron emission tomography. In the control task only go signals were presented. In the go/ no-go task, when compared with the control task, a signifcant increase in rCBF was noted in the following areas: (1) then principal sulci; (2) the anterodorsal frontal pole; (3) the anterior part of the inferior occipital sulcus which appeared to be the V4; and (4) then parieto- occipital region. The increase in the principal sulci was related to the the no-go decision and motor suppression in the area.

Lesion studies

Muir et al. (1996) investigated the dissociable effects of bilateral excitotoxic lesions of different regions of the rat neocortex on a five-choice-serial reaction time task. Whereas *medial prefrontal cortical lesions impaired performance of the task* as revealed by a reduction in choice accuracy, an increase in the latency to respond correctly to the visual target and enhanced perseverative responding, *lesions of the anterior cingulate cortex specifically increased premature responding*. By contrast, lateral frontal cortical lesions did not significantly disrupt baseline performance of the task, but rather increased the latency to respond correctly to the visual target during various manipulations, for example, when the length of the intertrial interval was varied unpredictably.*Lesions of the parietal cortex failed to disrupt any aspect of task performance* investigated.

In a second work Muir et al. (1996) investigated the role of the cholinergic innervation of the cingulate gyrus in visual attentional function and acquisition of a visual conditional discrimination task. After lesioning the vertical limb diagonal band of Broca (VDB) which provides the main cholinergic projection to cingulate cortex, animals were not significantly impaired on the 5-choice-serial reaction time task. This task has previously be shown to be sensitive to AMPA lesions of the nucleus basalis magnocellularis (NBM). Lesions of the VDB did significantly affect the acquisition of a visual conditional discrimination task. *These results suggest a role for the cholinergic innervation of the cingulate cortex in conditional learning* but not for continous attentional performance.

Rushworth et al. (1997) assessed the ability of monkeys with inferior convexity (IC) lesions to perform visual pattern association tasks and color-matching tasks, both with and without delay.These experiments showed that the lesions had no effect on the task. They therefore suggest that the IC may be more important for stimulius selection and attention.

Sasaki et al. (1994) got findings resulting from field potential analyses in the cerebral cortex of behaving monkeys and also on those with EEG and MEG studies in human subjects. They consider the prefrontal cortex to be essential for initiative and reasoning behaviour in humans, and delayed and discriminative movement tasks in monkeys. Several parts of the prefrontal cortex, especially the prearcuate cortex, are very active in initiating simple reaction-time hand movements in response to visual stimuli. In addition the prefrontal cortex can decide not to move and to subsequently suppress movement. They found a "no-go potential" in the prefrontal cortex of monkeys which is specific to the no-go reaction in the go/ no-go reaction time hand movement task with discrimination between different colours or sound stimuli. Similar no-go potentials were also recorded from the human scalp and their location in the cerebral hemispheres was examined by using magnetencephalography (MEG). Effects on cooling the prefrontal cortex of monkeys on visually initiated reaction time movements were consistent with the assumption that the prefrontal cortex is an important cortical area in the sending of motor commands for visually initiated reaction time movement in monkeys.

The no-go potentials of monkeys appears with a latency of 85–150 ms after the onset of the no-go stimulus in the dorsal bank of the principal sulcus and in the rostroventral corner of the prefrontal cortex *on both hemispheres contra- and ipsilateral to the operant hand.* When a monkey was first trained to use the right hand the no-go potential was more predominant in the contralateral (left) prefrontal loci than in the ipsilateral (right). After the operant hand was switched to the left, the no-go potential in the right hemisphere gradually increased in size and that in the left hemisphere remained much the same as before, or decreased slightly.

In humans the no-go potential appeared in dorsolateral part of the frontal lobe in both hemispheres at about 140 ms after the onset.

The timing of ERPs in cortical areas used by stimulus-response pathways in animals

Gemba et al. (1995) recorded the field potentials with electrodes implanted in various cortical areas while a naive monkey was learning reaction time hand movements with complex tone. When cortical surface-negative, depth-positive potential at a latency of about 80 ms after a stimulus onset appeared in the rostral bank of the inferior limb of the arcuate sulcus of the left cerebral hemisphere, and became gradually larger, the monkey began to respond to the stimulus with the movement. The authors therefore suggest that the prefrontal area, especially in the left hemisphere, plays a significant role for a monkey to associate a stimulus with appropriate motor execution.

The cortical areas used by stimulus-response pathways in patients with schizophrenia

Reaction times of patients with schizophrenia

Maruff et al. (1995) investigated the presence of attentional asymmetries in patients with schizophrenia with particular emphasis on the stage of disease, medication status and clinical symptom severity. The used a modified version of Posner's COVAT *and found reaction times to right visual field targets significanly slower than reaction times to left visual field targets when targets followed invalid spatial or non-spatial cues.* Attentional asymmetries partially resolved with brief periods of medication and completely resolved with long periods of medication. No assymmetries were found in controls nur in unmedicated subjects without schizophrenia. They noted the simularity in the COVAT performance of the acute schizophrenic subjects and that of normal subjects concurrently processing language and spatial information. The authors suggest that activation of the anterior cingulate cortex is common to tasks that involve selection of both visual and verbal information.

Carter et al. (1996) present evidence suggesting a left-hemisphere attentional deficit in schizophrenia. They say, that previous research has linked the discrimination of local targets to left hemispheric processes and the discrimination of global targets to right hemispheric processes. In a global/ local task they found impaired detection of local targets, consistent with al left hemispheric deficit.

Carnahan et al. (1997) examined schizophrenic and control right-handed males performing aiming movements with a mouse (controlled by either the left or right hand) on a graphics tablet towards targets of differing size and distances appearing on a computer screen. Results showed that, for reaction time, the controls were faster than the schizophrenics and the latter had a *left hand advantage* for movement preparation (reaction time) while the controls showed no such differences.

Bustillo et al. (199/) found nondeficit patients exhibiting a significant and abnormal asymmetry, with slower reaction time to targets presented in the right visual field than in the left visual field. This right visual field disadvantage was found only at the 100 ms cue-target interval. The deficit patients were slowest in overall reaction time but, similar to the normal subjects, showed no asymmetry. These results are consistent with slower visual information processing in the left compared to the right cerebral hemisphere in nondeficit schizophrenia.

rCBF, PET, MRSI

In a further study Carter et al. (1997) investigated the anterior cingulate dysfunction in PET and selective attention deficits in schizophrenia during single-trial Stroop task performance. Compared to 15 normal subjects the 14 patients with schizophrenia failed to activate the anterior cingulate gyrus during selective attention performance (while naming the color of color-incongruent stimuli).

Haznedar et al. (1997) studied fifty unmedicated male schizophrenic patients and 24 normal men with positron emissiont tomography and found lower relative metabolic rates in the anterior cingulate and higher rates in the posterior cingulate of the schizophrenic patients compared with the normal men.

Deicken et al. (1997) measured N-acetylaspartate (NAA), a putative neuronal marker in the anterior cingulate region of 26 schizophrenic patients and 16 control subjects using in vivo proton magnetic resonance spectroscopic imaging (1H MRSI). Relative to the control group, the patients with schizophrenia demonstrated signicantly lower NAA in both the right and left anterior cingulate regions. There was no association between NAA and duration of illness or medication dosage. The NAA findings provide support for either neuronal dysfunction or neuronal loss in the anterior cingulate region in schizophrenia.

Dolan et al. (1995) used positron emission tomography to examine the regulatory role of dopamine on cortical function in normal subjects and unmedicated schizophrenic patients. In these patients, relative to controls, an impaired cognitive activation of the anterior cingulate cortex was significantly modulated by a manipulation of dopaminergic transmission with apomorphine. After apomorphine, the schizophrenic subjects displayed a significantly enhanced cognitive activation of the anterior cingulate cortex relative to controls. These data provide in vivo evidence that an impaired cognitive task activation of the anterior cingulate cortex in schizophrenic patients can be significantly modulated by a dopaminergic manipulation.

Holocomb et al. (1996) and Holocomb and Lahti (1997) generated an rCBF comparison between schizophrenic patients and matched healthy control subjects while both groups are performing a practiced auditory recognition task. The task involves discriminating between two auditory tones. The sensorimotor control task involves repetitive tone and motor stimulation, there also is a rest condition task. Image analysis resulted in two subtracted group images: sensorimotor control task minus rest and auditory discrimination task minus sensorimotor control task. For the decision task, healthy control subjects activated the anterior cingulate, right insula, and right middle frontal cortex.

The schizophrenic subjects had considerable more activation in the sensorimotor control task, in the decision task they showed very little incremental increase in activation except for a mild flow increase in right frontal cortex. One might speculate that schizophrenic persons need already "full effort" to perform the easiest of tasks.

Evoked potentials

Egan et al. (1994) looked for event-related potential abnormalities correlating with structural brain alterations and clinical features in patients with chronic schizophrenia. They found significant correlations between the hippocampal area and the amplitude of the auditory and visual N200, between the *right* hippocampus and the *visual* P300, and between the *left* temporal lobe and the *auditory* P300.

Matsuoka et al. (1996) recorded visual event-related potentials (ERP) during a simple response task (SRT) and a discriminative response task (DRT) in remitted schizophrenic outpatients and age-matched controls to examine two endogenous negative potentials: NA and N2c. The NA potentials were derived by subtracting the ERPs for SRT from those for non-target stimuli in DRT. The N2c was calculated as the difference between ERPs for target and non-target stimuli in DRT. Schizophrenics showed retardation in NA and N2c peaks and degradation in N2c amplitude relative to controls. The NA peak latency increased as much as the latencies of N2c and reaction time for DRT in schizophrenia.

Dopamine receptors, GABAergic neurons in schizophrenia

Meador-Woodruff et al. (1997) quantified the levels of mRNA molecules encoding the five dopamine receptors in postmortem brain samples from 16 schizophrenic patients and 9 control subjects and found a dramatic *decrease of dopamine receptor transcripts in the prefrontal cortex but restricted to the D3 and D4 receptors and localized to Brodmann area 11 (orbitofrontal cortex)*. No differences were found in striatum or visual cortex.

Davis and Lewis (1995) found a selective increase in the density of a subpopulation of GABAergic local circuit neurons in the prefrontal cortex of patients with schizophrenia.

Kalus et al. (1997) showed an increase in a subpopulation of GABAergic local circuit neurons in the layers Va and Vb of the anterior cingulate cortex of patients with schizophrenia. The authors suggest this may indicate an increased inhibition of projection neurons, thus altering the neuronal output pattern of the anterior cingulate cortex in schizophrenia. Other authors could not replicate these findings (Woo et al. 1997).

Hemispheric asymmetries in healthy subjects

Martinez et al. (1997) used fMRI to explore the brain substrate associated with global and local processing of visuospatial patterns. Within the right hemisphere, the area of activation was greater under conditions of global processing than under local processing conditions. In the left hemisphere, activation to global and local input was comparable.

Burbaud et al. (1995) investigated the laterization of prefrontal cortex activity during internal mental calculation in 16 human volunteers. A clear lateralization was seen in right – but not left- handed subjects. In right-handed subjects, activation was clearly lateralized in the left dorsolateral prefrontal cortex, whereas a bilateral activation was found in left-handed subjects.

Casey examined the role of the anterior cingulate in the development of attention. They correlated attentional measures assessed with a *visual* discrimination paradigm with magnetic resonance imaging based measures of the anterior cingulate. There were significant correlations between attentional performance and *right, but not left* anterior cingulate measures. Performance was faster and more accurate during trials requiring predominantly controlled processes for those children with larger right anterior cingulate measures. The results are consistend with adult neuroimaging findings of activation in the right anterior cingulate during (visual) attention tasks and with lesion studies implicating greater right hemisphere involvement in (visual) attentional processes.

Bechara et al. (1998) tested the hypothesis that cognitive functions related to working memory (assessed with delay tasks) are distinct from those related to decision making (assessed with a gambling task), and that working memory and decision making depend in part on separate anatomical substrates. They compared normal controls, subjects with lesions in the ventromedial (VM) or dorsolateral/high mesial (DL/M) prefrontal cortices. VM subjects with more anterior lesions performed defectively on the gambling but not on the delay task. VM subjects with more posterior lesions were impaired on both tasks. Right DL/M subjects were impaired on the delay task but not the gambling task. Left DL/M subjects were not impaired on either task. These findings underscore the special importance of *the ventromedial prefrontal region in decision making and the right DL/M region in (visual) working memory.*

This localization of working memory was also identified by Kammer et al. (1997). The engagement of working memory in a visual task involving letter detection produced significant activation in the *dorsolateral prefrontal cortex* in *both* hemispheres.l

Coull et al. (1996) investigated the functional anatomy of the rapid visual information processing task (RVIP) using positron emission tomography in healthy volunteers. In RVIP single digits are presented in quick succession on a computer screen, and target sequences of numbers must be detected with a button press. Compared with a rest condition (eyes closed), the RVIP task increased rCBF *bilaterally* in the inferior frontal gyri, parietal cortex and fusiform gyrus, and also in the *right* frontal superior gyrus rostrally. The authors suggest that these data are consistent with the existence of a *right fronto-parietal network for sustained, and possibly selective, (visual) attention, and a left fronto-parietal network for the phonological loop component of working memory.*

Mohr et al. (1994) presented tachistocopically function words, content words and pronounceable non-words (pseudowords) either in the left or the right visual field or with identical copies flashed simultaneously in both visual half-fields. Their findings were consistent with the view that the neuronal counterparts of words are Hebbian cell assemblies consisting of strongly connected excitatory neurons of both hemispheres. Since function words show a right visual field advantage in addition to their Bi gain, their assemblies are likely to have most of their neurons located in the left hemisphere. Neuronal assemblies corresponding to content words may be less strongly lateralized.

Heider (1996) investigated hemispheric asymmetries in visuospatial functions in women and men with a tachistoscopic task using lateralized presentation of "Necker" cubes. The analysis of hits in the cube task showed *a right visual field advantage for women and a left visual field advantage for men*, but women had more false alarms in the right visual field, whereas men showed the reverse pattern again.

Podell et al. (1995) used the Cognitive Bias Task (CBT) to investigate the lateralized frontral lobe functions in males. Their findings suggest a dynamic balance between two *synergistic decision-making systems in the frontal lobes: context-dependent in the left hemisphere and context-invariant in the right.*

Knight (1994) reports that the effects of attention can onset as early as 25 ms after stimulation, indicating that humans are able to exert attention effects on inputs to the primary auditory cortex. Similar effects of selective attention have been reported in the visual and somatosensory modalities.

There are many clinical observations (e.g. neglect syndrome) supporting a right frontal dominance in attention capacity. An enlarged right frontal lobe may provide the underlying anatomical substrate for this attention asymmetry in humans.

In the cat, prefrontal cortex controls a thalamic gating mechanism which can produce modality-specific suppression of sensory inputs to primary cortical regions.

Reaction times of patients with depression

Elliot et al. (1997) scanned six patients with unipolar depression and six matched controls while performing easy and hard Tower of London problems using positron emission tomography. In normal subjects, the task engaged a network of prefrontal cortex, anterior cingulate, posterior cortical areas and subcortical structures including the striatum, thalamus and cerebellung. Depressive patients failed to show significant activation in the cingulate and striatum; activation in the other prefrontal and posterior cortical regions was significantly attenuated relative to controlls. Crucially, patients also failed to show the normal augmentation of activation in the caudate nucleus, anterior cingulate and right prefrontal cortex associated with increasing task difficulty.

Ferstl et al. (1994) ask whether the modality-shift effect is specific for schizophrenia patients. Several studies have found that the reaction times of schizophrenia patients is longer when successive imperative stimuli are of different modality (e.g. light followed by sound) than when they are identical (e.g. sound followed by sound). Their results indicated that a *shift from light to sound* stimuli lengthened the reaction time for schizophrenia patients and patients with mood disorders but not for patients with other diagnoses and healthy subjects. A modality shift from sound to light showed a difference only between schizophrenia patients and normal controls; there was no significant difference between schizophrenia patients and other patient groups.

Dopamine and reaction times

Baunez et al. (1995) trained rats to depress a layer and wait for the onset of a light stimulus, occuring after four equiprobable and variable intervals. At the stimulus onset, they had to release the lever within a reaction time limit for food reinforcement.

Following activation of dopaminergic transmission after *systemic injection of d-amphetamine* or *intrastriatal injection of dopamine* the rats showed premature responses and shortened reaction times. In contrast, *blocking D2 receptors with raclopride*, the rats showed delayed responses and lengthened reaction times.

These results indicate that a "critical level" of dopamine activity (neither too low nor too high) *in the striatum* is necessary for a correct execution of the movement in a conditioned motor task with temporal constraint.

In a second paper Baunez et al. (1995) produced dopamine depletion in rats by infusing the neurotoxin 6-hydroxydopamine (6-OHDA) bilaterally into the dorsal part of the *striatum*. They found an increase in the number of delayed responses and a lengthening of RTs. Lesions of the subthalamic nucleus with ibotenic acid induced the opposite: increase in the number of premature responses and a decrease of RTs.

Watanabe et al. (1997) measured prefrontal extracellular dopamine concentration using in vivo microdialysis in monkeys performing in a delayed alternation task as a typical working memory paradigm and in a sensory-guided control task. *They observed significant increases in dopamine concentration during both tasks* as compared with the basal resting level

Cai and Arnsten (1997) treated aged monkeys with dopamine D1 receptor agonists A77636 or SKF81297 and observed that low doses improved performance although higher doses impaired or had no effect on performance. The authors demonstrate that there is a narrow range of D1 receptor stimulation for optimal PFC cognitive function.

Murphy et al. (1997) had shown that FG7142 impairs spatial working memory in rats and monkeys through excessive DA receptor stimulation in the PFC. Lower clozapine doses (1–3 mg/kg p.o.) reversed the FG7142-induced spatial working memory deficits, whereas doses in the clinical range (6mg/kg p.o.) did not improve cognitive function in most animals.

Sesack et al. (1995) sought to determine the ultrastructural associations between dopamine terminals and local circuit neurons in the monkey prefrontal cortex. The dopamine terminals synaptically target the distal dendrites of both pyramidal cells and GABA interneurons.

Thierry et al. (1994) report that according to microiontophoretic studies, both dopamine (DA) and norepinephrine (NA) inhibit the spontaneous firing rate of PFC cells. Because NA was preserving or enhancing the evoked responses, it has been proposed that NA increases the signal to noise ratio. In contrast the activation of the DA pathway blocked these evoked responses. Preliminary data indicate that the excitatory responses result mainly from the activation of recurrent collaterals of efferent pyramidal cells. Therefore by suppressing these collateral excitations on neighbouring PFC cells, the activation of the mesocortical DA neurons could allow a spatial focalization of the cortical signal.

Robbins et al. (1994) report that prefrontal dopamine *depletion* facilitates attentional shifting .

Yudofsky and Hales (1997) say that dopamine has an effect on cognitive speed and reaction time. Patients with *untreated Parkinson's disease have slower cognitive speed* than patients on levodopa, even after controlling for motor dysfunction. This effect is probably mediated through mesencephalic-prefrontal dopaminergic pathways.

Hienz et al. (1997) found *cocaine reducing the reaction times* in two baboons and not affecting it in one baboon whereas quinpirole, a relatively selective *D2/D3 dopamine agonist, lengthened reaction time* in a dose dependend manner.

Nyberg et al. (1995) measured *no* prolonged reaction time in three healthy subjects after zuclopenthixol, a D2 receptor antagonist.

Moukhles et al. (1994) transplanted dopamine-rich ventral mesencephalic suspension in a partially dopamine-depleted striatum of rats performing a reaction-time motor task. In the lesion group premature responses and delayed responses were seen. In the grafted animals the premature responses improved totally, the number of delayed responses remained high.

Brockel and Fowler (1995) treated rats which were trained to react to visual stimuli with *haloperidol* for 3 months. The *reaction time increased.* Scopalamine, benztropine and d-amphetamine ameliorated haloperidol -induced reaction time slowing.

Hauber et al. (1994) found a *significant increase in RT after 6-hydroxydopamine (6-OHDA) lesion of the medial prefrontal rat cortex.*

Hauber (1996) examined the effects of dopamine D1 or D2 receptor blockade on initiation and execution of movements using a simple reaction time task for rats. The preferential D2 antagonist *haloperidol* caused a delayed movement initiation, as indicated by an *inrease in reaction time*. In addition, movement execution was slowed. He concluded that D1 and D2 receptors are both involved in movement initiation and execution processes.

Puumala and Sirvio (1998) investigated whether differences in the function of monoaminergic systems could account for the variability in attention and impulsive behaviour between rats tested in the five-choice serial reaction time task. Following training the rats were decapitated and the levels of dopamine, noradrenaline and other substances determined. Multivariate regression analysis revealed that the indices of utilization of *serotonin in the left frontal cortex and dopamine in the right frontal cortex* together accounted for 49% of the variability in attentional performance between subjects.

Dopamine and prefrontal cortex

Lidow et al. (1998) used in situ hybridization histochemistry to determine the laminar levels of the five distinct dopamine receptor mRNAs in the primate prefrontal cortex and to compare striatal and cortical levels. All five subtypes of dopamine receptor mRNA are present in both the monkey striatum and the cerebral cortex but in different proportions within each structure.

The levels of *D1 and D2 mRNAs are noticeable stronger in the striatum* than in the cortex, whereas *D4 and D5 expression is clearly higher in the cortex. The D3 transcripts appear nearly equivalent in the striatum and the cortex. Within the prefrontal cortex, mRNAs encoding all dopamine receptor subtypes are expressed most strongly in layer V*. Their prominence in layer V, which contains the corticostriatal and corticotectal projection neurons, provides a basis for dopaminergic regulation of these descending control systems.

Arnsten (1997) reviews evidence that either insufficient or excessive dopamine D1 receptor stimulation is detrimental to PFC function. *High levels of catecholamine release during stress may serve to take the PFC "off-line" to allow faster, more habitual responses mediated by the posterior and/or subcortical structures* to regulate behavior.

Wilkinson (1997) suggests that the prefrontal cortex and the striatal terminal fields are possibly linked in an "inverse" manner, whereby *a change in prefrontal dopamine transmission in one direction occasions an opposite change in dopamine function in striatal territories*.

Shim et al. (1996) made single unit recordings from DA neurons in control and PFC-lesioned rats. The number of spontaneously active DA neurons in the ventral tegmental area (VTA) was significantly decreased. In the substantia nigra the same lesions increased the firing rate. These results suggest that *PFC lesions alter the activity of DA neurons*.

Karreman and Moghaddam (1996) examined whether *the prefrontal cortex exerts a tonic control over the basal release of dopamine in the limbic striatum.* Using intracerebral microdialysis in freely moving rats, it was demonstrated that application of tetrodoxin in the contralateral PFC significantly decreased the release of dopamine in the medial striatum. Conversely, blockade of the tonic inhibitory GABAergic input in the PFC with bicuculline increased the release of dopamine in the medial striatum. This control is mediated by glutamatergic afferents to the dopamine cell body.

Law-Tho et al. (1994) studied the effects of dopamine on layer V pyramidal cells of the prefrontal cortex using intracellular recording in rat brain slices. Their findings indicate *that DA decreases both glutamatergic and gabaergic synaptic transmission in neurons located in layer V of rat prefrontal cortex.*

Jedema and Moghaddam (1994) used in vivo microdialysis to assess the hypothesis that the *stress-induced increase in dopamine release in the prefrontal cortex is mediated by stress-activated glutamate neurotransmission in this region* of the rat prefrontal cortex.

What is the junction between striatal dopamine and reaction time? Either the striatum plays a role in the RT pathway or changing the striatal dopamine elicits a compensating reaction in the cortex dopamine. With this hypothesis the contradictions of dopamine effect do not exist any more: *subcortical dopamine depletion and cortical dopamine activation leads to lengthened reaction times, subcortical dopamine activation and cortical dopamine depletion causes prolonged reaction times.*

Other substances and reaction times

Jin et al. (1997) investigated the effects of phencyclidine (PCP) on choice reaction in a 3-choice serial reaction task for studying attentional function. *PCP (3.2 mg/kg) significantly delayed choice reaction time and reduced choice accuracy. A novel sigma receptor antagonist antagonized both the delayed choice reaction time and the decreased choice accuracy elicited by PCP administration.* These findings indicated that PCP (3.2mg/kg) significantly induced attention deficit in a 3-choice serial reaction time task, and that this process may be mediated by sigma receptors.

Meyer-Lindenberg et al. (1997) evaluated the psychometric effects of sulpiride in comparison with placebo in 12 healthy volunteers using time estimation, critical flicker fusion, and choice reaction tasks at baseline and 4h after oral administration of either 300 mg of sulpiride or placebo. Critical flicker fusion frequency was lower *and choice-reaction decision time was prolonged under medication.* Choice reaction movement time was not altered.

Hasbroucq et al. (1997) administered a single oral dose of fluvoxamine (100 mg) or a placebo (randomized double blind, cross-over design) to eight healthy volunteers who performed a choice reaction task in which stimulus intensity, stimulus-response compatibility and response repertoire were manipulated. *Fluvoxamine shortened reaction time* without decreasing the accuracy of the responses. The authors suggest that fluvoxamine spares the processing stages of stimulus preprocessing and response selection.

Dehaene et al. (1999)

The experimental design of Dehaene et al. consisted in trials of four visual stimuli appearing successively: a random-letter string mask, a prime number which was presented for a very short duration (43ms), another mask, and a target number. The subjects werde not informed of the presence of the prime number. The subjects were asked to press a response key with one hand if the target number was larger than 5, and with the other hand if the target number was smaller than 5. Half of the trials were of the congruent type (prime and target number falling on the same side of 5), and half were incongruent. The authors established that prime-target congruity has a significant influence on behavioural, electrical and haemodynamic measures of brain function. They then showed that the interference between prime and target can be attributed to a covert, prime-induced activation of motor cortex. This indicated that the prime was unconsciously processed according to task instructions, all the way down to the motor system. The task of Dehaene compromised 9 stimuli (numbers 1-9), 2 keys, together 11 elements of interest (= alternative elements, set of alternatives). The proposed pathway of Dehaene's task:

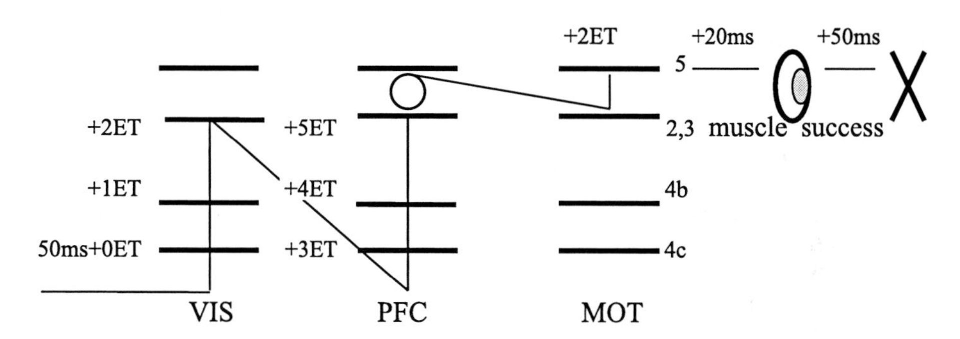

Fig. 187. The hypothetical pathway of Dehaene's task

The arrival of the information in 4c of VIS is about 50ms. The time it takes from the pyramidal neurons of MOT to success is 70 ms. The transduction time from PFC to MOT is 2ET. As the mean cycle time 25.6 ms is used (mean of 14 subjects in 1.3.2).

Therefore the whole time needed to respond to the target stimulus is:

$= 50 + 5ET + 11CT + 2ET + 70$
$= 50 + 5*12.8 + 11*25.6 + 2*12.8 + 70$
$= 50 + 64 + 281.6 + 25.6 + 70$
$= 491.2$ (ms)

The response times for arabic numbers lie between ~485 ms for congruent trials and ~ 505 ms for incongruent trials. The entire response time distribution (for all combinations of verbal/arabic presentation) was shifted by ~24 ms in incongrent trials compared with congruent trials. If Dehaene's conclusion that the prime stimulus and the target stimulus are processed in parallel, then there must be two simultaneous search and find processes upon the same set of alternatives in layer 2,3.

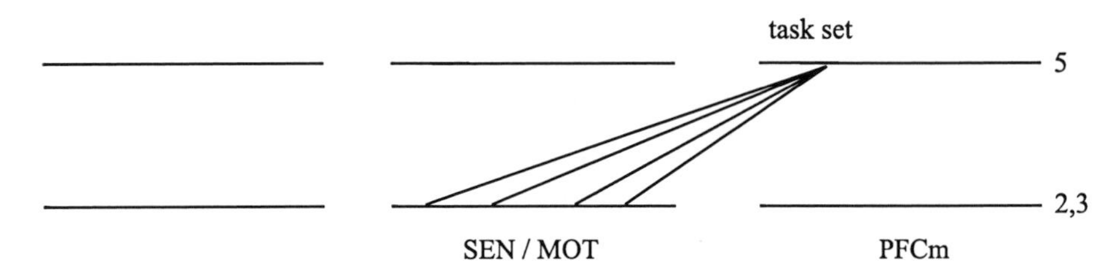

Fig. 188. The task set of Dehaene's task

The set of alternatives is called "task memory set" or "task set", the shortliving memory sets which saven the intermediate results of the search and find process are called "working memory sets" or "working sets". The results of Dehaene would imply simultaneous working sets for the prime and the target stimulus. In order to to remain isolated the two working sets could use different rhythms. What are the differences between the two working sets and why does the target working set go conscious, but the prime working set does not? The search and find process does not become conscious at all. The target stimulus and the target response do it. Obviously the prime stimulus cannot produce any subjective feeling.

The main difference between the two working sets is the duration of the prime stimulus (43ms). This means that the time to search for the stimulus is confined to 43 ms (during this time the stimulus remains active in the image layer of SEN). If we use the above value of 25.6 ms duration for one searching cycle, there is only one chance to find the prime stimulus. Either it is found at the first trial or the stimulus is gone at the second and all subsequent trials. The probabilty to find it first is 1/11 = 9.1%.

Perhaps a second trial is possible because this trial begins at 25.6 + 12.8 = 38.4 ms. So at 43 ms, the information is on it's way to the action layer (layer). The probabilty to find the prime stimulus within the first two trials is 1/11 + 1/10 = 19.1% (?). If the prime stimulus is not found within 43 ms, there is no chance of finding the prime response. But it's (empty) working set may stay active for some time.

Dehaene finds an influence of the congruency of the prime stimulus on the target processing. In order to get the knowledge of the congrency of the prime stimulus, the system has to through the complete prime processing. This is only possible in some 10 to 20%.

Supposedly only trials with finding the prime stimulus at first (= 9.1%) influence the target process by a delay of 24 ms. This would mean that one successful prime process delays the target process for 24*11 = 264 ms. In the whole, this delay would diminsh because of the rareness of the event.

BUT: The distribution in Dehaene Fig.2b contradicts this theory. If the theory would be correct there should be some very high reaction times (a peak at the right end of the distribution). This is not the case. The congruent and the incongruent distribution differ for all reaction times from each other.

This is the precise structure of Dehaene's task:

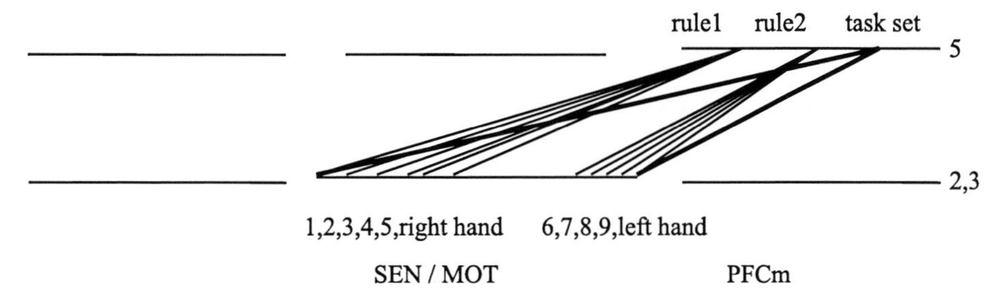

Fig. 189. The precise structure of Dehaene's task

The distribution of reaction times (Fig.2b of Dehaene) argues that in *all* trials the prime processing is present. And it is striking, that the delay in incongruent trials is about one cyle time (25.6 ms).That means in *all* trials the prime processing can only perform one full searching cycle during the 43 ms of prime stimulus activity. When the target process arrives it has to start new, abandoning the prime process after it's first searching cycle. This could be achieved by lateral inhibition of the top elements of rule1 and rule2. If the target process uses the same rule as the prime process, no change of the rule is necessary and the search

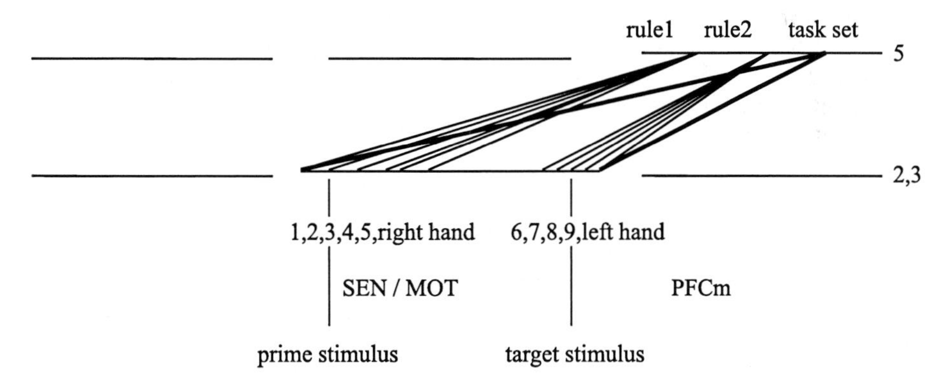

Fig. 190. The prime stimulus and the target stimulus in Dehaene's task

The search affects *all* stimulus and response elements. How does the prime stimulus influence the target process in *all* trials? The system has the time to activate at least one memory set due to the prime stimulus, probable it's rule:

At the momement the target stimulus appears, the memory of a congruent or incongruent prime stimulus has to remain vivid, because it differently influences the target process. How can this be achieved?

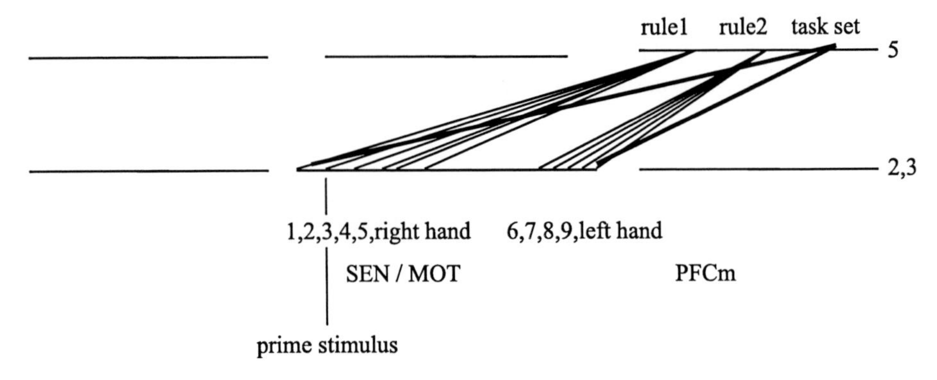

Fig. 191. The prime stimulus is active first in Dehaene's task.

The prime stimulus activates it's task set (searching set) and it's rule set. These sets remain active for some time even when the prime stimulus has gone. Therefore the information of congruency is still active within the rule set.

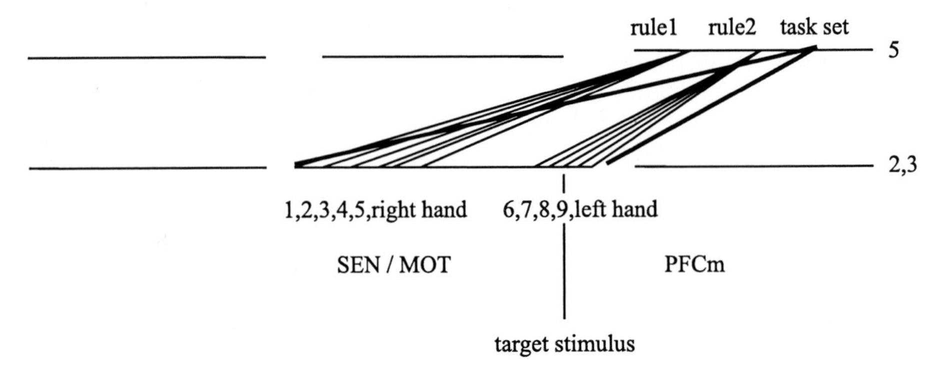

Fig. 192. The target stimulus becomes active in Dehaene's task

The remnant activity of rule1 causes a delay of (incongruent) trial. The system does not have to process the prime fully, the activity of it's rule may make the difference between congruent and incongruent target processing. The consequence is: the prime stimulus remains unconscious because it only manages to activate it's rule but no longer activation of it's task set with finding of the stimulus by the task set is possible. Though the authors used only four numbers (1,4,6, and 9) the subjects expected 9 numbers and two responses. They therefore had to perform the 11 searching actions on average. Dehaene et al. propose that the slower responses in incongruent trials might be due to response competition. They have proven that the rule1 interferes with the target processing. If the rule sets use symmetric relations one has to explain, why they do not participate in the selection process.

References

Alexander GE, Crutcher MD, De Long MR (1990) Basal ganglia – thalamocortical circuits: parallel substrates for motor, oculomotor, "prefrontal" and "limbic" functions. Prog Brain Res 85: 119–146

Arnsten AF (1997) Catecholamine regulation of the prefrontal cortex. J Psychopharmacol 11(2): 151–62

Baker SC, Rogers RD, Owen AM, Frith CD, Dolan RJ, Frackowiak RS, Robbins TW (1996) Neural systems engaged by planning: a PET study of the Tower of London task. Neuropsychologia 34(6): 515–526

Baunez C, Nieoullon A, Amalric M (1995) Dopamine and complex sensorimotor integration: further studies in a conditioned motor task in the rat. Neuroscience 65(2): 375–84

Baunez C, Nieoullon A, Amalric M (1995) In a rat model of parkinsonism, lesions of the subthalamic nucleus reverse increases of reaction time but induce a dramatic prematur responding deficit. J Neurosci 15(10): 6531–41

Bechara A, Damasio H, Tranel D, Anderson SW (1998) Dissociation of working memory from decision making within the human prefrontal cortex. J Neurosci 18(1): 428–437

Belin TR, Rubin DB (1995) The analysis of repeated-measures data on schizophrenic reaction time using mixture models. Stat Med Apr 30; 14(8): 747–768

Bench CJ, Friston KJ, Brown RG et al (1993) Regional cerebral blood flow in depression measured by positron emission tomography: the relationship with clinical dimensions. Psychol Med 23: 579–590

Brockel BJ, Fowler SC (1995) Effects of chronic haloperidol on reaction time and errors in a sustained attention task: partial reversal by anticholinergics and by amphetamine. J Pharmacol Exp Ther 275(3): 1090–1098

Burbaud P, Degreze P, Lafon P, Franconi JM, Bouligand B, Bioulac B, Caille JM, Allard M (1995) Lateralization of prefrontal activation during mental calculation: a functional magnetic resonance imaging study. J Neurophysiol 74(5): 2194–2200

Bustillo, JR, Thaker G, Buchanan RW, Moran M, Kirkpatrick B, Carpenter Jr WT (1997) Visual information-processing impairments in deficit and nondeficit schizophrenia. Am J Psychiatry 154(5): 647–654

Cai JX, Arnsten AF (1997) Dose-dependent effects of the dopamine D1 receptor agonists A77636 or H251581297 on spatial working memory in aged monkeys. J Pharmacol Exp Ther 283(1): 21–38

Carnahan H, Aguilar O, Malla A, Norman R (1997) An investigation into movement planning and execution deficits in individuals with schizophrenia. Schizophr Res 23(3): 213–21

Carter CS, Robertson LC, Nordahl TE, Chaderjian M, Oshara-Celaya L (1996) Perceptual and attentional asymmetries in schizophrenia: further evidence for a left-hemispheric deficit. Psychiatry Res 62(2): 111–119

Carter CS, Mintun M, Nichols T, Cohen JD (1997) Anterior cingulate gyrus dysfunction and selective attention deficit in schizophrenia: [15O]H2O PET study during single-trial Stroop task performance. Am J Psychiatry 154(12): 1670–1675
Casey BJ, Trainor R, Giedd J, Vauss Y, Vaituzis CK, Hamburger S, Kozuch P, Rapoport JL (1997) The role of the anterior cingulate in automatic and controlled processes: a developmental neuroanatomical study. Dev Psychobiol 30(1): 61–69
Cohen A, Semple WE, Gross M et al (1988) Functional localization of sustained attention: comparison to sensory stimulation in the absence of instruction. Neuropsychiatry Neuropsychol Behav Neurol 1: 3–20
Coull JT, Frith CD, Frackowiak RS, Grasby PM (1996) A fronto-parietal network for rapid visual information processing: a PET study of sustained attention and working memory. Neuropsychologia 34(11): 1085–1095
Davis SR, Lewis DA (1995) Local circuit neurons of the prefrontal cortex in schizophrenia: selective increase in the density of calbindin-immunoreactive neurons. Psychiatry Res 59(1–2): 81–96
Dee HL, Van Allen MW (1971) Simple and choice reaction time and motor strength in unilateral cerebral disease. Acta Psychiatry Scand 47: 315–323
Dehaene S, Naccache L, LeClec G, Koechlin E, Mueller M, Dehaene-Lambertz G, van den Moortele P, LeBihan D (1999) Imaging unconscious semantic priming. Nature 395: 597– 600
Dehaene S, Jonides J, Smith EE, Spitzer M (1999) Thinking and problem solving. In: Zigmond MJ, Bloom FE, Landis SC, Roberts JL, Squire LR (eds) Fundamental neuroscience. Academic Press, San Diego
Deiber MP, Pasingham RE, Colebatch JG, Friston KJ, Nixon PD, Frackowiak RSJ (1991) Cortical areas and the selection of movement: a study with positron emission tomography, Exp.Brain Res 84: 393–402
Deicken RF, Zhou L, Schuff N, Weiner MW (1997) Proton magnetic resonance spectroscopy of the anterior cingulate region in schizophrenia. Schizophr Res 27(1): 65–71
Dolan RJ, Bench CJ, Liddle PF et al (1993) Dorsolateral prefrontal cortex dysfunction in the major psychoses: symptom or disease specifity? J Neurol Neurosurg Psychiatry 56: 1290–1294
Dolan RJ, Fletcher P, Frith CD, Friston KJ, Frackowiak RS, Grasby PM (1995) Dopaminergic modulation of impaired cognitive activation in the anterior cingulate cortex in schizophrenia. Nature 378(6553): 180–182
Donders FC (1868) Die Schnelligkeit psychischer Prozesse. Arch Anat Phys 657–681
Drevets WC, Videen TO, Price JL et al (1992) A functional anatomical study of unipolar depression. J Neurosci 12: 3628–3641
Egan MF, Duncan CC, Suddath RL, Kirch DG, Mirsky AF, Wyatt RJ (1994) Event-related potential abnormalities correlate with structural brain alterations and clinical features in patients with chronic schizophrenia. Schizopr Res 11(3): 259–271
Elliot R, Baker SC, Rogers RD, O'Leary DA, Paykel ES, Frith CD, Dolan RJ, Sahakian BJ (1997) Prefrontal dysfunction in depressed patients performing a complex planning task: a study using positron emission tomography. Psychol Med 27(4): 931–942
Ferstl R, Hanewinkel R, Krag P (1994) Is the modality-shift effect specific for schizophrenia patients? Schizophrenia Bulletin 20(2): 367–373
Freedman R, Adler LE, Waldo MC et al (1983) Neurophysiological evidence for a defect in inhibitory pathways in schizophrenia: comparison of medicated and drug-free patients. Biol Psychiatry 18: 537–551
Frith CD, Friston K, Liddle PF and Frackowiak RSJ (1991) Willed action and the prefrontal cortex in man: a study with PET, Proc R Soc Lond 244: 241–246
Gemba H, Mike N, Sasaki K (1995) Field potential change in the prefrontal cortex of the left hemisphere during learning processes of reaction time hand movement with complex tone in the monkey. Neurosci Lett 190(2): 93–96
Godefroy O, Rousseaux M (1995) Binary choice in patients with prefrontal or posterior brain damage. A relative judgement theory analysis. Neuropsychologia 34(10): 1029–1038

Godefroy O, Rousseaux M (1997) Novel decision making in patients with prefrontal or posterior brain damage. Neurology 49(3): 695–701
Green DM, Smith AF, Gierke SM (1983) Choice reaction time with a random foreperiod. Perception Psychophysics 34: 195–208
Hasbroucq T, Rihet P, Blin O, Possamai CA (1997) Serotonin and human information processing: fluvoxamine can improve reaction time performance. Neurosci Lett 229(3): 204–208
Hanewinkel R, Ferstl R (1996) Effects of modality shift and motor response shift on simple reaction time in schizophrenia patients. J Abnorm Psychol 105(3): 459–463
Hauber W, Bubser M, Schmidt WJ (1994) 6-Hydroxydopamine lesion of the rat prefrontal cortex impairs motor initiation but not motor execution. Exp Brain Res 99(3): 524–528
Hauber W (1996) Impairments of movement initiation and execution induced by a blockade of dopamine D1 or D2 receptors are reversed by a blockade of N-methyl-D-aspartate receptors. Neuroscience 73(1): 121–130
Haznedar, MM, Buchsbaum MS, Luu C, Hazlett EA, Siegel BV Jr, Lohr J, Wu J, Haier RJ, Bunney Jr WE (1997) Decreased anterior cingulate gyrus metabolic rate in schizophrenia. Am J Psychiatry 154(5): 682–684
Heider B (1996) Sex-related engagement of the hemispheres in visuospatial processing. Int J Neurosci 88(1–2): 83–95
Hick WE (1952) On the rate of gain of information. J Exp Psychol 4: 11–26
Hienz RD, Zarcone TJ, Pyle DA, Brady JV (1997) Effects of cocaine and quinpirole on perceptual and motor functions in baboons. Psychopharmacology 134(1): 38–45
Holocomb HH, Caudill PJ, Zhao Z et al (1996) Brain metabolism patterns and sensitivity to attentional effort associated with tone recognition task. Biol Psychiatry 39: 1013–1022
Holocomb HH, Cascella NG, Thaker GK et al (1996b) Functional sites of neuroleptic drug action in human brain: PET/FDG studies with and without haloperidol. Am J Psychiatry 153: 41–49
Holocomb HH, Lahti A (1997) Images of regional cerebral blood flow elevations. In: Yudofsky SC, Hales RE (eds) The American psychatric press textbook of neuropsychiatry. 3rd edn. American Psychiatric Press, Washington DC, London, p 869
Hudspeth AJ. Hearing (2000) In Kandel ER, Schwartz JH, Jessell TM (eds) Principles of neural science. 4th edn. McGraw-Hill, New York
Huey ED, Wexler BD (1994) Abnormalities in rapid, automatic aspects of attention in schizophrenia: blunted inhibition of return. Schizophr Res 14(1): 57–63
Hyman R (1953) Stimulus information as a determinant of reaction time. J Exp Psychol 45: 188–196
Jedema HP, Moghaddam B (1994) Glutamatergic control of dopamine release during stress in the rat prefrontal cortex. J Neurochem 63(2): 785–788
Jin J, Yamamoto T, Watanabe S (1997) The involvement of sigma receptors in the choice reaction performance deficits induced by phencyclidine. Eur J Pharmacol 319(2–3): 147–152
Judd LL, McAdams L, Budnick B et al (1992) Sensory gating deficits in schizophrenia: new results. Am J Psychiatry 149: 488–493
Kalus P, Senitz D, Beckmann H (1997) Altered distribution of parvalbumin-immunoreactive local circuit neurons in the anterior cingulate cortex of schizophrenic patients. Psychiatry Res 75(1): 49–59
Kandel ER, Schwartz JH, Jessell TM (eds) Principles of neural science. 4th edn. McGraw-Hill, New York
Karreman M, Moghaddam B (1996) The prefrontal cortex regulates the basal release of dopamine in the limbic striatum: an effect mediated by ventral tegmental area. J Neurochem 66(2): 589–98
Kawashima R, Satoh K, Itoh H, Ono S, Furumoto S, Gotoh R, Koyoma M, Yoshioka S, Takahashi T, Takahashi K, Yanagisawa T, Fukuda H (1996) Functional anatomy of GO/

NO-GO discrimination and response selection – a PET study in man. Brain Res 728(1): 79–89
Kim S-G, Ashe J, Hendrich K et al (1993) Functional magnetic resonance imaging of motor cortex: hemispheric asymmetry and handedness. Science 261: 615–617
Klingberg T, Roland PE (1997) Interference between two concurrent tasks is associated with activation of overlapping fields in the cortex. Brain Res Cogn Brain Res 6(1): 1–8
Link H27, Heath RA. A sequential theory of psychological discrimination. Psychometrika 40: 77–105
Knight RT (1994) Attention regulation and human prefrontal cortex. In: Thierry A-M, Glowinski J, Goldman-Rakic PS, Christen Y (eds) Motor and cognitive functions of the prefrontal cortex. Springer, Berlin Heidelberg New York Tokyo
Kosslyn SM, Gazzaniga MS, Galaburda AM, Rabin C (1999) Hemispheric Specialization. In: Zigmond MJ, Bloom FE, Landis SC, Roberts JL, Squire LR (eds) Fundamental neuroscience. Academic Press, San Diego
Kotrla KJ. Functional neuroimaging in neuropsychiatry. In: Yudofsky SC, Hales RE (eds) The American psychatric press textbook of neuropsychiatry. 3rd edn. American Psychiatric Press, Washington DC, London
Lahti AC, Holocomb HH, Medoff DR et al (1995) Ketamine activates psychosis and alters limbic blood flow in schizophrenia. Neuroreport 6: 868–869
Law-Tho D, Hirsch JC, Crepel F (1994) Dopamine modulation of synaptic transmission in rat prefrontal cortex: an in vitro electrophysiological study. Neurosci Res 21(2): 151–160
Liddle PF, Friston KJ, Frith CD, Hirsch SR, Jones T, Frackowiak RS (1992) Patterns of cerebral blood flow in Schizophrenia. Br J Psychiatry 160: 179–186
Lidow MS, Wang F, Cao Y, Goldman-Rakic PS (1998) Layer V neurons bear the majority of mRNAs encoding the five distinct dopamine receptor subtypes in the primate prefrontal cortex. Synapse 28(1): 10–20
Maier W, Franke P, Kopp B, Hardt J, Hain C, Rist F (1994) Reaction time paradigms in subjects at rist for schizophrenia. Schizophr Res 13(1): 35–43
Manoach DS, SchlaugG, Siewert B et al (1997) Profrontal cortex fMRI signal changes are correlated with working memory load. Neuroreport 8(2): 545–549
Massman PJ, Levin BE, Delis DC et al (1992) Heterogeneity of memory impairment in Parkinson's disease: evidence for three major profile subtypes. J Clin Exp Neuropsychol 14: 81
Martinez A, Moses P, Frank L, Buxton R, Wong E, Stiles J (1997) Hemispheric asymmetries in global and local processing: evidence from fMRI. Neuroreport 8(7): 1685–1689
Maruff P, Hay D, Malone V, Currie J (1995) Asymmetries in the covert orienting of visual spatial attention in schizophrenia. Neuropsychologica 33: 1205–1223
Matsuoka H, Saito H, Ueno T, Sato M (1996) Altered endogenous negativities of the visual event-related potential in remitted schizophrenia. Electroencephalogr Clin Neurophysiol 100(1): 18–24
Mayberg HS (1994) Frontal lobe dysfunction in secondary depression. J Neuropsychiatry Clin Neurosci 6: 428–442
Mayberg HS, Lewis PJ, Regenold W et al (1994) Paralimbic hypoperfusion in unipolar depression. J Nucl Med 35: 929–934
Mayberg HS, Mahurin RK, Brannan SK (1997) Neuropsychiatric aspects of mood and affective disorders. In: Yudofsky SC, Hales RE (eds) The American psychatric press textbook of neuropsychiatry. 3rd edn. American Psychiatric Press, Washington DC, London
Meador-Woodruff JH, Haroutunian V, Powchik P, Davidson M, Davis KL, Watson SJ (1997) Dopamine receptor transcript expression in striatum and prefrontal and occipital cortex. Focal abnormalities in orbitofrontal cortex in schizophrenia. Arch Gen Psychiatry 54(12): 1089–1095
Merkel J (1885) Die zeitlichen Verhältnisse der Willensthätigkeit. Phil Stud 2: 73–127

Meyer-Lindenberg A, Rammseyer T, Ulferts J, Gallhofer B (1997) The effects of sulpiride on psychomotor performance and subjective tolerance. Eur Neuropsychopharmacol 7(3): 219–223

Mohr B, Pulvermuller F, Zaidel E (1994) Lexical decision after left, right and bilateral presentation of function words, content words and non-words: evidence for interhemispheric interaction. Neuropsychologia 32(1): 105–124

Moukhles H, Amalric M, Nieoullon A, Daszuta A (1994) Behavioural recovery of rats grafted with dopamine cells after partial striatal dopaminergic depletion in a conditioned reaction-time task. Neuroscience 63(1): 73–84

Muir JL, Everitt BJ, Robbins TW (1996) The cerebral cortex of the rat and visual attentional function: dissociable effects of mediofrontal, cingulate, anterior dorsolateral, and parietal cortex lesions on a five-choice serial reaction time task. Cereb Cortex 6(3): 470–481

Muir JL, Bussey TJ, Everitt BJ, Robbins TW (1996) Dissociable effects of AMPA-induced lesions of the vertical limb diagonal band of Broca on performance of the 5-choice serial reaction task and on acquisition of a conditional visual discrimination. Behav Brain Res 82(1): 31–44

Murphy BL, Roth RH, Arnsten AF (1997) Clozapine reverses the spatial working memory deficits induced by FG7142 in monkeys. Neuropsychopharmacology 16(6): 433–437

Neylan TC, Reynolds CF III, Kupfer DJ (1997) Electrodiagnostic techniques in neuropsychiatry. In: Yudofsky SC, Hales RE (eds) The American psychatric press textbook of neuropsychiatry. 3rd edn. American Psychiatric Press, Washington DC, London

Nyberg S, Farde L, Bartfai A, Halldin C (1995) Central D2 receptor occupancy and effects of zuclopenthixol acetate in humans. Int Clin Psychopharmacol 10(4): 221–227

Osmon DC, http://www.uwm.edu/~neuropsych/RT.html

O'Sullivan BT, Roland PE, Kawashima R (1994) A PET study of somatosensory discrimation in man. Microgeometry versus macrogeometry. Eur J Neurosci 6: 137–148

Paus T, Petrides M, Evans AC, Meyer E (1993) Role of the human anterior cingulate cortex in the control of oculomotor, manual, and speech responses: a positron emission tomography study, J Neurophysiol 70: 453–469

Podell K, Lovell M, Zimmermann M, Goldberg E (1995) The cognitive bias task and lateralized frontal lobe functions in males. J Neuropsychiatry Clin Neurosci 7(4): 491–501

Posner MI, Raichle ME (1994) Images of mind. Scientific American Library, New York

Pratt J, Klingstone A, Khoe W (1997) Inhibition of return in location- and identity-based choice decision tasks. Percept Psychophys 59(6): 964–971

Pugh KR, Offywith BA, Shaywitz SE et al (1996) Auditory selective attention: an fMRI investigation. Neuroimage 4(3Pt1): 159–173

Puumula T, Sirvio J (1998) Changes in activities of dopamine and serotonin systems in the frontal cortex underlie poor choice accuracy and impulsivity of rats in an attention task. Neuroscience 83(2): 489–499

Rao SM, Binder JR, Bandettini PA et al (1993) Functional magnetic resonance imaging of complex human movements. Neurology 43: 2311–2318

Reid RC (1999) Vision. In: Zigmond MJ, Bloom FE, Landis SC, Roberts JL, Squire LR (eds) Fundamental neuroscience. Academic Press, San Diego

Remy P, Zilbovicius M, Leroy-Willig A, Syrota A, Samson Y (1994) Movement- and task-related activations of motor cortical areas: a positron emission tomographic study. Ann Neurol 36(1): 19–26

Richter W, Andersen PM, Georgopoulos AP, Kim SG (1997) Sequential activity in human motor areas during a delayed cued finger movement task studied by time- resolved fMRI. Neuroreport 8(5): 1257–1261

Robbins TW, Roberts AC, Owen AM, Sahakian BJ, Everitt BJ, Wilkinson L, Muir J, De Salvia M, Tovée M (1994) Monoaminergic-dependent cognitive functions of the prefrontal cortex in monkey and man. In: Thierry A-M, Glowinski J, Goldman-Rakic PS, Christen Y (eds) Motor and cognitive functions of the prefrontal cortex. Springer, Berlin Heidelberg New York Tokyo

Roland PE (1993) Brain activation. Wiley-Liss, New York

Rushworth MF, Nixon PD, Eacott MJ, Passingham RE (1997) Ventral prefrontal cortex is not essential for working memory. J Neurosci 17(12): 4829–4838

Salisbury DF, O'Donnel BF, McCarley RW, Nestor PG, Faux SF, Smith RS (1994) Parametric manipulations of auditory stimuli differentially affect P3 amplitude in schizophrenics and controls. Psychophysiology 31(1): 29–36

Salisbury DF, O'Donnel BF, McCarley RW, Shenton ME, Benavage A (1994) The N2 event-related potential reflects attention deficit in schizophrenia. Biol Psychol 39(1): 1–13

Sasaki K, Gemba H, Nambu A, Matsuzake R (1994) Acitivity of the prefrontal cortex on no-go decision and motor suppression. In: Thierry A-M, Glowinski J, Goldman-Rakic PS, Christen Y (eds) Motor and cognitive functions of the prefrontal cortex. Springer, Berlin Heidelberg New York Tokyo

Sasaki K, Nambu A, Tsujimoto T, Matsuzaki R, Kyuhou S, Gemba H (1996) Studies on integrative functions of the human frontal association cortex with MEG. Brain Res Cogn Brain Res 5(1–2): 165–174

Schluter ND, Rushworth MF, Passingham RE, Mills KR (1998) Temporary interference in human lateral premotor cortex suggests dominance for the selection of movements. A study using transcranial magnetic stimulation. Brain 121(Pt5): 785–799

Schroeder MM, Lipton RB, Ritter W, Giesser BS, Vaughan Jr HG (1995) Event-related potential correlates of early processing in normal aging. J Neurosci 80: 371–382

Sesack SR, Bressler CN, Lewis DA (1995) Ultrastructural associations between dopamine terminals and local circuit neurons in the monkey prefrontal cortex: a study of calretinin-immunoreactive cells. Neurosci Lett 200(1): 9–12

Shim SS, Bunney BS, Shi WX (1996) Effects of lesions in the medial prefrontal cortex on the activity of midbrain dopamine neurons. Neuropsychopharmacology 15(5): 437–441

Siegel C, Waldo M, Mizner G et al (1984) Deficits in sensory gating in schizophrenic patients and their relatives. Arch Gen Psychiatry 41: 607–612

Silbersweig DA, Stern E, Frith C et al (1995) A functional neuroanatomy of hallucinations in schizophrenia. Nature 378: 176–179

Simpson GV, Belliveau JW, Ilmoniemi RJ et al (1995) Multi-modal spatiotemporal mapping of brain activity during visual spatial attention. Human Brain Mapping [Suppl] 1: 193

Spitzer M (1999) The mind within the net: models of learning, thinking, and acting. MIT Press, Cambridge

Sternberg S (1966) High- speed scanning in human memory. Science 153: 652–654

Sternberg S (1967) Two operations in character recognition: Some evidence from reaction – time measurement. Perception Psychophysics 2: 45–53

Sternberg S (1969) The discovery of processing stages: Extension of Donders' method. In WG Koster (ed) Attention and performance II. North Holland Publishing Company, Amsterdam, 276–315

Sternberg S (1969) Memory-scanning: mental processes revealed by reaction-time experiments. Amer Scientist 57: 421–457

Sternberg S (1971) Decomposing mental processes with reacion-time data. Invited address to the Annual Meeting of the Midwestern Psychological Association Detroit

Sternberg S (1973) Evidence against self-terminating memory search from properties of RT distributions. Annual Meeting of the Psychomemic Society St. Louis

Sternberg S (1975) Memory scanning: new findings and current controversies. Quart J Exper Psychol 27: 1–32

Stöhr M, Dichgans J, Diener HC, Buettner UW (1989) Evozierte Potentiale. 2nd edn. Springer, Berlin Heidelberg New York Tokyo

Strandburg RJ, Marsh JT, Brown WS, Asarnow RF, Guthrie D, Higa J, Yee-Bradbury CM, Nuechterlein KH (1994) Reduced attention-related negative potentials in schizophrenic adults. Psychophysiology 31(3): 272–281

Tamminga CA, Thaker GK, Buchanan R et al (1992) Limbic system abnormalities identified in schizophrenia using positron emission tomography with fluorodeoxyglucose and neocortical alterations with deficit syndrome. Arch Gen Psychiatry 49: 522–530
Tamminga CA (1997) Neuropsychiatric aspects of schizophrenia. In: Yudofsky SC, Hales RE (eds) The American psychatric press textbook of neuropsychiatry, 3rd edn. American Psychiatric Press, Washington DC, London
Taylor JL, McCloskey DI (1996) Selection of motor responses on the basis of unperceived stimuli. Exp Brain Res 110(1): 62–66
Thierry AM, Jay TM, Pirot JS, Mantz J, Godbaut R, Glowinski J (1994) Influence of afferent systems on the activity of the rat prefrontal cortex: electrophysiological and pharmacological characterization. In: Thierry A-M, Glowinski J, Goldman-Rakic PS, Christen Y (eds) Motor and cognitive functions of the prefrontal cortex. Springer, Berlin Heidelberg New York Tokyo
Tsujimoto T, Ogawa M, Nishikawa S, Tsukuda H, Kakiuchi T, Sasaki K (1997) Activation of the prefrontal, occipital and parietal cortices during go/ no-go discrimination tasks in the monkey as revealed by positron emission tomography. Neurosci Lett 224(2): 111–114
Villringer K, Minoshima S, Hock C, Obrig H, Ziegler S, Dirnagl U, Schwaiger M, Villringer A (1997) Assessment of local brain activation. A simultaneous PET and near-infrared spectroscopy study. Adv Exp Med Biol 413: 149–153
Volz H-P, Gaser C, Häger F, Rzanny R, Mentzel H-J, Kreitschmann-Andermahr I, Kaiser WA, Sauer H (1997) Brain activation during cognitive stimulation with the Wisconsin Card Sorting Test – a functional MRI study on healthy volunteers and schizophrenics. Psychiatry Res 75: 145–157
Watanabe M, Kodama T, Hikosaka K (1997) Increase of extracellular dopamine in primate prefrontal cortex during a working memory task. J Neurophysiol 78(5): 2795–2798
Wilkinson LS (1997) The nature of interactions involving prefrontal and striatal dopamine systems. J Psychopharmacol 11(2): 143–50
Woo TU, Miller JL, Lewis DA (1997) Schizophrenia and the parvalbumin-containing class of cortical local circuit neurons. Am J Psychiatry 154(7): 1013–1015
Wundt W (1896) Vorlesungen über die Menschen- und Thierseele
Yudofsky SC, Hales RE (eds) The American psychatric press textbook of neuropsychiatry, 3rd edn. American Psychiatric Press, Washington DC, London
Zigmond MJ, Bloom FE, Landis SC, Roberts JL, Squire LR (eds) Fundamental neuroscience. Academic Press, San Diego London Boston New York Sydney Tokyo Toronto

Summary

In the first part of the book, a number of preparatory studies, empirical methods and first hypothetical models are developed.

At first, bihemispheric reaction tasks were considered. There was evidence that there existed a linear relation between reaction time and the number of alternatives in choice reaction tasks. The number of alternative stimuli plus the number of alternative responses and the reaction times are strongly related. The precise type of this relation is questionable. The law of Hick which suggests a logarithmic model has dominated this discussion for a long time. The question arises whether new data fit Hick's law or any other law. The results of the preparatory studies of this chapter fit better to a linear relation between reaction time and the number of alternatives. But this is only valid when the number of alternatives is not too high. At higher numbers, the relation is clearly non-linear.

Then bihemispheric visual median finger reaction times were investigated.

In this chapter there is the astonishing finding that each finger alone performing the task v11 has nearly the same reaction time, however many fingers working together (mostly from v55 to v1010) show a different median finger reaction time for each finger. There are two questions to be answered: is the cycle time or the the cycle number responsible for the different reaction times and secondly what could be the cause of this differentiation? This chapter shows that the cycle times of the different fingers of the left hand are rather constant and the cycle number is the cause of the increased reaction time of single fingers.

The second question for the cause of this sudden differentiation may be answered by many contradictions against lateral inhibition as the cause of different median finger reaction times. Rather differentiating influences (from layer 5) are supposed to be the cause.

The next two chapters deal with monohemispheric visual and auditory reaction tasks.The theme of this chapter is looking for an asymmetry between the right and left hemispheric visual pathways of the task v22. A righthanded subject has a slight left hand (i.e. right hemisphere) advantage of cycle time and a right hand (i.e. left hemisphere) advantage of cycle number. It has been said previously that the cycle number depends on the fingers' pre-activation but the cycle time does not. Perhaps a right hemisphere advantage in visual cycle time can be explained by a general right hemisphere visual processing advantage.

The auditory pathways of the tasks a22l and a22r are the objects of this chapter. Similar to the visual case, a possible asymmetry of cycle time and cycle number between the two hemispheres has to be examined. The cycle times and the cycle numbers are again directly observed. Of special interest is the computed cycle number and its relative interindividual constancy.

Now the intraindividual variability of cycle time was investigated. It is known, that the reaction time depends on various factors like smoking, menstrual period, sleep and so on. The question is whether the individual cycle time depends on the same factors or whether it is more stable than the reaction time.

The next chapters make an effort to apply the methods of monohemispheric reaction tasks to patients.

The cycle time, cycle number and intercepts in patients with monohemispheric brain damage (The Pathways of Patients with Monohemispheric Brain Lesions) are investigated. Patients with a defined (by CT or MRI) monohemispheric brain damage have a known cause. Therefore, they are well suited for testing the measures proposed by this work.

Correspondingly, the cycle times and cycle numbers of patients with schizophrenia are described. It deals with the search strategies of patients with schizophrenia.The reaction tasks of schizophrenics are altered in some characteristic ways. There is a long known increase of reaction time and effekts like inhibition of return, the modality shift effect and the crossover effect. Some authors have examined the relationship between reaction tasks and evoked potentials, some the saccadic reaction time.

In the next chapter, the event-related potentials of monohemispheric tasks are used to give further evidence for the hitherto acquired knowledge.

It treats the evoked potentials of auditory reaction tasks. Is it possible to observe the empirical correlation of the hypothetical periods of a reaction task? To answer this question, event-related potentials were recorded and evaluated from some of the subjects of the previous work. There are some important differences between the healthy subjects and the patients with schizophrenia which make it necessary to write this special chapter. Many patients show a lateral asymmetry between daCTl and daCTr larger than in healthy subjects. The second finding is an increased cycle number vCN in some patients.

Now, the actual *second part* of the work starts with the spatiotemporal structure of stimulus-response pathways.The basis of the whole work is the measurement of the elementary times. In order to achieve this goal, different methods have been developed which should give convergent evidence.

Because the chronophoretic results are often ambiguous, further methods to measure the basic parameters have been developed. One of these is the "first peak method".

The program FPM is described. It uses a procedure called "NESTLE" which calculates the elementary times from the distribution of reaction times. FPM is the abbreviation for "first peak method" in order to underscore the eminent role of the first peak as the minimal useful reaction time which must have the simplest pathway (FPM31e uses 5 ms intervals, FPM26f58 uses 1ms intervals.

Then, the program SINGLE is used for the same purpose. This method can be named chronophoresis because the elementary times are represented by plots in a time figure.

The elementary times are the fundamental parameters. Their mesurement is crucial to all the following studies. A special program (SINGLE) has been developed to get these parameters from the measured reaction times of a task. In both the chronophoresis and the FPM only the responses of the indexfinger are evaluated. Now the results of the chronophoresis are compared to the results of the FPM program. A convergence table compares all the results (SINGLE100n, SINGLE100r,u, (median – fp)/n method, (fp – con)/m method and NESTLE index and takes as the elementary time the result which fits the most methods.

The next chapter is about the attributes of elementary times. They show a remarkable intraindividual stability at different times and a high symmetry between the two hemispheres. Some of the subjects have been asked to repeat the tasks some days later. These repetitions replicate the elementary times of a subject but not the number of elementary times, which are necessary to perform a single trial of the task. These changes are due to implicit learning.

Of special interest were the symmetries between the auditory or visual elementary times of the right and left hemisphere.

With the elementary times as the components of the pathways, it is possible to investigate the time structure of the stimulus-response pathways.

The distribution pattern of the reaction times is used by the program FPM in order to calculate the length of the linear and the cyclical portions of the pathway of the tasks a11r, a11l, v11r, v11l, a22r, a22l, v22r, and v22l.

In a next step, the constant value (70 for auditory tasks, 120 ms for visual tasks), the first peak, and the median reaction time are used to compute the spatiotemporal structure of the special pathway.

The pathways of the individual subjects show certain variations: there are a number of variations of the linear portion and a number of variations of the cyclical portion of the pathway. The variability of the linear pathway is the theme of the following chapter. Of special interest are the minimal pathway and the median pathway.

Subsequently, the variability of the cyclical pathway is treated. The random search, the fast mode, the slow mode, the double search and the triple search are presented as variations of the search process.

The concluding chapter presents the discussion of this whole section. The items are critical reconsidered and possible errors discussed.

In the *third part* of this book, the methods of measuring the elementary times (FPM 31e, FPM 26f58, SINGLE106n, and SINGLE104r) and the methods of determining the length of the linear and the cyclical pathway (FPM31e) are used to get an equation system for each subject measured. For each of the eight tasks (a11r, a11l, v11r, v11l, a22r, a22l, v22r, and v22l) a single equation describes the spatiotemporal structure of the relevant pathway. In some subjects, additional tasks (a33r, a33l, v33r, v33l) were measured and in some subjects the tasks and the calculations were replicated in order to prove the validity of the methods.

In this part of the work, three groups are examined with approximately 20 subjects.each. The first group consists of healhty subjects, the second of patients with monohemispheric brain lesions (tumors, stroke etc.).In the third group, patients with schizophrenia are assembled.

By searching for deviations of the pathways in the two groups with diseases compared with the healthy group, some applications of the stimulus-response pathway structures in neurology and psychiatry are introduced.

The initial chapters presents the convergence tables of the FPM results for each of the healthy subjects, and the collection of equations for this control group. Then, the statistics of these equations is presented with the symmetries of elementary times and pathways, the implicit learning axiom and variations of the searching process like double search and triple search.

The next chapter presents the convergence tables for the patients with monohemispheric brain lesions. Patients with a defined (by CT or MRI) monohemispheric brain are well suited for testing the equations proposed by this work and their asymmetries caused by the brain lesion.

It is followed by the collection of equations for the pathways of these patients. The patients with brain lesions were all different. It cannot be expected that each of these patients must show disturbed stimulus response pathways at all or the same disturbance as the other ones.The question therefore is: which localisations of brain lesions do produce disorders of xNNy pathways?

This is the theme of the subsequent *chapter* where the deviations from the normal pathways of healthy subjects are presented: many patients show asymmetrically increased elementary times at the side of the brain lesion. Altogether there exist the following patterns of disturbation of mental pathways in patients with monohemispheric brain lesions.

There are three possibilities:

- An increased elementary time can be found in both sensory modalities in the lesioned hemisphere. This pattern is can be caused by a frontal monohemispheric lesion.
- The elementary time of only one sensory modality is increased. This pattern can be caused by a monohemispheric lesion in the non-frontal brain.
- The lesion does not affect the elementary times of the mental pathways.

In this chapter, it is attempted to correlate the asymmetrically increasesd elementary times with the localisation of the stimulus-response pathways

The next three chapters of this section deal with the pathways of patients with schizophrenia.

First, the convergence tables with the FPM results are presented and then the collection of equations for this group.

Next, the deviations from the normal pathways of healthy subjects are described. Some patients show asymmetric elementary times, many patients show prolonged linear pathways in one or two of the auditory tasks a22r and/or a22l. The specificity and sensitivity is the same as in other findings for patients with schizophrenia: there are patients who do not show this finding and there are healthy subjects who have the same finding as the patients.

Lastly, the results of this part are critically discussed: Why do the tasks a33y still show a greater variation in patients with schizöphrenia? What is the cause of the prolongation of the linear pathways in this group?

The *fourth part* of the book presents a critical evaluation of the model and the results.

The event-related potentials are used to confirm the elementary times and pathway structures of healthy subjects. The same method is used to confirm the findings of patients with schizophrenia.

The next chapters present a comprehensive view of the model. Here, the memory sets and the set systems are defined, and the neural structure of the x11y, x22y, and x33y pathways is discussed. Now the individual pathway (with its individual linear and cyclical composition) is interpreted as a complex spatiotemporal process.Then these spatiotemporal processes are embedded into the neural structure of the human cortex.

It is followed by the simulation of set systems and its contribution to the validation and understanding of the model. Having these spatiotemporal structures of the stimulus- response pathways, it is possible to emulate them in a computer programm. This simuluation produces the same distributions as the subjects, using the equations of the subjects.

Here the complete model is presented with memory sets, set systems, and large memory sets. The memory sets can exhibit a slow, spontaneous, asynchronous activity or a fast, stimulated synchronous activity. Set systems which show stimulated activity compete with each other. Competitive tasks (e.g. finger tapping) may slow the reaction time of one task. The arguments are gathered why this model is more plausible than that of the preparatory studies. At last, the pathology of memory sets is discussed.

Are there experiments which can disprove the presented model?

The last chapter is about the unsolved problems, the limitations of the model and the results: which are the possible errors, which are the unprecise items, which are the open questions the modell is leaving? The pathways of mind are described by their length and their duration. The living mind which uses this pathways cannot be explained by these models. The only theme of this book is how the mind behaves, not what the mind is.

Next are the references for the themes of this book. The results are compared to the results of other researchers who use brain mapping techniques like functional magnetic resonance imaging and other methods to investigate the stimulus response pathways within the human brain.

In the last, *fifth part* of the book, the literature and a summary are included.